Digital Signal Processing

Konstantinos N. Plataniotis · Anastasios N. Venetsanopoulos

Color Image Processing and Applications

Springer
Berlin
Heidelberg
New York
Barcelona
Hong Kong
London
Milano
Paris
Singapore
Tokyo

Konstantinos N. Plataniotis
Anastasios N. Venetsanopoulos

Color Image Processing and Applications

With 100 Figures

Springer

Series Editors

Prof. Dr.-Ing. Arild Lacroix
Johann-Wolfgang-Goethe-Universität
Institut für Angewandte Physik
Robert-Mayer-Str. 2-4
D-60325 Frankfurt

Prof. Anastasios N. Venetsanopoulos
University of Toronto
Department of Electrical & Computer Engineering
10 King's College Road
M5S 3G4 Toronto, Ontario
Canada

Authors

Ph. D. Konstantinos N. Plataniotis
Prof. Anastasios N. Venetsanopoulos
University of Toronto
Department of Electrical & Computer Engineering
10 King's College Road
M5S 3G4 Toronto, Ontario
Canada

e-mails: kostas@dsp.toronto.edu
anv@dsp.toronto.edu

ISBN 978-3-642-08626-7

Library of Congress Cataloging-in-Publication Data

Plataniotis, Konstantinos N.:

Color Image Processing and Applications / Konstantinos N. Plataniotis; Anastasios N. Venetsanopoulos. - Berlin; Heidelberg; New York; Barcelona; Hong Kong; London; Milano; Paris; Singapore; Tokyo: Springer 2000

Softcover version of the original hardcover edition 2000

Cover-Design: de'blik, Berlin

Preface

The perception of color is of paramount importance to humans since they routinely use color features to sense the environment, recognize objects and convey information. Color image processing and analysis is concerned with the manipulation of digital color images on a computer utilizing digital signal processing techniques. Like most advanced signal processing techniques, it was, until recently, confined to academic institutions and research laboratories that could afford the expensive image processing hardware needed to handle the processing overhead required to process large numbers of color images. However, with the advent of powerful desktop computers and the proliferation of image collection devices, such as digital cameras and scanners, color image processing techniques are now within the grasp of the general public.

This book is aimed at researchers and practitioners that work in the area of color image processing. Its purpose is to fill an existing gap in scientific literature by presenting the state of the art research in the area. It is written at a level which can be easily understood by a graduate student in an Electrical and Computer Engineering or Computer Science program. Therefore, it can be used as a textbook that covers part of a modern graduate course in digital image processing or multimedia systems. It can also be used as a textbook for a graduate course on digital signal processing since it contains algorithms, design criteria and architectures for processing and analysis systems.

The book is structured into four parts. The first, Chapter 1, deals with color principles and is aimed at readers who have very little prior knowledge of color science. Readers interested in color image processing may read the second part of the book (Chapters 2-5). It covers the major, although somewhat mature, fields of color image processing. Color image processing is characterized by a large number of algorithms that are specific solutions to specific problems, for example vector median filters have been developed to remove impulsive noise from images. Some of them are mathematical or content independent operations that are applied to each and every pixel, such as morphological operators. Others are algorithmic in nature, in the sense that a recursive strategy may be necessary to find edge pixels in an image.

The third part of the book, Chapters 6-7, deals with color image analysis and coding techniques. The ultimate goal of color image analysis is to enhance human-computer interaction. Recent applications of image analysis includes compression of color images either for transmission across the internetwork or coding of video images for video conferencing. Finally, the fourth part (Chapter 8) covers emerging applications of color image processing. Color is useful for accessing multimedia databases. Local color information, for example in the form of color histograms, can be used to index and retrieve images from the database. Color features can also be used to identify objects of interest, such as human faces and hand areas, for applications ranging from video conferencing, to perceptual interfaces and virtual environments. Because of the dual nature of this investigation, processing and analysis, the logical dependence of the chapters is somewhat unusual. The following diagram can help the reader chart the course.

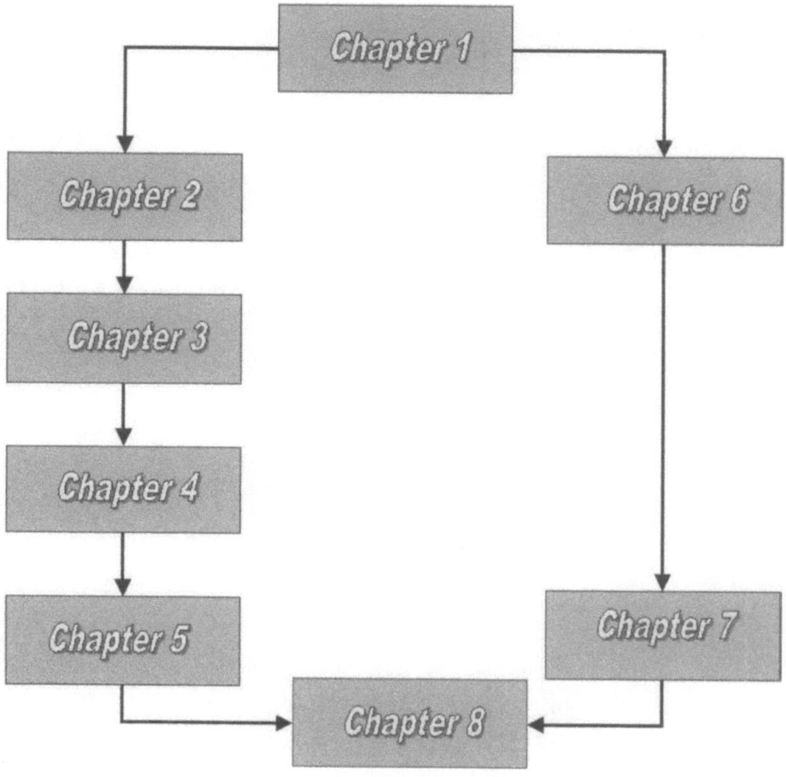

Logical dependence between chapters

Acknowledgment

We acknowledge a number of individuals who have contributed in different ways to the preparation of this book. In particular, we wish to extend our appreciation to Prof. M. Zervakis for contributing the image restoration section, and to Dr. N. Herodotou for his informative inputs and valuable suggestions in the emerging applications chapter. Three graduate students of ours also merit special thanks. Shu Yu Zhu for her input and high quality figures included in the color edge detection chapter, Ido Rabinovitch for his contribution to the color image coding section and Nicolaos Ikonomakis for his valuable contribution in the color segmentation chapter. We also thank Nicolaos for reviewing the chapters of the book and helping with the Latex formating of the manuscript. We also grateful to Terri Vlassopoulos for proofreading the manuscript, and Frank Holzwarth of Springer Verlag for his help during the preparation of the book. Finally, we are indebted to Peter Androutsos who helped us tremendously on the development of the companion software.

Contents

List of Figures

List of Tables

1. Color Spaces

1.1 Basics of Color Vision

Color is a sensation created in response to excitation of our visual system by electromagnetic radiation known as light [1], [2], [3]. More specific, color is the perceptual result of light in the visible region of the electromagnetic spectrum, having wavelengths in the region of 400nm to 700nm, incident upon the retina of the human eye. Physical power or radiance of the incident light is in a spectral power distribution (SPD), often divided into 31 components each representing a 10nm band [4]-[13].

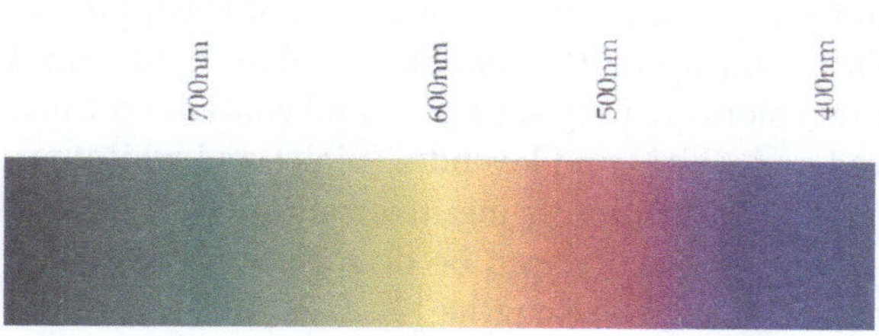

Fig. 1.1. The visible light spectrum

The human retina has three types of color photo-receptor cells, called cones , which respond to radiation with somewhat different spectral response curves [4]-[5]. A fourth type of photo-receptor cells, called roads , are also present in the retina. These are effective only at extremely low light levels, for example during night vision. Although rods are important for vision, they play no role in image reproduction [14], [15].

The branch of color science concerned with the appropriate description and specification of a color is called colorimetry [5], [10]. Since there are exactly three types of color photo-receptor cone cells, three numerical components are necessary and sufficient to describe a color, providing that appropriate spectral weighting functions are used. Therefore, a color can be specified by a tri-component vector. The set of all colors form a vector space called color space or color model. The three components of a color can be defined in many different ways leading to various color spaces [5], [9].

Before proceeding with color specification systems (color spaces), it is appropriate to define a few terms: Intensity (usually denoted I), brightness

(Br), luminance (Y), lightness (L^*), hue (H) and saturation (S), which are often confused or misused in the literature. The intensity (I) is a measure, over some interval of the electromagnetic spectrum, of the flow of power that is radiated from, or incident on a surface and expressed in units of watts per square meter [4], [18], [16]. The intensity (I) is often called a linear light measure and thus is expressed in units, such as watts per square meter [4], [5]. The brightness (Br) is defined as the attribute of a visual sensation according to which an area appears to emit more or less light [5]. Since brightness perception is very complex, the Commission Internationale de L'Eclairage (CIE) defined another quantity luminance (Y) which is radiant power weighted by a spectral sensitivity function that is characteristic of human vision [5]. Human vision has a nonlinear perceptual response to luminance which is called lightness (L^*). The nonlinearity is roughly logarithmic [4].

Humans interpret a color based on its lightness (L^*), hue (H) and saturation (S) [5]. Hue is a color attribute associated with the dominant wavelength in a mixture of light waves. Thus hue represents the dominant color as perceived by an observer; when an object is said to be red, orange, or yellow the hue is being specified. In other words, it is the attribute of a visual sensation according to which an area appears to be similar to one of the perceived colors: red, yellow, green and blue, or a combination of two of them [4], [5]. Saturation refers to the relative purity or the amount of white light mixed with a hue. The pure spectrum colors are fully saturated and contain no white light. Colors such as pink (red and white) and lavender (violet and white) are less saturated, with the degree of saturation being inversely proportional to the amount of white light added [1]. A color can be de-saturated by adding white light that contains power at all wavelengths [4]. Hue and saturation together describe the chrominance. The perception of color is basically determined by luminance and chrominance [1].

To utilize color as a visual cue in multimedia, image processing, graphics and computer vision applications, an appropriate method for representing the color signal is needed. The different color specification systems or color models (color spaces or solids) address this need. Color spaces provide a rational method to specify, order, manipulate and effectively display the object colors taken into consideration. A well chosen representation preserves essential information and provides insight to the visual operation needed. Thus, the selected color model should be well suited to address the problem's statement and solution. The process of selecting the best color representation involves knowing how color signals are generated and what information is needed from these signals. Although color spaces impose constraints on color perception and representation they also help humans perform important tasks. In particular, the color models may be used to define colors, discriminate between colors, judge similarity between color and identify color categories for a number of applications [12], [13].

Color model literature can be found in the domain of modern sciences, such as physics, engineering, artificial intelligence, computer science, psychology and philosophy. In the literature four basic color model families can be distinguished [14]:

1. **Colorimetric color models**, which are based on physical measurements of spectral reflectance. Three primary color filters and a photo-meter, such as the CIE *chromaticity diagram* usually serve as the initial points for such models.
2. **Psychophysical color models**, which are based on the human perception of color. Such models are either based on subjective observation criteria and comparative references (e.g. Munsell color model) or are built through experimentation to comply with the human perception of color (e.g. Hue, Saturation and Lightness model).
3. **Physiologically inspired color models**, which are based on the three *primaries*, the three types of *cones* in the human *retina*. The Red-Green-Blue (RGB) color space used in computer hardware is the best known example of a physiologically inspired color model.
4. **Opponent color models**, which are based on perception experiments, utilizing mainly pairwise opponent primary colors, such as the Yellow-Blue and Red-Green color pairs.

In image processing applications, color models can alternatively be divided into three categories. Namely:

1. **Device-oriented color models**, which are associated with input, processing and output signal devices. Such spaces are of paramount importance in modern applications, where there is a need to specify color in a way that is compatible with the hardware tools used to provide, manipulate or receive the color signals.
2. **User-oriented color models**, which are utilized as a bridge between the human operators and the hardware used to manipulate the color information. Such models allow the user to specify color in terms of perceptual attributes and they can be considered an experimental approximation of the human perception of color.
3. **Device-independent color models**, which are used to specify color signals independently of the characteristics of a given device or application. Such models are of importance in applications, where color comparisons and transmission of visual information over networks connecting different hardware platforms are required.

In 1931, the Commission Internationale de L'Eclairage (CIE) adopted standard color curves for a hypothetical standard observer. These color curves specify how a specific spectral power distribution (SPD) of an *external stimulus* (visible radiant light incident on the eye) can be transformed into a set of three numbers that specify the color. The CIE color specification system is based on the description of color as the luminance component Y and two additional components X and Z [5]. The spectral weighting curves of X and Z have been standardized by the CIE based on statistics from experiments involving human observers [5]. The CIE XYZ ***tristimulus values*** can be used to describe any color. The corresponding color space is called the CIE XYZ color space. The XYZ model is a device independent color space that is useful in applications where consistent color representation across devices with different characteristics is important. Thus, it is exceptionally useful for color management purposes.

The CIE XYZ space is perceptually highly non uniform [4]. Therefore, it is not appropriate for quantitative manipulations involving color perception and is seldom used in image processing applications [4], [10]. Traditionally, color images have been specified by the non-linear red (R'), green (G') and blue (B') tristimulus values where color image storage, processing and analysis is done in this non-linear RGB (R'G'B') color space. The red, green and blue components are called the primary colors . In general, hardware devices such as video cameras, color image scanners and computer monitors process the color information based on these primary colors. Other popular color spaces in image processing are the YIQ (North American TV standard), the HSI (Hue, Saturation and Intensity), and the HSV (Hue, Saturation, Value) color spaces used in computer graphics.

Although XYZ is used only indirectly it has a significant role in image processing since other color spaces can be derived from it through mathematical transforms. For example, the linear RGB color space can be transformed to and from the CIE XYZ color space using a simple linear three-by-three matrix transform. Similarly, other color spaces, such as non-linear RGB, YIQ and HSI can be transformed to and from the CIE XYZ space, but might require complex and non-linear computations. The CIE have also derived and standardized two other color spaces, called L*u*v* and L*a*b*, from the CIE XYZ color space which are perceptually uniform [5].

The rest of this chapter is devoted to the analysis of the different color spaces in use today. The different color representation models are discussed and analyzed in detail with emphasis placed on motivation and design characteristics.

1.2 The CIE Chromaticity-based Models

Over the years, the CIE committee has sponsored the research of color perception. This has lead to a class of widely used mathematical color models. The derivation of these models has been based on a number of color matching experiments, where an observer judges whether two parts of a visual stimulus match in appearance. Since the colorimetry experiments are based on a matching procedure in which the human observer judges the visual similarity of two areas the theoretical model predicts only matching and not perceived colors. Through these experiments it was found that light of almost any spectral composition can be matched by mixtures of only three primaries (lights of a single wavelength). The CIE had defined a number of standard observer color matching functions by compiling experiments with different observers, different light sources and with various power and spectral compositions. Based on the experiments performed by CIE early in this century, it was determined that these three primary colors can be broadly chosen, provided that they are independent.

The CIE's experimental matching laws allow for the representation of colors as vectors in a three-dimensional space defined by the three primary colors. In this way, changes between color spaces can be accomplished easily. The next few paragraphs will briefly outline how such a task can be accomplished.

According to experiments conducted by Thomas Young in the nineteenth century [19], and later validated by other researchers [20], there are three different types of cones in the human retina, each with different absorption spectra: $S_1(\lambda)$, $S_2(\lambda)$, $S_3(\lambda)$, where $380 \leq \lambda \leq 780$ (nm). These approximately peak in the yellow-green, green and blue regions of the electromagnetic spectrum with significant overlap between S_1 and S_2. For each wavelength the absorption spectra provides the weight with which light of a given spectral distribution (SPD) contributes to the cone's output. Based on Young's theory, the color sensation that is produced by a light having SPD $C(\lambda)$ can be defined as:

$$\alpha_i(C) = \int_{\lambda_1}^{\lambda_2} S_i(\lambda)C(\lambda)\, d\lambda \tag{1.1}$$

for $i = 1,2,3$. According to (1.1) any two colors $C_1(\lambda)$, $C_2(\lambda)$ such that $\alpha_i(C_1) = \alpha_i(C_2)$, $i = 1,2,3$ will be perceived to be identical even if $C_1(\lambda)$ and $C_2(\lambda)$ are different. This well known phenomenon of spectrally different stimuli that are indistinguishable to a human observer is called *metamers* [14] and constitutes a rather dramatic illustration of the perceptual nature of color and the limitations of the color modeling process. Assume that three primary colors C_k , $k = 1,2,3$ with SPD $C_k(\lambda)$ are available and let

$$\int C_k(\lambda)\, d\lambda = 1 \tag{1.2}$$

To match a color C with spectral energy distribution $C(\lambda)$, the three primaries are mixed in proportions of β_k, $k = 1,2,3$. Their linear combination $\sum_{k=1}^{3} \beta_k C_k(\lambda)$ should be perceived as $C(\lambda)$. Substituting this into (1.1) leads to:

$$\alpha_i(C) = \int (\sum_{k=1}^{3} \beta_k C_k(\lambda)) S_i(\lambda)\, d\lambda = \sum_{k=1}^{3} \beta_k \int S_i(\lambda) C_k(\lambda)\, d\lambda \tag{1.3}$$

for $i = 1,2,3$.

The quantity $\int S_i(\lambda)C_k(\lambda)\,d\lambda$ can be interpreted as the ith, $i = 1,2,3$ cone response generated by one unit of the kth primary color:

$$\alpha_{i,k} = \alpha_i(C_k) = \int S_i(\lambda)C_k(\lambda)\,d\lambda \tag{1.4}$$

Therefore, the color matching equations are:

$$\sum_{k=1}^{3} \beta_k \alpha_{i,k} = \alpha_i(C) = \int S_i(\lambda)C(\lambda)\,d\lambda \tag{1.5}$$

assuming a certain set of primary colors $C_k(\lambda)$ and spectral sensitivity curves $S_i(\lambda)$. For a given arbitrary color, β_k can be found by simply solving (1.4) and (1.5).

Following the same approach w_k can be defined as the amount of the kth primary required to match the reference white, providing that there is available a reference white light source with known energy distribution $w(\lambda)$. In such a case, the values obtained through

$$T_k(C) = \frac{\beta_k}{w_k} \tag{1.6}$$

for $k = 1,2,3$ are called *tristimulus* values of the color C, and determine the relative amounts of primitives required to match that color. The tristimulus values of any given color $C(\lambda)$ can be obtained given the spectral tristimulus values $T_k(\lambda)$, which are defined as the tristimulus values of unit energy spectral color at wavelength λ. The spectral tristimulus $T_k(\lambda)$ provide the so-called *spectral matching curves* which are obtained by setting $C(\lambda) = \delta(\lambda - \lambda^*)$ in (1.5).

The spectral matching curves for a particular choice of color primaries with an approximately red, green and blue appearance were defined in the CIE 1931 standard [9]. A set of pure monochromatic primaries are used, blue ($435.8nm$), green ($546.1nm$) and red ($700nm$). In Figures 1.2 and 1.3 the Y-axis indicates the relative amount of each primary needed to match a stimulus of the wavelength reported on the X-axis. It can be seen that some of the values are negative. Negative numbers require that the primary in question be added to the opposite side of the original stimulus. Since negative sources are not physically realizable it can be concluded that the arbitrary set of three primary sources cannot match all the visible colors. However, for any given color a suitable set of three primary colors can be found.

Based on the assumption that the human visual system behaves linearly, the CIE had defined spectral matching curves in terms of virtual primaries. This constitutes a linear transformation such that the spectral matching curves are all positive and thus immediately applicable for a range of practical situations. The end results are referred to as the CIE 1931 standard observer matching curves and the individual curves (functions) are labeled

$\bar{x}$, $\bar{y}$, $\bar{z}$ respectively. In the CIE 1931 standard the matching curves were selected so that $\bar{y}$ was proportional to the human luminosity function, which was an experimentally determined measure of the perceived brightness of monochromatic light.

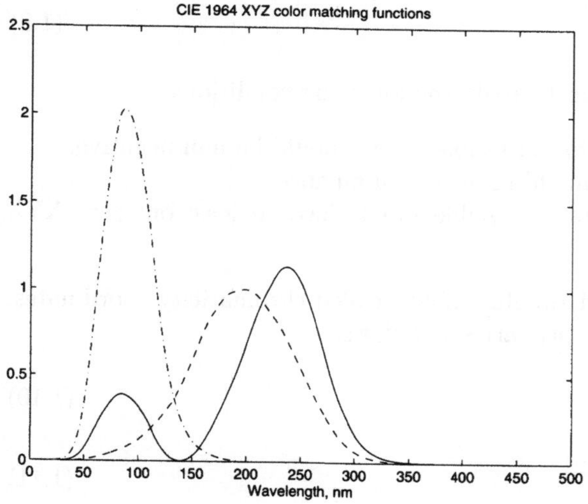

Fig. 1.2. The CIE XYZ color matching functions

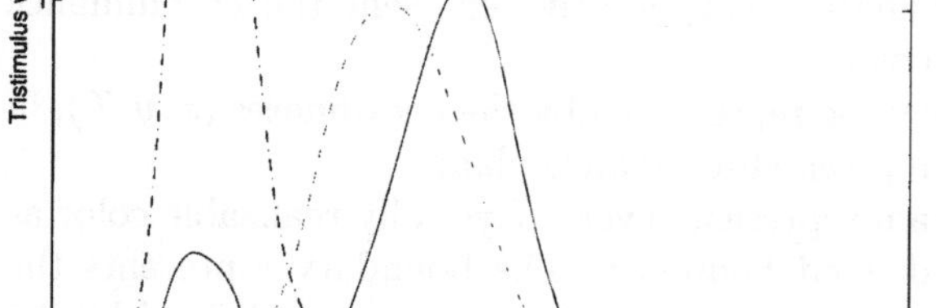

Fig. 1.3. The CIE RGB color matching functions

If the spectral energy distribution $C(\lambda)$ of a stimulus is given, then the chromaticity coordinates can be determined in two stages. First, the tristimulus values X, Y, Z are calculated as follows:

$$X = \int \bar{x}(\lambda)C(\lambda)\,d\lambda \tag{1.7}$$

$$Y = \int \bar{y}(\lambda)C(\lambda)\,d\lambda \tag{1.8}$$

$$Z = \int \bar{z}(\lambda)C(\lambda)\,d\lambda \tag{1.9}$$

The new set of primaries must satisfy the following conditions:

1. The XYZ components for all visible colors should be non-negative.
2. Two of the primaries should have zero luminance.
3. As many spectral colors as possible should have at least one zero XYZ component.

Secondly, normalized tristimulus values, called chromaticity coordinates, are calculated based on the primaries as follows:

$$x = \frac{X}{X+Y+Z} \tag{1.10}$$

$$y = \frac{Y}{X+Y+Z} \tag{1.11}$$

$$z = \frac{Z}{X+Y+Z} \tag{1.12}$$

Clearly $z = 1 - (x + y)$ and hence only two coordinates are necessary to describe a color match. Therefore, the chromaticity coordinates project the $3-D$ color solid on a plane, and they are usually plotted as a parametric $x-y$ plot with z implicitly evaluated as $z = 1 - (x + y)$. This diagram is known as the chromaticity diagram and has a number of interesting properties that are used extensively in image processing. In particular,

1. The chromaticity coordinates (x, y) jointly represent the chrominance components of a given color.
2. The entire color space can be represented by the coordinates (x, y, T), in which $T = \mathit{constant}$ is a given chrominance plane.
3. The chromaticity diagram represents every physically realizable color as a point within a well defined boundary. The boundary represents the primary sources. The boundary vertices have coordinates defined by the chromaticities of the primaries.
4. A white point is located in the center of the chromaticity diagram. More saturated colors radiate outwards from white. Complementary pure colors can easily be determined from the diagram.
5. In the chromaticity diagram, the color perception obtained through the superposition of light coming from two different sources, lies on a straight line between the points representing the component lights in the diagram.

6. Since the chromaticity diagram reveals the range of all colors which can be produced by means of the three primaries (gamut), it can be used to guide the selection of primaries subject to design constraints and technical specifications.
7. The chromaticity diagram can be utilized to determine the hue and saturation of a given color since it represents chrominance by eliminating luminance. Based on the initial objectives set out by CIE, two of the primaries, X and Z, have zero luminance while the primary Y is the luminance indicator determined by the light-efficiency function $V(\lambda)$ at the spectral matching curve $\bar{y}$. Thus, in the chromaticity diagram the dominant wavelength (hue) can be defined as the intersection between a line drawn from the reference white through the given color to the boundaries of the diagram. Once the hue has been determined, then the purity of a given color can be found as the ratio $r = \frac{wc}{wp}$ of the line segments that connect the reference white with the color (wc) to the line segment between the reference white and the dominant wavelength/hue (wp).

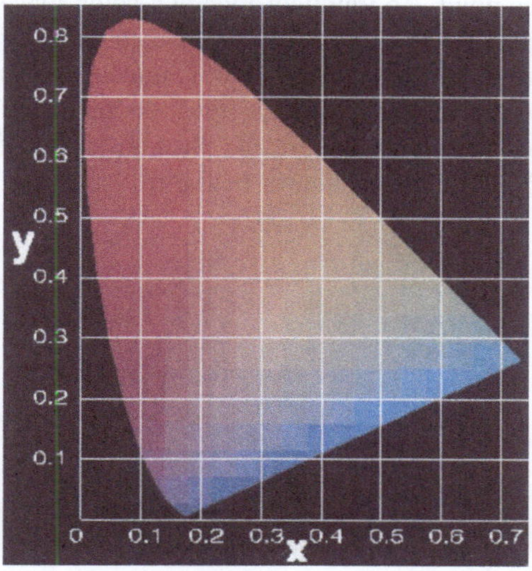

Fig. 1.4. The chromaticity diagram

1.3 The CIE-RGB Color Model

The fundamental assumption behind modern colorimetry theory, as it applies to image processing tasks, is that the initial basis for color vision lies in the different excitation of three classes of photo-receptor cones in the retina. These include the red, green and blue receptors, which define a trichromatic space whose basis of primaries are pure colors in the short, medium and high portions of the visible spectrum [4], [5], [10].

As a result of the assumed linear nature of light, and due to the principle of superposition, the colors of a mixture are a function of the primaries and the fraction of each primary that is mixed. Throughout this analysis, the primaries need not be known, just their tristimulus values. This principle is called additive reproduction. It is employed in image and video devices used today where the color spectra from red, green and blue light beams are physically summed at the surface of the projection screen. Direct view color CRT's (cathode ray tube) also utilize additive reproduction. In particular, the CRT's screen consists of small dots which produce red, green and blue light. When the screen is viewed from a distance the spectra of these dots add up in the retina of the observer. In practice, it is possible to reproduce a large number of colors by additive reproduction using the three primaries: red, green and blue. The colors that result from additive reproduction are completely determined by the three primaries.

The video projectors and the color CRT's in use today utilize a color space collectively known under the name RGB, which is based on the red, green and blue primaries and a white reference point. To uniquely specify a color space based on the three primary colors the chromaticity values of each primary color and a white reference point need to be specified. The gamut of colors which can be mixed from the set of the RGB primaries is given in the (x, y) chromaticity diagram by a triangle whose vertices are the chromaticities of the primaries (Maxwell triangle) [5], [20]. This is shown in Figure 1.5.

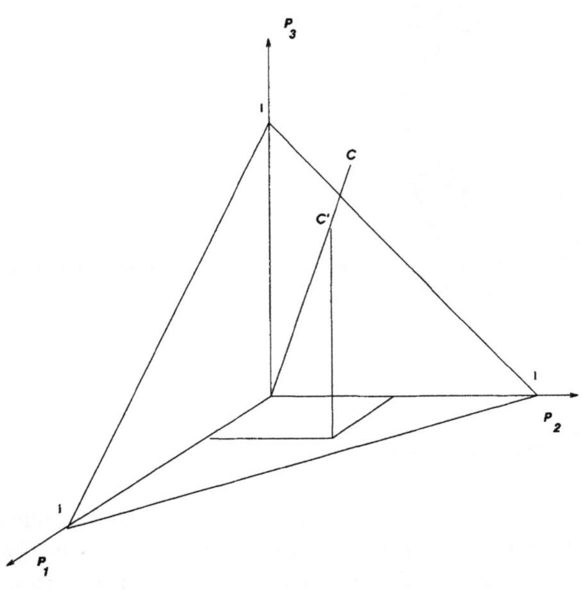

Fig. 1.5. The Maxwell triangle

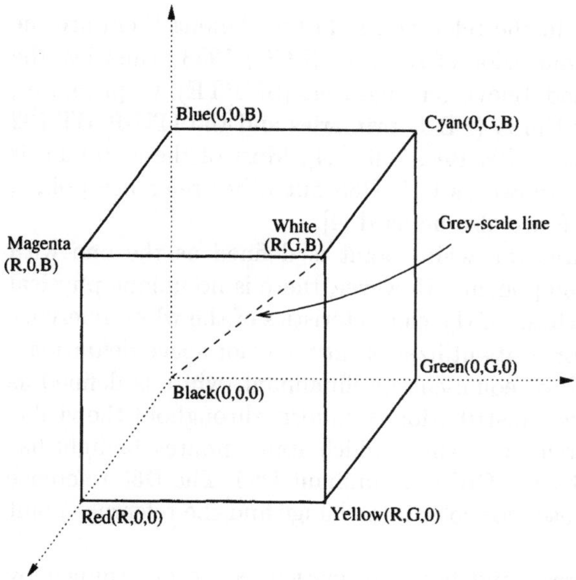

Fig. 1.6. The RGB color model

In the red, green and blue system the color solid generated is a bounded subset of the space generated by each primary. Using an appropriate scale along each primary axis, the space can normalized, so that the maximum is 1. Therefore, as can be seen in Figure 1.6 the RGB color solid is a cube, called the RGB cube. The origin of the cube, defined as $(0, 0, 0)$ corresponds to black and the point with coordinates $(1, 1, 1)$ corresponds to the system's brightest white.

In image processing, computer graphics and multimedia systems the RGB representation is the most often used. A digital color image is represented by a two dimensional array of three variate vectors which are comprised of the pixel's red, green and blue values. However, these pixel values are relative to the three primary colors which form the color space. As it was mentioned earlier, to uniquely define a color space, the chromaticities of the three primary colors and the reference white must be specified. If these are not specified within the chromaticity diagram, the pixel values which are used in the digital representation of the color image are meaningless [16].

In practice, although a number of RGB space variants have been defined and are in use today, their exact specifications are usually not available to the end-user. Multimedia users assume that all digital images are represented in the same RGB space and thus use, compare or manipulate them directly no matter where these images are from. If a color digital image is represented in the RGB system and no information about its chromaticity characteristics is available, the user cannot accurately reproduce or manipulate the image.

Although in computing and multimedia systems there are no standard primaries or white point chromaticities, a number of color space standards have been defined and used in the television industry. Among them are the Federal Communication Commission of America (FCC) 1953 primaries, the Society of Motion Picture and Television Engineers (SMPTE) 'C' primaries, the European Broadcasting Union (EBU) primaries and the ITU-R BT.709 standard (formerly known as CCIR Rec. 709) [24]. Most of these standards use a white reference point known as CIE D65 but other reference points, such as the CIE illuminant E are also be used [4].

In additive color mixtures the white point is defined as the one with equal red, green and blue components. However, there is no unique physical or perceptual definition of white, so the characteristics of the white reference point should be defined prior to its utilization in the color space definition.

In the CIE illuminant E, or equal-energy illuminant, white is defined as the point whose spectral power distribution is uniform throughout the visible spectrum. A more realistic reference white, which approximates daylight has been specified numerically by the CIE as illuminant D65. The D65 reference white is the one most often used for color interchange and the reference point used throughout this work.

The appropriate red, green and blue chromaticities are determined by the technology employed, such as the sensors in the cameras, the phosphors within the CTR's and the illuminants used. The standards are an attempt to quantify the industry's practice. For example, in the FCC-NTSC standard, the set of primaries and specified white reference point were representative of the phosphors used in color CRTs of a certain era.

Although the sensor technology has changed over the years in response to market demands for brighter television receivers, the standards remain the same. To alleviate this problem, the European Broadcasting Union (EBU) has established a new standard (EBU Tech 3213). It is defined in Table 1.1.

Table 1.1. EBU Tech 3213 Primaries

Colorimetry	Red	Green	Blue	White D65
x	0.640	0.290	0.150	0.3127
y	0.330	0.600	0.060	0.3290
z	0.030	0.110	0.790	0.3582

Recently, an international agreement has finally been reached on the primaries for the High Definition Television (HDTV) specification. These primaries are representative of contemporary monitors in computing, computer graphics and studio video production. The standard is known as ITU-R BT.709 and its primaries along with the D65 reference white are defined in Table 1.2.

The different RGB systems can be converted amongst each other using a linear transformation assuming that the white references values being used are known. As an example, if it is assumed that the D65 is used in both

Table 1.2. EBU Tech 3213 Primaries

Colorimetry	Red	Green	Blue	White D65
x	0.640	0.300	0.150	0.3127
y	0.330	0.600	0.060	0.3290
z	0.030	0.100	0.790	0.3582

systems, then the conversion between the ITU-R BT.709 and SMPTE 'C' primaries is defined by the following matrix transformation:

$$\begin{bmatrix} R_{709} \\ G_{709} \\ B_{709} \end{bmatrix} = \begin{bmatrix} 0.939555 & 0.050173 & 0.010272 \\ 0.017775 & 0.9655795 & 0.016430 \\ -0.001622 & -0.004371 & 1.005993 \end{bmatrix} \begin{bmatrix} R_c \\ G_c \\ B_c \end{bmatrix} \tag{1.13}$$

where R_{709}, G_{709}, B_{709} are the linear red, green and blue components of the ITU-R BT.709 and R_c, G_c, B_c are the linear components in the SMPTE 'C' system. The conversion should be carried out in the linear voltage domain, where the pixel values must first be converted into linear voltages. This is achieved by applying the so-called *gamma correction*.

1.4 Gamma Correction

In image processing, computer graphics, digital video and photography, the symbol γ represents a numerical parameter which describes the nonlinearity of the intensity reproduction. The cathode-ray tube (CRT) employed in modern computing systems is nonlinear in the sense that the intensity of light reproduced at the screen of a CRT monitor is a nonlinear function of the voltage input. A CRT has a power law response to applied voltage. The light intensity produced on the display is proportional to the applied voltage raised to a power denoted by γ [4], [16], [17]. Thus, the produced intensity by the CRT and the voltage applied on the CRT have the following relationship:

$$I_{int} = (v')^{\gamma} \tag{1.14}$$

The relationship which is called the 'five-halves' power law is dictated by the physics of the CRT electron gun. The above function applies to a single electron gun of a gray-scale CRT or each of the three red, green and blue electron guns of a color CRT. The functions associated with the three guns on a color CRT are very similar to each other but not necessarily identical. The actual value of γ for a particular CRT may range from about 2.3 to 2.6 although most practitioners frequently claim values lower than 2.2 for video monitors.

The process of pre-computing for the nonlinearity by computing a voltage signal from an intensity value is called *gamma correction*. The function required is approximately a 0.45 power function. In image processing applications, gamma correction is accomplished by analog circuits at the camera.

In computer graphics, gamma correction is usually accomplished by incorporating the function into a frame buffer lookup table. Although in image processing systems gamma was originally used to refer to the nonlinearity of the CRT, it is generalized to refer to the nonlinearity of an entire image processing system. The γ value of an image or an image processing system can be calculated by multiplying the γ's of its individual components from the image capture stage to the display.

The model used in (1.14) can cause wide variability in the value of gamma mainly due to the black level errors since it forces the zero voltage to map to zero intensity for any value of gamma. A slightly different model can be used in order to resolve the black level error. The modified model is given as:

$$I_{int} = (voltage + \epsilon)^{2.5} \tag{1.15}$$

By fixing the exponent of the power function at 2.5 and using the single parameter to accommodate black level errors the modified model fits the observed nonlinearity much better than the variable gamma model in (1.14).

The voltage-to-intensity function defined in (1.15) is nearly the inverse of the luminance-to-brightness relationship of human vision. Human vision defines luminance as a weighted mixture of the spectral energy where the weights are determined by the characteristics of the human retina. The CIE has standardized a weighting function which relates spectral power to luminance. In this standardized function, the perceived luminance by humans relates to the physical luminance (proportional to intensity) by the following equation:

$$L^* = \begin{cases} 116(\frac{Y}{Y_n})^{\frac{1}{3}} - 16 & \text{if } \frac{Y}{Y_n} > 0.008856 \\ 903.3(\frac{Y}{Y_n})^{\frac{1}{3}} & \text{if } \frac{Y}{Y_n} \leq 0.008856 \end{cases} \tag{1.16}$$

where Y_n is the luminance of the reference white, usually normalized either to 1.0 or 100. Thus, the lightness perceived by humans is, approximately, the cubic root of the luminance. The lightness sensation can be computed as intensity raised, approximately to the third power. Thus, the entire image processing system can be considered linear or almost linear.

To compensate for the nonlinearity of the display (CRT), gamma correction with a power of $(\frac{1}{\gamma})$ can be used so that the overall system γ is approximately 1.

In a video system, the gamma correction is applied to the camera for precomputing the nonlinearity of the display. The gamma correction performs the following transfer function:

$$voltage' = (voltage)^{\frac{1}{\gamma}} \tag{1.17}$$

where *voltage* is the voltage generated by the camera sensors. The gamma corrected value is the reciprocal of the gamma resulting in a transfer function with unit power exponent.

To achieve subjectively pleasing images, the end-to-end power function of the overall imaging system should be around 1.1 or 1.2 instead of the mathematically correct linear system.

The REC 709 specifies a power exponent of 0.45 at the camera which, in conjunction with the 2.5 exponent at the display, results in an overall exponent value of about 1.13. If the γ value is greater than 1, the image appears sharper but the scene contrast range, which can be reproduced, is reduced. On the other hand, reducing the γ value has a tendency to make the image appear soft and washed out.

For color images, the linear values R, G, and B values should be converted into nonlinear voltages R', G' and B' through the application of the gamma correction process. The color CRT will then convert R', G' and B' into linear red, green and blue light to reproduce the original color.

The ITU-R BT.709 standard recommends a gamma exponent value of 0.45 for the High Definition Television. In practical systems, such as TV cameras, certain modifications are required to ensure proper operation near the dark regions of an image, where the slope of a pure power function is infinite at zero. The red tristimulus (linear light) component may be gamma-corrected at the camera by applying the following convention:

$$R'_{709} = \begin{cases} 4.5R & \text{if } R \leq 0.018 \\ 1.099R^{0.45} - 0.099 & \text{if } 0.018 < R \end{cases} \tag{1.18}$$

with R denoting the linear light and R'_{709} the resulting gamma corrected value. The computations are identical for the G and B components.

The linear R, G, and B are normally in the range $[0, 1]$ when color images are used in digital form. The software library translates these floating point values to 8-bit integers in the range of 0 to 255 for use by the graphics hardware. Thus, the gamma corrected value should be:

$$R' = 255R^{\frac{1}{\gamma}} \tag{1.19}$$

The constant 255 in (1.19) is added during the A/D process. However, gamma correction is usually performed in cameras, and thus, pixel values are in most cases nonlinear voltages. Thus, intensity values stored in the framebuffer of the computing device are gamma corrected on-the-fly by hardware look up tables on their way to the computer monitor display. Modern image processing systems utilize a wide variety of sources of color images, such as images captured by digital cameras, scanned images, digitized video frames and computer generated images. Digitized video frames usually have a gamma correction value between 0.5 and 0.45. Digital scanners assume an output gamma in the range of 1.4 to 2.2 and they perform their gamma correction accordingly. For computer generated images the gamma correction value is usually unknown. In the absence of the actual gamma value the recommended gamma correction is 0.45.

In summary, pixel values alone cannot specify the actual color. The gamma correction value used for capturing or generating the color image is needed. Thus, two images which have been captured with two cameras operating under different gamma correction values will represent colors differently even if the same primaries and the same white reference point are used.

1.5 Linear and Non-linear RGB Color Spaces

The image processing literature rarely discriminates between linear RGB and non-linear (R′G′B′) gamma corrected values. For example, in the JPEG and MPEG standards and in image filtering, non-linear RGB (R′G′B′) color values are implicit. Unacceptable results are obtained when JPEG or MPEG schemes are applied to linear RGB image data [4]. On the other hand, in computer graphics, linear RGB values are implicitly used [4]. Therefore, it is very important to understand the difference between linear and non-linear RGB values and be aware of which values are used in an image processing application. Hereafter, the notation R′G′B′ will be used for non-linear RGB values so that they can be clearly distinguished from the linear RGB values.

1.5.1 Linear RGB Color Space

As mentioned earlier, intensity is a measure, over some interval of the electromagnetic spectrum, of the flow of power that is radiated from an object. Intensity is often called a linear light measure. The linear R value is proportional to the intensity of the physical power that is radiated from an object around the 700 nm band of the visible spectrum. Similarly, a linear G value corresponds to the 546.1 nm band and a linear B value corresponds to the 435.8 nm band. As a result the linear RGB space is device independent and used in some color management systems to achieve color consistency across diverse devices.

The linear RGB values in the range [0, 1] can be converted to the corresponding CIE XYZ values in the range [0, 1] using the following matrix transformation [4]:

$$\begin{bmatrix} X \\ Y \\ Z \end{bmatrix} = \begin{bmatrix} 0.4125 & 0.3576 & 0.1804 \\ 0.2127 & 0.7152 & 0.0722 \\ 0.0193 & 0.1192 & 0.9502 \end{bmatrix} \begin{bmatrix} R \\ G \\ B \end{bmatrix} \tag{1.20}$$

The transformation from CIE XYZ values in the range [0, 1] to RGB values in the range [0, 1] is defined by:

$$\begin{bmatrix} R \\ G \\ B \end{bmatrix} = \begin{bmatrix} 3.2405 & -1.5372 & -0.4985 \\ -0.9693 & 1.8760 & 0.0416 \\ 0.0556 & -0.2040 & 1.0573 \end{bmatrix} \begin{bmatrix} X \\ Y \\ Z \end{bmatrix} \tag{1.21}$$

Alternatively, tristimulus XYZ values can be obtained from the linear RGB values through the following matrix [5]:

$$\begin{bmatrix} X \\ Y \\ Z \end{bmatrix} = \begin{bmatrix} 0.490 & 0.310 & 0.200 \\ 0.117 & 0.812 & 0.011 \\ 0.000 & 0.010 & 0.990 \end{bmatrix} \begin{bmatrix} R \\ G \\ B \end{bmatrix} \tag{1.22}$$

The linear RGB values are a physical representation of the chromatic light radiated from an object. However, the perceptual response of the human visual system to radiate red, green, and blue intensities is non-linear and more complex. The linear RGB space is, perceptually, highly non-uniform and not suitable for numerical analysis of the perceptual attributes. Thus, the linear RGB values are very rarely used to represent an image. On the contrary, non-linear R′G′B′ values are traditionally used in image processing applications such as filtering.

1.5.2 Non-linear RGB Color Space

When an image acquisition system, e.g. a video camera, is used to capture the image of an object, the camera is exposed to the linear light radiated from the object. The linear RGB intensities incident on the camera are transformed to non-linear RGB signals using gamma correction. The transformation to non-linear R′G′B′ values in the range [0, 1] from linear RGB values in the range [0, 1] is defined by:

$$\begin{aligned} R' &= \begin{cases} 4.5R, & \text{if } R \leq 0.018 \\ 1.099R^{\frac{1}{\gamma_C}} - 0.099, & \text{otherwise} \end{cases} \\ G' &= \begin{cases} 4.5G, & \text{if } G \leq 0.018 \\ 1.099G^{\frac{1}{\gamma_C}} - 0.099, & \text{otherwise} \end{cases} \\ B' &= \begin{cases} 4.5B, & \text{if } B \leq 0.018 \\ 1.099B^{\frac{1}{\gamma_C}} - 0.099, & \text{otherwise} \end{cases} \end{aligned} \tag{1.23}$$

where γ_C is known as the *gamma factor* of the camera or the acquisition device. The value of γ_C that is commonly used in video cameras is $\frac{1}{0.45}$ ($\simeq 2.22$) [4]. The above transformation is graphically depicted in Figure 1.7. The linear segment near low intensities minimizes the effect of sensor noise in practical cameras and scanners.

Thus, the digital values of the image pixels acquired from the object and stored within a camera or a scanner are the R′G′B′ values usually converted to the range of 0 to 255. Three bytes are then required to represent the three components, R', G', and B' of a color image pixel with one byte for each component. It is these non-linear R′G′B′ values that are stored as image data files in computers and are used in image processing applications. The RGB symbol used in image processing literature usually refers to the R′G′B′

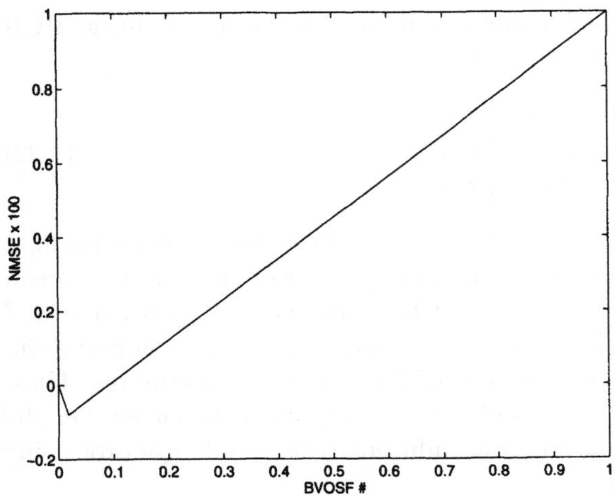

Fig. 1.7. Linear to Non-linear Light Transformation

values and, therefore, care must be taken in color space conversions and other relevant calculations.

Suppose the acquired image of an object needs to be displayed in a display device such as a computer monitor. Ideally, a user would like to see (perceive) the exact reproduction of the object. As pointed out, the image data is in R′G′B′ values. Signals (usually voltage) proportional to the R′G′B′ values will be applied to the red, green, and blue guns of the CRT (Cathode Ray Tube) respectively. The intensity of the red, green, and blue lights generated by the CRT is a non-linear function of the applied signal. The non-linearity of the CRT is a function of the electrostatics of the cathode and the grid of the electron gun. In order to achieve correct reproduction of intensities, an ideal monitor should invert the transformation at the acquisition device (camera) so that the intensities generated are identical to the linear RGB intensities that were radiated from the object and incident in the acquisition device. Only then will the perception of the displayed image be identical to the perceived object.

A conventional CRT has a power-law response, as depicted in Figure 1.8. This power-law response, which inverts the non-linear (R′G′B′) values in the range [0, 1] back to linear RGB values in the range [0, 1], is defined by the following power function [4]:

$$R = \begin{cases} \frac{R'}{4.5}, & \text{if } R' \leq 0.018 \\ \left(\frac{R' + 0.099}{1.099}\right)^{\gamma_D} & \text{otherwise} \end{cases}$$

$$G = \begin{cases} \frac{G'}{4.5}, & \text{if } G' \leq 0.018 \\ \left(\frac{G' + 0.099}{1.099}\right)^{\gamma_D} & \text{otherwise} \end{cases} \qquad (1.24)$$

$$B = \begin{cases} \frac{B'}{4.5}, & \text{if } B' \leq 0.018 \\ \left(\frac{B' + 0.099}{1.099}\right)^{\gamma_D} & \text{otherwise} \end{cases}$$

The value of the power function, γ_D, is known as the *gamma* factor of the display device or CRT. Normal display devices have γ_D in the range of 2.2 to 2.45. For exact reproduction of the intensities, *gamma* factor of the display device must be equal to the *gamma* factor of the acquisition device ($\gamma_C = \gamma_D$). Therefore, a CRT with a *gamma* factor of 2.2 should correctly reproduce the intensities.

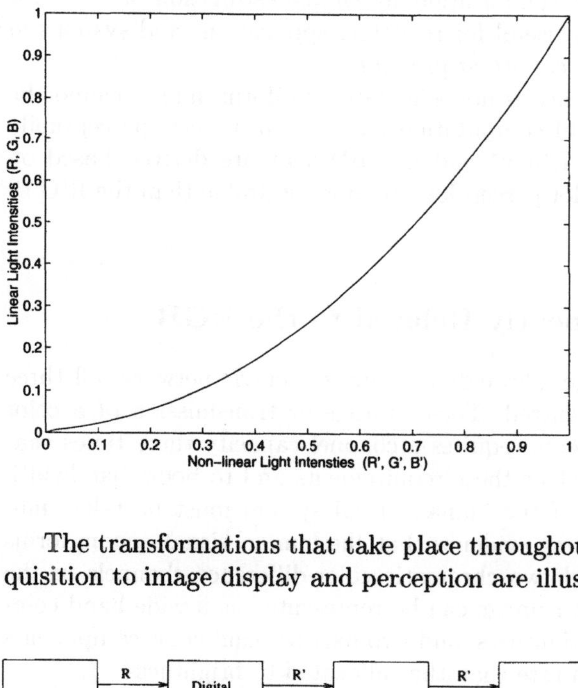

Fig. 1.8. Non-linear to linear Light Transformation

The transformations that take place throughout the process of image acquisition to image display and perception are illustrated in Figure 1.9.

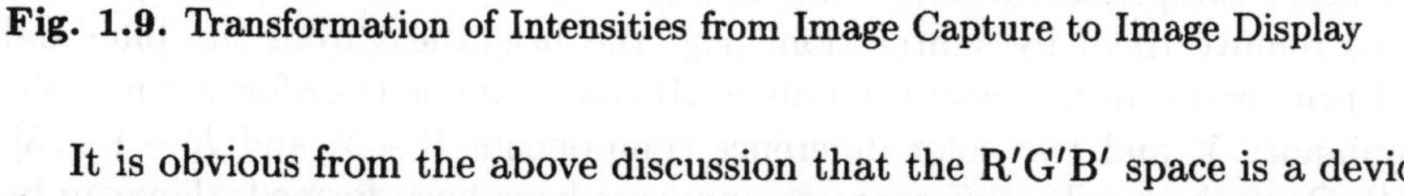

Fig. 1.9. Transformation of Intensities from Image Capture to Image Display

It is obvious from the above discussion that the R′G′B′ space is a device dependent space. Suppose a color image, represented in the R′G′B′ space, is displayed on two computer monitors having different *gamma* factors. The red, green, and blue intensities produced by the monitors will not be identical and the displayed images might have different appearances. Device dependent spaces cannot be used if color consistency across various devices, such as display devices, printers, etc., is of primary concern. However, similar devices (e.g. two computer monitors) usually have similar *gamma* factors and in such cases device dependency might not be an important issue.

As mentioned before, the human visual system has a non-linear perceptual response to intensity, which is roughly logarithmic and is, approximately, the inverse of a conventional CRT's non-linearity [4]. In other words, the perceived red, green, and blue intensities are approximately related to the R′G′B′ values. Due to this fact, computations involving R′G′B′ values have an approximate relation to the human color perception and the R′G′B′ space is less perceptually non-uniform relative to the CIE XYZ and linear RGB spaces [4]. Hence, distance measures defined between the R′G′B′ values of two color vectors provide a computationally simple estimation of the error between them. This is very useful for real-time applications and systems in which computational resources are at premium.

However, the R′G′B′ space is not adequately uniform, and it cannot be used for accurate perceptual computations. In such instances, perceptually uniform color spaces (e.g. L*u*v* and L*a*b*) that are derived based on the attributes of human color perception are more desirable than the R′G′B′ space [4].

1.6 Color Spaces Linearly Related to the RGB

In transmitting color images through a computer-centric network, all three primaries should be transmitted. Thus, storage or transmission of a color image using RGB components requires a channel capacity three times that of gray scale images. To reduce these requirements and to boost bandwidth utilization, the properties of the human visual system must be taken into consideration. There is strong evidence that the human visual system forms an achromatic channel and two chromatic color-difference channels in the retina. Consequently, a color image can be represented as a wide band component corresponding to brightness, and two narrow band color components with considerably less data rate than that allocated to brightness.

Since the large percentage (around 60%) of brightness is attributed to the green primary, then it is advantageous to base the color components on the other two primaries. The simplest way to form the two color components is to remove them by subtraction, (e.g. the brightness from the blue and red primaries). In this way the unit RGB color cube is transformed into the luminance Y and two color difference components $B - Y$ and $R - Y$ [33], [34]. Once these color difference components have been formed, they can be sub-sampled to reduce the bandwidth or data capacity without any visible degradation in performance. The color difference components are calculated from nonlinear gamma corrected values R',G',B' rather than the tristimulus (linear voltage) R, G, B primary components.

According to the CIE standards the color imaging system should operate similarly to a gray scale system, with a CIE luminance component Y formed as a weighted sum of RGB tristimulus values. The coefficients in the weighted sum correspond to the sensitivity of the human visual system to each of the RGB primaries. The coefficients are also a function of the chromaticity of the white reference point used. International agreement on the REC. 709 standard provides a value for the luminance component based on the REC. 709 primaries [24]. Thus, the Y'_{709} luminance equation is:

$$Y'_{709} = 0.2125R'_{709} + 0.7154G'_{709} + 0.0721B'_{709} \tag{1.25}$$

where R'_{709}, B'_{709} and G'_{709} are the gamma-corrected (nonlinear) values of the three primaries. The two color difference components $B'_{709} - Y'_{709}$ and $R'_{709} - Y'_{709}$ can be formed on the basis of the above equation.

Various scale factors are applied to the basic color difference components for different applications. For example, the $Y'P_RP_B$ is used for component analog video, such as BetaCam, and $Y'C_BC_R$ for component digital video, such as studio video, JPEG and MPEG. Kodak's YCC (PhotoCD model) uses scale factors optimized for the gamut of film colors [31]. All these systems utilize different versions of the $(Y'_{709}, B'_{709}-Y'_{709}, R'_{709}-Y'_{709})$ which are scaled to place the extrema of the component signals at more convenient values.

In particular, the $Y'P_RP_B$ system used in component analog equipment is defined by the following set:

$$\begin{bmatrix} Y'_{601} \\ P_B \\ P_R \end{bmatrix} = \begin{bmatrix} 0.299 & 0.587 & 0.114 \\ -0.168736 & -0.331264 & 0.5 \\ 0.5 & -0.418686 & -0.081312 \end{bmatrix} \begin{bmatrix} R' \\ G' \\ B' \end{bmatrix} \tag{1.26}$$

and

$$\begin{bmatrix} R' \\ G' \\ B' \end{bmatrix} = \begin{bmatrix} 1. & 0. & 1.402 \\ 1. & -0.344136 & -0.714136 \\ 1. & 1.772 & 0. \end{bmatrix} \begin{bmatrix} Y'_{601} \\ P_B \\ P_R \end{bmatrix} \tag{1.27}$$

The first row comprises the luminance coefficients which sum to unity. For each of the other two columns the coefficients sum to zero, a necessity for color difference formulas. The 0.5 weights reflect the maximum excursion of P_B and P_R for the blue and the red primaries.

The Y'C_BC_R is the Rec ITU-R BT. 601-4 international standard for studio quality component digital video. The luminance signal is coded in 8 bits. The Y' has an excursion of 219 with an offset of 16, with the black point coded at 16 and the white at code 235. Color differences are also coded in 8-bit forms with excursions of 112 and offset of 128 for a range of 16 through 240 inclusive.

To compute Y'C_BC_R from nonlinear R'G'B' in the range of $[0, 1]$ the following set should be used:

$$\begin{bmatrix} Y'_{601} \\ C_B \\ C_R \end{bmatrix} = \begin{bmatrix} 16 \\ 128 \\ 128 \end{bmatrix} + \begin{bmatrix} 65.481 & 128.553 & 24.966 \\ -37.797 & -74.203 & 112.0 \\ 112.0 & -93.786 & -18.214 \end{bmatrix} \begin{bmatrix} R' \\ G' \\ B' \end{bmatrix} \tag{1.28}$$

with the inverse transform

$$\begin{bmatrix} R' \\ G' \\ B' \end{bmatrix} = \begin{bmatrix} 0.00456821 & 0.0 & 0.00625893 \\ 0.00456621 & -0.00153632 & -0.00318811 \\ 0.00456621 & 0.00791071 & 0.0 \end{bmatrix} \cdot \left(\begin{bmatrix} Y'_{601} \\ P_B \\ P_R \end{bmatrix} - \begin{bmatrix} 16 \\ 128 \\ 128 \end{bmatrix} \right) \tag{1.29}$$

When 8-bit R′G′B′ are used, black is coded at 0 and white is at 255. To encode $Y'C_BC_R$ from R′G′B′ in the range of $[0, 255]$ using 8-bit binary arithmetic the transformation matrix should be scaled by $\frac{256}{255}$. The resulting transformation pair is as follows:

$$\begin{bmatrix} Y'_{601} \\ P_B \\ P_R \end{bmatrix} = \begin{bmatrix} 16 \\ 128 \\ 128 \end{bmatrix} + \frac{1}{256} \begin{bmatrix} 65.481 & 128.553 & 24.966 \\ -37.797 & -74.203 & 112.0 \\ 112.0 & -93.786 & -18.214 \end{bmatrix} \begin{bmatrix} R'_{255} \\ G'_{255} \\ B'_{255} \end{bmatrix} \tag{1.30}$$

where R'_{255} is the gamma-corrected value, using a gamma-correction lookup table for $\frac{1}{\gamma}$. This yields the RGB intensity values with integer components between 0 and 255 which are gamma-corrected by the hardware. To obtain R′G′B′ values in the range $[0, 255]$ from $Y'C_BC_R$ using 8-bit arithmetic the following transformation should be used:

$$\begin{bmatrix} R' \\ G' \\ B' \end{bmatrix} = \frac{1}{256} \begin{bmatrix} 0.00456821 & 0.0 & 0.00625893 \\ 0.00456621 & -0.00153632 & -0.00318811 \\ 0.00456621 & 0.00791071 & 0.0 \end{bmatrix} \left(\begin{bmatrix} Y'_{601} \\ P_B \\ P_R \end{bmatrix} - \begin{bmatrix} 16 \\ 128 \\ 128 \end{bmatrix} \right) \tag{1.31}$$

Some of the coefficients when scaled by $\frac{1}{256}$ may be larger than unity and, thus some clipping may be required so that they fall within the acceptable RGB range.

The Kodak YCC color space is another example of a predistorted color space, which has been designed for the storage of still color images on the Photo-CD. It is derived from the predistorted (gamma-corrected) R′G′B′ values using the ITU-R BT.709 recommended white reference point, primaries, and gamma correction values. The YCC space is similar to the $Y'C_BC_R$ discussed, although scaling of $B' - Y'$ and $R' - Y'$ is asymmetrical in order to accommodate a wide color gamut, similar to that of a photographic film. In particular the following relationship holds for Photo-CD compressed formats:

$$Y' = \frac{255}{1.402} Y'_{601} \tag{1.32}$$

$$C_1 = 156 + 111.40(B' - Y') \tag{1.33}$$

$$C_2 = 137 + 135.64(R' - Y') \tag{1.34}$$

The two chrominance components are compressed by factors of 2 both horizontally and vertically. To reproduce predistorted R′G′B′ values in the range of [0, 1] from integer PhotoYCC components the following transform is applied:

$$\begin{bmatrix} R' \\ G' \\ B' \end{bmatrix} = \frac{1}{256} \begin{bmatrix} 0.00549804 & 0.0 & 0.0051681 \\ 0.00549804 & -0.0015446 & -0.0026325 \\ 0.00549804 & 0.0079533 & 0.0 \end{bmatrix} \left(\begin{bmatrix} Y' \\ C_1 \\ C_2 \end{bmatrix} - \begin{bmatrix} 0 \\ 156 \\ 137 \end{bmatrix} \right) \tag{1.35}$$

The $B' - Y'$ and $R' - Y'$ components can be converted into polar coordinates to represent the perceptual attributes of hue and saturation. The values can be computed using the following formulas [34]:

$$H = tan^{-1}(\frac{B' - Y'}{R' - Y'}) \tag{1.36}$$

$$S = ((B' - Y')^2 + (R' - Y')^2)\frac{1}{2} \tag{1.37}$$

where the saturation S is the length of the vector from the origin of the chromatic plane to the specific color and the hue H is the angle between the $R' - Y'$ axis and the saturation component [33].

1.7 The YIQ Color Space

The YIQ color specification system, used in commercial color TV broadcasting and video systems, is based upon the color television standard that was adopted in the 1950s by the National Television Standard committee (NTSC) [10], [1], [27], [28]. Basically, YIQ is a recoding of non-linear R′G′B′ for transmission efficiency and for maintaining compatibility with monochrome TV standards [1], [4]. In fact, the Y component of the YIQ system provides all the video information required by a monochrome television system.

The YIQ model was designed to take advantage of the human visual system's greater sensitivity to change in luminance than to changes in hue or saturation [1]. Due to these characteristics of the human visual system, it is useful in a video system to specify a color with a component representative of luminance Y and two other components: the in-phase I, an orange-cyan axis, and the quadrature Q component, the magenta-green axis. The two chrominance components are used to jointly represent hue and saturation .

With this model, it is possible to convey the component representative of luminance Y in such a way that noise (or quantization) introduced in transmission, processing and storage is minimal and has a perceptually similar effect across the entire tone scale from black to white [4]. This is done by allowing more bandwidth (bits) to code the luminance (Y) and less bandwidth (bits) to code the chrominance (I and Q) for efficient transmission and storage purposes without introducing large perceptual errors due to quantization [1]. Another implication is that the luminance (Y) component of an image can be processed without affecting its chrominance (color content). For instance, histogram equalization to a color image represented in YIQ format can be done simply by applying histogram equalization to its Y component [1]. The relative colors in the image are not affected by this process.

The ideal way to accomplish these goals would be to form a luminance component (Y) by applying a matrix transform to the linear RGB components and then subjecting the luminance (Y) to a non-linear transfer function to achieve a component similar to lightness L^*. However, there are practical reasons in a video system why these operations are performed in the opposite order [4]. First, gamma correction is applied to each of the linear RGB. Then, a weighted sum of the nonlinear components is computed to form a component representative of luminance Y. The resulting component (luma) is related to luminance but is not the same as the CIE luminance Y although the same symbol is used for both of them.

The nonlinear RGB to YIQ conversion is defined by the following matrix transformation [4], [1]:

$$\begin{bmatrix} Y \\ I \\ Q \end{bmatrix} = \begin{bmatrix} 0.299 & 0.587 & 0.114 \\ 0.596 & -0.275 & -0.321 \\ 0.212 & -0.523 & 0.311 \end{bmatrix} \begin{bmatrix} R' \\ G' \\ B' \end{bmatrix} \tag{1.38}$$

As can be seen from the above transformation, the blue component has a small contribution to the brightness sensation (luma Y) despite the fact that human vision has extraordinarily good color discrimination capability in the blue color [4]. The inverse matrix transformation is performed to convert YIQ to nonlinear R′G′B′.

Introducing a cylindrical coordinate transformation, numerical values for hue and saturation can be calculated as follows:

$$H_{YIQ} = tan^{-1}(\frac{Q}{I}) \tag{1.39}$$

$$S_{YIQ} = (I^2 + Q^2)^{\frac{1}{2}} \tag{1.40}$$

As described it, the YIQ model is developed from a perceptual point of view and provides several advantages in image coding and communications applications by decoupling the luma (Y) and chrominance components (I and Q). Nevertheless, YIQ is a perceptually non-uniform color space and thus not appropriate for perceptual color difference quantification. For example, the Euclidean distance is not capable of accurately measuring the perceptual color distance in the perceptually non-uniform YIQ color space. Therefore, YIQ is not the best color space for quantitative computations involving human color perception.

1.8 The HSI Family of Color Models

In image processing systems, it is often convenient to specify colors in a way that is compatible with the hardware used. The different variants of the RGB monitor model address that need. Although these systems are computationally practical, they are not useful for user specification and recognition of colors. The user cannot easily specify a desired color in the RGB model. On the other hand, perceptual features, such as perceived luminance (intensity), saturation and hue correlate well with the human perception of color . Therefore, a color model in which these color attributes form the basis of the space is preferable from the users point of view. Models based on lightness, hue and saturation are considered to be better suited for human interaction. The analysis of the user-oriented color spaces starts by introducing the family of intensity, hue and saturation (HSI) models [28], [29]. This family of models is used primarily in computer graphics to specify colors using the artistic notion of tints, shades and tones. However, all the HSI models are derived from the RGB color space by coordinate transformations. In a computer centered image processing system, it is necessary to transform the color coordinates to RGB for display and vice versa for color manipulation within the selected space.

The HSI family of color models use approximately cylindrical coordinates. The saturation (S) is proportional to radial distance, and the hue (H) is a function of the angle in the polar coordinate system. The intensity (I) or lightness (L) is the distance along the axis perpendicular to the polar coordinate plane. The dominant factor in selecting a particular HSI model is the definition of the lightness, which determines the constant-lightness surfaces, and thus, the shape of the color solid that represents the model. In the cylindrical models, the set of color pixels in the RGB cube which are assigned a common lightness value (L) form a constant-lightness surface. Any line parallel to the main diagonal of the color RGB cube meets the constant-lightness surface at most in one point.

The HSI color space was developed to specify, numerically, the values of hue, saturation, and intensity of a color [4]. The HSI color model is depicted in Figure 1.10. The hue (H) is measured by the angle around the vertical axis and has a range of values between 0 and 360 degrees beginning with red at 0°. It gives a measure of the spectral composition of a color. The saturation (S) is a ratio that ranges from 0 (i.e. on the I axis), extending radially outwards to a maximum value of 1 on the surface of the cone. This component refers to the proportion of pure light of the dominant wavelength and indicates how far a color is from a gray of equal brightness. The intensity (I) also ranges between 0 and 1 and is a measure of the relative brightness. At the top and bottom of the cone, where $I = 0$ and 1 respectively, H and S are undefined and meaningless. At any point along the I axis the Saturation component is zero and the hue is undefined. This singularity occurs whenever $R = G = B$.

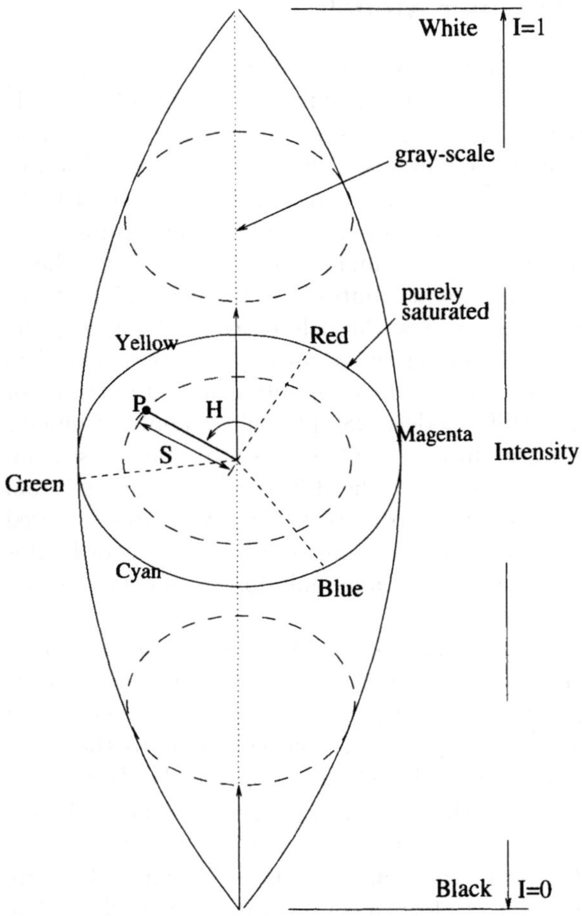

Fig. 1.10. The HSI Color Space

The HSI color model owes its usefulness to two principal facts [1], [28]. First, like in the YIQ model, the intensity component I is decoupled from the chrominance information represented as hue H and saturation S. Second, the hue (H) and saturation (S) components are intimately related to the way in which humans perceive chrominance [1]. Hence, these features make the HSI an ideal color model for image processing applications where the chrominance is of importance rather than the overall color perception (which is determined by both luminance and chrominance). One example of the usefulness of the

HSI model is in the design of imaging systems that automatically determine the ripeness of fruits and vegetables [1]. Another application is color image histogram equalization performed in the HSI space to avoid undesirable shifts in image hue [10].

The simplest way to choose constant-lightness surfaces is to define them as planes. A simplified definition of the perceived lightness in terms of the R,G,B values is $L = \frac{R'+G'+B'}{3}$, where the normalization is used to control the range of lightness values. The different constant-lightness surfaces are perpendicular to the main diagonal of the RGB cube and parallel to each other. The shape of a constant lightness surface is a triangle for $0 \leq L \leq \frac{M}{3}$ and $\frac{2M}{3} \leq L \leq M$ with $L \in [0, M]$ and where M is a given lightness threshold.

The theory underlying the derivation of conversion formulas between the RGB space and HSI space is described in detail in [1], [28]. The image processing literature on HSI does not clearly indicate whether the linear or the non-linear RGB is used in these conversions [4]. Thus the non-linear (R′G′B′), which is implicit in traditional image processing, shall be used. But this ambiguity must be noted.

The conversion from R′G′B′ (range [0, 1]) to HSI (range [0, 1]) is highly nonlinear and considerably complicated:

$$H = \cos^{-1}\left[\frac{\frac{1}{2}[(R' - G') + (R' - B')]}{[(R' - G')^2 + (R' - B')(G' - B')]^{\frac{1}{2}}}\right] \tag{1.41}$$

$$S = 1 - \frac{3}{(R' + G' + B')}[\min(R', G', B')] \tag{1.42}$$

$$I = \frac{1}{3}(R' + G' + B') \tag{1.43}$$

where $H = 360° - H$, if $(B'/I) > (G'/I)$. Hue is normalized to the range [0, 1] by letting $H = H/360°$. Hue (H) is not defined when the saturation (S) is zero. Similarly, saturation (S) is undefined if intensity (I) is zero.

To transform the HSI values (range [0, 1]) back to the R′G′B′ values (range [0, 1]), then the H values in [0, 1] range must first be converted back to the un-normalized $[0^o, 360^o]$ range by letting $H = 360°(H)$. For the $R'G'$ (red and green) sector $(0° < H \leq 120°)$, the conversion is:

$$B' = I\,(1 - S) \tag{1.44}$$

$$R' = I\left[1 + \frac{S\cos H}{\cos(60° - H)}\right] \tag{1.45}$$

$$G' = 3I - (R' + B') \tag{1.46}$$

The conversion for the $G'B'$ (green and blue) sector $(120° < H \leq 240°)$ is given by:

$$H = H - 120° \tag{1.47}$$

$$R' = I\,(1 - S) \tag{1.48}$$

$$G' = I\left[1 + \frac{S\,\cos H}{\cos\,(60^\circ - H)}\right] \tag{1.49}$$

$$B' = 3I - (R' + G') \tag{1.50}$$

Finally, for the $B'R'$ (blue and red) sector ($240^\circ < H \leq 360^\circ$), the corresponding equations are:

$$H = H - 240^\circ \tag{1.51}$$

$$G' = I\,(1 - S) \tag{1.52}$$

$$B' = I\left[1 + \frac{S\,\cos H}{\cos\,(60^\circ - H)}\right] \tag{1.53}$$

$$R' = 3I - (G' + B') \tag{1.54}$$

Fast versions of the transformation, containing fewer multiplications and avoiding square roots, are often used in hue calculations. Also, formulas without trigonometric functions can be used. For example, hue can be evaluated using the following formula [44]:

1. If $B' = \min(R',\ G',\ B')$ then

$$H = \frac{G' - B'}{3(R' + G' - 2B'} \tag{1.55}$$

2. If $R' = \min(R',\ G',\ B')$ then

$$H = \frac{B' - R'}{R' + G' - 2B'} + \frac{1}{3} \tag{1.56}$$

3. If $G' = \min(R',\ G',\ B')$ then

$$H = \frac{B' - R'}{R' + G' - 2B'} + \frac{2}{3} \tag{1.57}$$

Although the HSI model is useful in some image processing applications, the formulation of it is flawed with respect to the properties of color vision. The usual formulation makes no clear reference to the linearity or nonlinearity of the underlying RGB and to the lightness perception of human vision [4]. It computes the brightness as $(R' + G' + B')\,/3$ and assigns the name intensity I. Recall that the brightness perception is related to luminance Y. Thus, this computation conflicts with the properties of color vision [4].

In addition to this, there is a discontinuity in the hue at 360° and thus, the formulation introduces visible discontinuities in the color space. Another major disadvantage of the HSI space is that it is not perceptually uniform.

Consequently, the HSI model is not very useful for perceptual image computation and for conveyance of accurate color information. As such, distance measures, such as the Euclidean distance, cannot estimate adequately the perceptual color distance in this space.

The model discussed above is not the only member of the family. In particular, the double hexcone HLS model can be defined by simply modifying the constant-lightness surface. It is depicted in Figure 1.11. In the HLS model the lightness is defined as:

$$L = \frac{\max(R', G', B') + \min(R', G', B')}{2} \tag{1.58}$$

If the maximum and the minimum value coincide then $S = 0$ and the hue is undefined. Otherwise based on the lightness value, saturation is defined as follows:

1. If $L \leq 0.5$ then $S = \frac{(Max-Min)}{(Max+Min)}$
2. If $L > 0.5$ then $S = \frac{(Max-Min)}{(2-Max-Min)}$

where $Max = \max(R', G', B')$ and $Min = \min(R', G', B')$ respectively.

Similarly, hue is calculated according to:

1. If $R' = Max$ then

$$H = \frac{G' - B'}{Max - Min} \tag{1.59}$$

2. If $G' = Max$ then

$$H = \frac{B' - R'}{Max - Min} \tag{1.60}$$

3. If $B' = Max$ then

$$H = 4 + \frac{R' - G'}{Max - Min} \tag{1.61}$$

The backward transform starts by rescaling the hue angles into the range $[0, 6]$. Then, the following cases are considered:

1. If $S = 0$, hue is undefined and $(R', G', B') = (L, L, L)$
2. Otherwise, $i = Floor(H)$ (the $Floor(X)$ function returns the largest integer which is not greater than X), in which i is the sector number of the hue and $f = H - i$ is the hue value in each sector. The following cases are considered:

- if $L \leq L_{critical} = \frac{255}{2}$ then

$$Max = L(1 + S) \tag{1.62}$$

$$Mid1 = L(2fS + 1 - S) \tag{1.63}$$

$$Mid2 = L(2(1-f)S + 1 - S) \tag{1.64}$$

$$Min = L(1-S) \tag{1.65}$$

- if $L > L_{critical} = \frac{255}{2}$ then

$$Max = L(1-S) + 255S \tag{1.66}$$

$$Mid1 = 2((1-f)S - (0.5-f)Max) \tag{1.67}$$

$$Mid2 = 2(fL - (f-0.5)Max) \tag{1.68}$$

$$Min = L(1+S) - 255S \tag{1.69}$$

Based on these intermediate values the following assignments should be made:

1. if $i = 0$ then $(R', G', B') = (Max, Mid1, Min)$
2. if $i = 1$ then $(R', G', B') = (Mid2, Max, Min)$
3. if $i = 2$ then $(R', G', B') = (Min, Max, Mid1)$
4. if $i = 3$ then $(R', G', B') = (Min, Mid2, Max)$
5. if $i = 4$ then $(R', G', B') = (Mid1, Min, Max)$
6. if $i = 5$ then $(R', G', B') = (Max, Min, Mid2)$

The HSV (hue, saturation, value) color model also belongs to this group of hue-oriented color coordinate systems which correspond more closely to the human perception of color. This user-oriented color space is based on the intuitive appeal of the artist's tint, shade, and tone. The HSV coordinate system, proposed originally in Smith [36], is cylindrical and is conveniently represented by the hexcone model shown in Figure 1.12 [23], [27]. The set of equations below can be used to transform a point in the RGB coordinate system to the appropriate value in the HSV space.

$$H_1 = \cos^{-1}\left\{\frac{\frac{1}{2}[(R-G)+(R-B)]}{\sqrt{(R-G)^2+(R-B)(G-B)}}\right\} \tag{1.70}$$

$$H = H_1 \qquad , \text{ if } \quad B \leq G \tag{1.71}$$

$$H = 360^\circ - H_1 \qquad , \text{ if } \quad B > G \tag{1.72}$$

$$S = \frac{\max(R,G,B) - \min(R,G,B)}{\max(R,G,B)} \tag{1.73}$$

$$V = \frac{\max(R,G,B)}{255} \tag{1.74}$$

Here the RGB values are between 0 and 255. A fast algorithm used here to convert the set of RGB values to the HSV color space is provided in [23].

The important advantages of the HSI family of color spaces over other color spaces are:

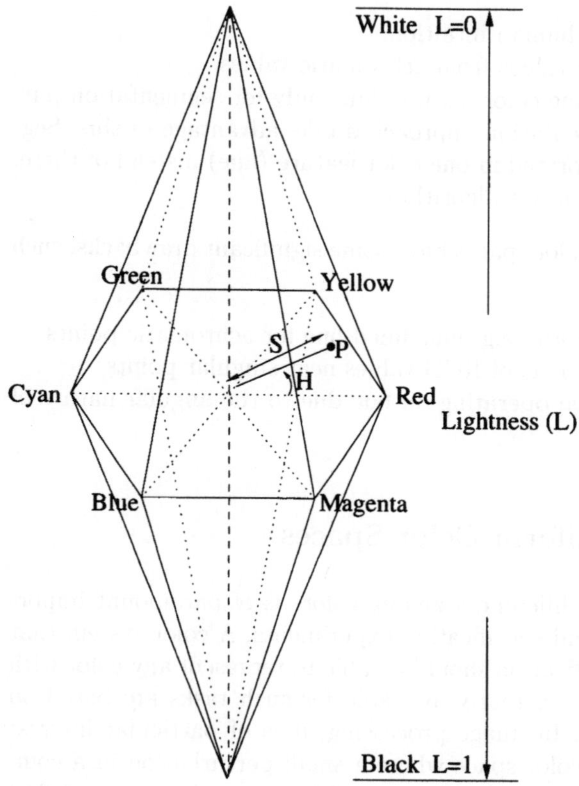

Fig. 1.11. The HLS Color Space

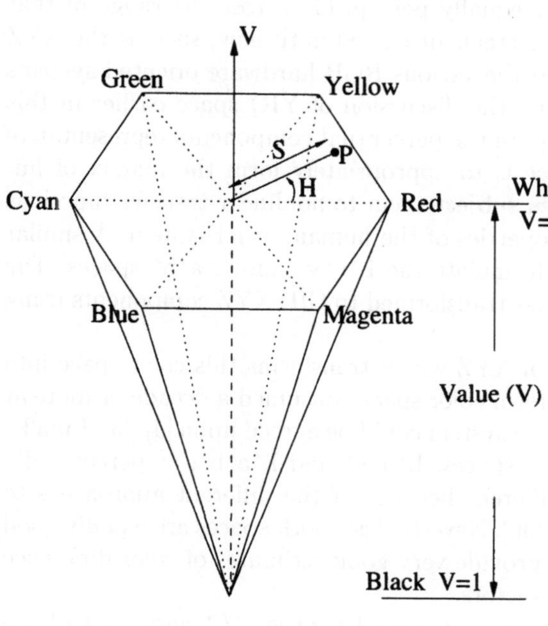

Fig. 1.12. The HSV Color Space

- Good compatibility with human intuition.
- Separability of chromatic values from achromatic values.
- The possibility of using one color feature, hue, only for segmentation purposes. Many image segmentation approaches take advantage of this. Segmentation is usually performed in one color feature (hue) instead of three, allowing the use of much faster algorithms.

However, hue-oriented color spaces have some significant drawbacks, such as:

- singularities in the transform, e.g. undefined hue for achromatic points
- sensitivity to small deviations of RGB values near singular points
- numerical instability when operating on hue due to the angular nature of the feature.

1.9 Perceptually Uniform Color Spaces

Visual sensitivity to small differences among colors is of paramount importance in color perception and specification experiments. A color system that is to be used for color specification should be able to represent any color with high precision. All systems currently available for such tasks are based on the CIE XYZ color model. In image processing, it is of particular interest in a perceptually uniform color space where a small perturbation in a component value is approximately equally perceptible across the range of that value. The color specification systems discussed until now, such as the XYZ or RGB tristimulus values and the various RGB hardware oriented systems are far from uniform. Recalling the discussion of YIQ space earlier in this chapter, the ideal way to compute the perceptual components representative of luminance and chrominance is to appropriately form the matrix of linear RGB components and then subject them to nonlinear transfer functions based on the color sensing properties of the human visual system. A similar procedure is used by CIE to formulate the L*u*v* and L*a*b* spaces. The linear RGB components are first transformed to CIE XYZ components using the appropriate matrix.

Finding a transformation of XYZ which transforms this color space into a reasonably perceptually uniform color space consumed a decade or more at the CIE and in the end, no single system could be agreed upon [4], [5]. Finally, in 1976, CIE standardized two spaces, L*u*v* and L*a*b*, as perceptually uniform. They are slightly different because of the different approaches to their formulation [4], [5], [25], [30]. Nevertheless, both spaces are equally good in perceptual uniformity and provide very good estimates of color difference (distance) between two color vectors.

Both systems are based on the perceived lightness L^* and a set of opponent color axes, approximately red-green versus yellow-blue. According to the CIE 1976 standard, the perceived lightness of a standard observer is assumed to follow the physical luminance (a quantity proportional to intensity) according to a cubic root law. Therefore, the lightness L^* is defined by the CIE as:

$$L^* = \begin{cases} 116(\frac{Y}{Y_n})^{\frac{1}{3}} - 16 \text{ if } \frac{Y}{Y_n} > 0.008856 \\ 903.3(\frac{Y}{Y_n})^{\frac{1}{3}} \quad \text{if } \frac{Y}{Y_n} \leq 0.008856 \end{cases} \tag{1.75}$$

where Y_n is the physical luminance of the white reference point . The range of values for L^* is from 0 to 100 representing a black and a reference white respectively. A difference of unity between two L^* values, the so-called ΔL^* is the threshold of discrimination.

This standard function relates perceived lightness to linear light luminance. Luminance can be computed as a weighted sum of red, green and blue components. If three sources appear red, green and blue and have the same power in the visible spectrum, the green will appear the brightest of the three because the luminous efficiency function peaks in the green region of the spectrum. Thus, the coefficients that correspond to contemporary CRT displays (ITU-R BT. 709 recommendation) [24] reflect that fact, when using the following equation for the calculation of the luminance:

$$Y_{709} = 0.2125R + 0.7154G + 0.0721B \tag{1.76}$$

The u^* and v^* components in L*u*v* space and the the a^* and b^* components in L*a*b* space are representative of chrominance. In addition, both are device independent color spaces. Both these color spaces are, however, computationally intensive to transform to and from the linear as well as nonlinear RGB spaces. This is a disadvantage if real-time processing is required or if computational resources are at a premium.

1.9.1 The CIE L*u*v* Color Space

The first uniform color space standardized by CIE is the L*u*v* illustrated in Figure 1.13. It is derived based on the CIE XYZ space and white reference point [4], [5]. The white reference point $[X_n, Y_n, Z_n]$ is the linear $RGB = [1, 1, 1]$ values converted to the XYZ values using the following transformation:

$$\begin{bmatrix} X_n \\ Y_n \\ Z_n \end{bmatrix} = \begin{bmatrix} 0.4125 & 0.3576 & 0.1804 \\ 0.2127 & 0.7152 & 0.0722 \\ 0.0193 & 0.1192 & 0.9502 \end{bmatrix} \begin{bmatrix} 1 \\ 1 \\ 1 \end{bmatrix} \tag{1.77}$$

Alternatively, white reference points can be defined based on the Federal Communications Commission (FCC) or the European Broadcasting Union (EBU) RGB values using the following transformations respectively [35]:

$$\begin{bmatrix} X_n \\ Y_n \\ Z_n \end{bmatrix} = \begin{bmatrix} 0.607 & 0.174 & 0.200 \\ 0.299 & 0.587 & 0.114 \\ 0.000 & 0.066 & 1.116 \end{bmatrix} \begin{bmatrix} 1 \\ 1 \\ 1 \end{bmatrix} \tag{1.78}$$

$$\begin{bmatrix} X_n \\ Y_n \\ Z_n \end{bmatrix} = \begin{bmatrix} 0.430 & 0.342 & 0.178 \\ 0.222 & 0.702 & 0.071 \\ 0.020 & 0.130 & 0.939 \end{bmatrix} \begin{bmatrix} 1 \\ 1 \\ 1 \end{bmatrix} \tag{1.79}$$

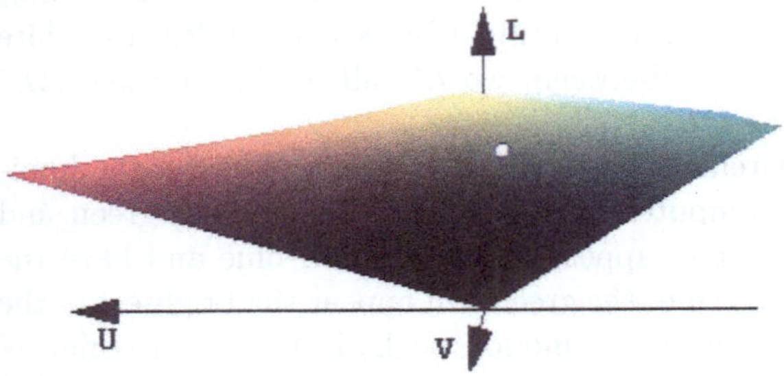

Fig. 1.13. The L*u*v* Color Space

The lightness component L^* is defined by the CIE as a modified cube root of luminance Y [4], [31], [37], [32]:

$$L^* = \begin{cases} 116\left(\frac{Y}{Y_n}\right)^{\frac{1}{3}} - 16 & \text{if } \frac{Y}{Y_n} > 0.008856 \\ 903.3\left(\frac{Y}{Y_n}\right) & \textit{Eotherwise} \end{cases} \tag{1.80}$$

The CIE definition of L^* applies a linear segment near black for $(Y/Y_n) \leq 0.008856$. This linear segment is unimportant for practical purposes [4]. L^* has a range [0, 100], and a L^* of unity is roughly the threshold of visibility [4].

Computation of u^* and v^* involves intermediate u', v', u'_n, and v'_n quantities defined as:

$$u' = \frac{4X}{X + 15Y + 3Z} \quad v' = \frac{9Y}{X + 15Y + 3Z} \tag{1.81}$$

$$u'_n = \frac{4X_n}{X_n + 15Y_n + 3Z_n} \quad v'_n = \frac{9Y_n}{X_n + 15Y_n + 3Z_n} \tag{1.82}$$

with the CIE XYZ values computed through (1.20) and (1.21).

Finally, u^* and v^* are computed as:

$$u^* = 13L^*(u' - u'_n) \tag{1.83}$$

$$v^* = 13L^*(v' - v'_n) \tag{1.84}$$

Conversion from L*u*v* to XYZ is accomplished by ignoring the linear segment of L^*. In particular, the linear segment can be ignored if the luminance variable Y is represented with eight bits of precision or less.

Then, the luminance Y is given by:

$$Y = \left(\frac{L^* + 16}{116}\right)^3 Y_n \quad (1.85)$$

To compute X and Z, first compute u' and v' as:

$$u' = \frac{u^*}{13L^*} + u'_n \quad v' = \frac{v^*}{13L^*} + v'_n \quad (1.86)$$

Finally, X and Z are given by:

$$X = \frac{1}{4}\left(\frac{u'(9.0 - 15.0\,v')\,Y}{v'} + 15.0\,u'\,Y\right) \quad (1.87)$$

$$Z = \frac{1}{3}\left(\frac{(9.0 - 15.0\,v')\,Y}{v'} - X\right) \quad (1.88)$$

Consider two color vectors $\mathbf{x}_{L^*u^*v^*}$ and $\mathbf{y}_{L^*u^*v^*}$ in the L*u*v* space represented as:

$$\mathbf{x}_{L^*u^*v^*} = [x_{L^*}, x_{u^*}, x_{v^*}]^T \text{ and } \mathbf{y}_{L^*u^*v^*} = [y_{L^*}, y_{u^*}, y_{v^*}]^T \quad (1.89)$$

The perceptual color distance in the L*u*v* space, called the total color difference ΔE^*_{uv} in [5], is defined as the Euclidean distance (L_2 norm) between the two color vectors $\mathbf{x}_{L^*u^*v^*}$ and $\mathbf{y}_{L^*u^*v^*}$:

$$\Delta E^*_{uv} = ||\mathbf{x}_{L^*u^*v^*} - \mathbf{y}_{L^*u^*v^*}||_{L_2}$$

$$\Delta E^*_{uv} = \left[(x_{L^*} - y_{L^*})^2 + (x_{u^*} - y_{u^*})^2 + (x_{v^*} - y_{v^*})^2\right]^{\frac{1}{2}} \quad (1.90)$$

It should be mentioned that in a perceptually uniform space, the Euclidean distance is an accurate measure of the perceptual color difference [5]. As such, the color difference formula ΔE^*_{uv} is widely used for the evaluation of color reproduction quality in an image processing system, such as color coding systems.

1.9.2 The CIE L*a*b* Color Space

The L*a*b* color space is the second uniform color space standardized by CIE. It is also derived based on the CIE XYZ space and white reference point [5], [37].

The lightness L^* component is the same as in the L*u*v* space. The L^*, a^* and b^* components are given by:

$$L^* = 116\left(\frac{Y}{Y_n}\right)^{\frac{1}{3}} - 16 \quad (1.91)$$

$$a^* = 500\left[\left(\frac{X}{X_n}\right)^{\frac{1}{3}} - \left(\frac{Y}{Y_n}\right)^{\frac{1}{3}}\right] \quad (1.92)$$

$$b^* = 200\left[\left(\frac{Y}{Y_n}\right)^{\frac{1}{3}} - \left(\frac{Z}{Z_n}\right)^{\frac{1}{3}}\right] \tag{1.93}$$

with the constraint that $\frac{X}{X_n}, \frac{Y}{Y_n}, \frac{Z}{Z_n} > 0.01$. This constraint will be satisfied for most practical purposes [4]. Hence, the modified formulae described in [5] for cases that do not not satisfy this constraint can be ignored in practice [4], [10].

The back conversion to the XYZ space from the L*a*b* space is done by first computing the luminance Y, as described in the back conversion of L*u*v*, followed by the computation of X and Z:

$$Y = \left(\frac{L^* + 16}{116}\right)^3 Y_n \tag{1.94}$$

$$X = \left(\frac{a^*}{500} + \left(\frac{Y}{Y_n}\right)^{\frac{1}{3}}\right)^3 X_n \tag{1.95}$$

$$Z = \left(-\frac{b^*}{200} + \left(\frac{Y}{Y_n}\right)^{\frac{1}{3}}\right)^3 Z_n \tag{1.96}$$

The perceptual color distance in the L*a*b* is similar to the one in the L*u*v*. The two color vectors $\mathbf{x}_{L^*a^*b^*}$ and $\mathbf{y}_{L^*a^*b^*}$ in the L*a*b* space can be represented as:

$$\mathbf{x}_{L^*a^*b^*} = [x_{L^*}, x_{a^*}, x_{b^*}]^T \quad \text{and} \quad \mathbf{y}_{L^*a^*b^*} = [y_{L^*}, y_{a^*}, y_{b^*}]^T \tag{1.97}$$

The perceptual color distance (or total color difference) in the L*a*b* space, ΔE^*_{ab}, between two color vectors $\mathbf{x}_{L^*u^*v^*}$ and $\mathbf{y}_{L^*u^*v^*}$ is given by the Euclidean distance (L_2 norm):

$$\Delta E^*_{ab} = ||\mathbf{x}_{L^*a^*b^*} - \mathbf{y}_{L^*a^*b^*}||_{L_2}$$

$$= \left[(x_{L^*} - y_{L^*})^2 + (x_{a^*} - y_{a^*})^2 + (x_{b^*} - y_{b^*})^2\right]^{\frac{1}{2}} \tag{1.98}$$

The color difference formula ΔE^*_{ab} is applicable to the observing conditions normally found in practice, as in the case of ΔE^*_{uv}. However, this simple difference formula values color differences too strongly when compared to experimental results. To correct the problem a new difference formula was recommended in 1994 by CIE [25], [31]. The new formula is as follows:

$$\Delta E^*_{ab_{94}} = [\frac{(x_{L^*} - y_{L^*})^2}{K_L S_L} + \frac{(x_{a^*} - y_{a^*})^2}{K_c S_c} + \frac{(x_{b^*} - y_{b^*})^2}{K_H S_H}]^{\frac{1}{2}} \tag{1.99}$$

where the factors K_L, K_c, K_H are factors to match the perception of the background conditions, and S_L, S_c, S_H are linear functions of the differences in chroma. Standard reference values for the calculation for $\Delta E^*_{ab_{94}}$ have been specified by the CIE. Namely, the values most often in use are $K_L = K_c = K_H = 1$, $S_L = 1$, $S_c = 1 + 0.045((x_{a^*} - y_{a^*})$ and $S_H = 1 + 0.015((x_{b^*} - y_{b^*})$ respectively. The parametric values may be modified to correspond to typical experimental conditions. As an example, for the textile industry, the K_L factor should be 2, and the K_c and K_H factors should be 1. For all other applications a value of 1 is recommended for all parametric factors [38].

1.9.3 Cylindrical L*u*v* and L*a*b* Color Space

Any color expressed in the rectangular coordinate system of axes L*u*v* or L*a*b* can also be expressed in terms of cylindrical coordinates with the perceived lightness L^* and the psychometric correlates of chroma and hue [37]. The chroma in the L*u*v* space is denoted as C^*_{uv} and that in the L*a*b* space C^*_{ab}. They are defined as [5]:

$$C^*_{uv} = \left((u^*)^2 + (v^*)^2\right)^{\frac{1}{2}} \tag{1.100}$$

$$C^*_{ab} = \left((a^*)^2 + (b^*)^2\right)^{\frac{1}{2}} \tag{1.101}$$

The hue angles are useful quantities in specifying hue numerically [5], [37]. Hue angle h_{uv} in the L*u*v* space and h_{ab} in the L*a*b* space are defined as [5]:

$$h_{uv} = \arctan\left(\frac{v^*}{u^*}\right) \tag{1.102}$$

$$h_{ab} = \arctan\left(\frac{b^*}{a^*}\right) \tag{1.103}$$

The saturation s^*_{uv} in the L*u*v* space is given by:

$$s^*_{uv} = \frac{C^*_{uv}}{L^*} \tag{1.104}$$

1.9.4 Applications of L*u*v* and L*a*b* spaces

The L*u*v* and L*a*b* spaces are very useful in applications where precise quantification of perceptual distance between two colors is necessary [5]. For example in the realization of perceptual based vector order statistics filters. If a degraded color image has to be filtered so that it closely resembles, in perception, the un-degraded original image, then a good criterion to optimize is the perceptual error between the output image and the un-degraded original image. Also, they are very useful for evaluation of perceptual closeness or perceptual error between two color images [4]. Precise evaluation of perceptual closeness between two colors is also essential in color matching systems used in various applications such as multimedia products, image arts, entertainment, and advertisements [6], [14], [22].

$L^*u^*v^*$ and $L^*a^*b^*$ color spaces are extremely useful in imaging systems where exact perceptual reproduction of color images (color consistency) across the entire system is of primary concern rather than real-time or simple computing. Applications include advertising, graphic arts, digitized or animated paintings etc. Suppose, an imaging system consists of various color devices, for example video camera/digital scanner, display device, and printer. A painting has to be digitized, displayed, and printed. The displayed and printed versions of the painting must appear as close as possible to the original image. $L^*u^*v^*$ and $L^*a^*b^*$ color spaces are the best to work with in such cases. Both these systems have been successfully applied to image coding for printing [4], [16].

Color calibration is another important process related to color consistency. It basically equalizes an image to be viewed under different illumination or viewing conditions. For instance, an image of a target object can only be taken under a specific lighting condition in a laboratory. But the appearance of this target object under normal viewing conditions, say in ambient light, has to be known. Suppose, there is a sample object whose image under ambient light is available. Then the solution is to obtain the image of the sample object under the same specific lighting condition in the laboratory. Then a correction formula can be formulated based on the images of the sample object obtained and these can be used to correct the target object for the ambient light [14]. Perceptual based color spaces, such as $L^*a^*b^*$, are very useful for computations in such problems [31], [37]. An instance, where such calibration techniques have great potential, is medical imaging in dentistry.

Perceptually uniform color spaces, with the Euclidean metric to quantify color distances, are particularly useful in color image segmentation of natural scenes using histogram-based or clustering techniques.

A method of detecting clusters by fitting to them some circular-cylindrical decision elements in the $L^*a^*b^*$ uniform color coordinate system was proposed in [39], [40]. The method estimates the clusters' color distributions without imposing any constraints on their forms. Boundaries of the decision elements are formed with constant lightness and constant chromaticity loci. Each boundary is obtained using only 1-D histograms of the $L^*H^\circ C^*$ cylindrical coordinates of the image data. The cylindrical coordinates $L^*H^\circ C^*$ [30] of the $L^*a^*b^*$ color space known as lightness, hue, and chroma, are given by:

$$L^* = L^* \tag{1.105}$$

$$H^\circ = \arctan(b^*/a^*) \tag{1.106}$$

$$C^* = (a^{*2} + b^{*2})^{1/2} \tag{1.107}$$

The $L^*a^*b^*$ space is often used in *color management systems* (CMS). A color management system handles the color calibration and color consistency issues. It is a layer of software resident on a computer that negotiates color reproduction between the application and color devices. Color management systems perform the color transformations necessary to exchange accurate color between diverse devices [4], [43]. A uniform, based on CIE L*u*v*, color space named TekHVC was proposed by Tektronix as part of its commercially available CMS [45].

1.10 The Munsell Color Space

The Munsell color space represents the earliest attempt to organize color perception into a color space [5], [14], [46]. The Munsell space is defined as a comparative reference for artists. Its general shape is that of a cylindrical representation with three dimensions roughly corresponding to the perceived lightness, hue and saturation. However, contrary to the HSV or HSI color models where the color solids were parameterized by hue, saturation and perceived lightness, the Munsell space uses the method of the *color atlas*, where the perception attributes are used for sampling.

The fundamental principle behind the Munsell color space is that of equality of visual spacing between each of the three attributes. Hue is scaled according to some uniquely identifiable color. It is represented by a circular band divided into ten sections. The sections are defined as red, yellow-red, yellow, green-yellow, green, blue-green, blue, purple-blue, purple and red-purple. Each section can be further divided into ten subsections if finer divisions of hue are necessary. A chromatic hue is described according to its resemblance to one or two adjacent hues. Value in the Munsell color space refers to a color's lightness or darkness and is divided into eleven sections numbered zero to ten. Value zero represents black while a value of ten represent white. The chroma defines the color's strength. It is measured in numbered steps starting at one with weak colors having low chroma values. The maximum possible chroma depends on the hue and the value being used. As can be seen in Fig. (1.14), the vertical axis of the Munsell color solid is the line of V values ranging from black to white. Hue changes along each of the circles perpendicular to the vertical axis. Finally, chroma starts at zero on the V axis and changes along the radius of each circle.

The Munsell space is comprised of a set of 1200 color chips each assigned a unique hue, value and chroma component. These chips are grouped in such a way that they form a three dimensional solid, which resembles a warped sphere [5]. There are different editions of the basic Munsell book of colors, with different finishes (glossy or matte), different sample sizes and a different number of samples. The glossy finish collection displays color point chips arranged on 40 constant-hue charts. On each constant-hue chart the chips are arranged in rows and columns. In this edition the colors progress from light at the top of each chart to very dark at the bottom by steps which are intended to be perceptually equal. They also progress from achromatic colors, such as white and gray at the inside edge of the chart, to chromatic colors at the outside edge of the chart by steps that are also intended to be perceptually equal. All the charts together make up the color atlas, which is the color solid of the Munsell system.

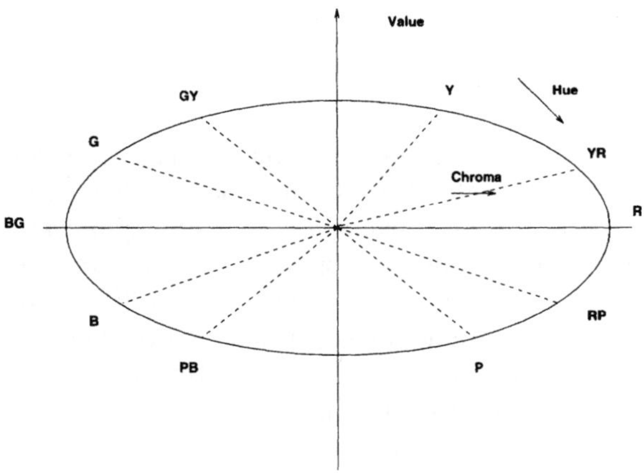

Fig. 1.14. The Munsell color system

Although the Munsell book of colors can be used to define or name colors, in practice is not used directly for image processing applications. Usually stored image data, most often in RGB format, are converted to the Munsell coordinates using either lookup tables or closed formulas prior to the actual application. The conversion from the RGB components to the Munsell hue (H), value (V) corresponding to luminance and chroma (C) corresponding to saturation, can be achieved by using the following mathematical algorithm [47]:

$$x = 0.620R + 0.178G + 0.204B$$

$$y = 0.299R + 0.587G + 0.144B$$

$$z = 0.056G + 0.942B \qquad (1.108)$$

A nonlinear transformation is applied to the intermediate values as follows:

$$p = f(x) - f(y) \qquad (1.109)$$

$$q = 0.4(f(z) - f(y)) \qquad (1.110)$$

where $f(r) = 11.6r^{\frac{1}{3}} - 1.6$. Further the new variables are transformed to:

$$s = (a + bcos(\theta))p \qquad (1.111)$$

$$t = (c + dsin(\theta))q \qquad (1.112)$$

where $\theta = tan^{-1}(\frac{p}{q})$, $a = 8.880$, $b = 0.966$, $c = 8.025$ and $d = 2.558$. Finally, the requested values are obtained as:

$$H = \arctan(\frac{s}{t}) \qquad (1.113)$$

$$V = f(y) \tag{1.114}$$

and

$$C = (s^2 + t^2)^{\frac{1}{2}} \tag{1.115}$$

Alternatively, conversion from RGB, or other color spaces, to the Munsell color space can be achieved through look-up tables and published charts [5].

In summary, the Munsell color system is an attempt to define color in terms of hue, chroma and lightness parameters based on subjective observations rather than direct measurements or controlled perceptual experiments. Although it has been found that the Munsell space is not as perceptually uniform as originally claimed and, despite the fact that it cannot directly integrate with additive color schemes, it is still in use today despite attempts to introduce colorimetric models for its replacement.

1.11 The Opponent Color Space

The opponent color space family is a set of physiologically motivated color spaces inspired by the physiology of the human visual system. According to the theory of color vision discussed in [48] the human vision system can be expressed in terms of opponent hues, yellow and blue on one hand and green and red on the other, which cancel each other when superimposed. In [49] an experimental procedure was developed which allowed researchers to quantitatively express the amounts of each of the basic hues present in any spectral stimulus. The color model of [50], [51], [52], [44] suggests the transformation of the RGB 'cone' signals to three channels, one achromatic channel (I) and two opponent color channels (RG, YB) according to:

$$RG = R - G \tag{1.116}$$

$$YB = 2B - R - G \tag{1.117}$$

$$I = R + G + B \tag{1.118}$$

At the same time a set of effective color features was derived by systematic experiments of region segmentation [53]. According to the segmentation procedure of [53] the color which has the deep valleys on its histogram and has the largest discriminant power to separate the color clusters in a given region need not be the R, G, and B color features. Since a feature is said to have large discriminant power if its variance is large, color features with large discriminant power were derived by utilizing the Karhunen-Loeve (KL) transformation. At every step of segmenting a region, calculation of the new color features is done for the pixels in that region by the KL transform of R, G, and B signals. Based on extensive experiments [53], it was concluded

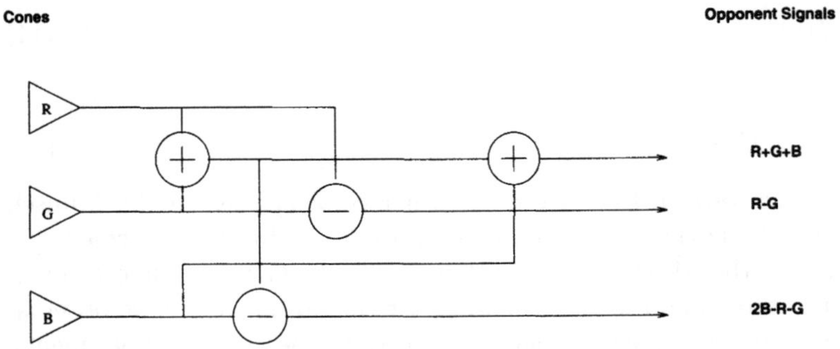

Fig. 1.15. The Opponent color stage of the human visual system

that three color features constitute an effective set of features for segmenting color images, [54], [55]:

$$I1 = \frac{(R+G+B)}{3} \tag{1.119}$$

$$I2 = (R-B) \tag{1.120}$$

$$I3 = \frac{(2G-R-B)}{2} \tag{1.121}$$

In the opponent color space hue could be coded in a circular format ranging through blue, green, yellow, red and black to white. Saturation is defined as distance from the hue circle making hue and saturation speciable with in color categories. Therefore, although opponent representation are often thought as a linear transforms of RGB space, the opponent representation is much more suitable for modeling perceived color than RGB is [14].

1.12 New Trends

The plethora of color models available poses application difficulties. Since most of them are designed to perform well in a specific application, their performance deteriorates rapidly under different operating conditions. Therefore, there is a need to merge the different (mainly device dependent) color spaces into a single standard space. The differences between the monitor RGB space and device independent spaces, such as the HVS and the CIE L*a*b* spaces impose problems in applications, such as multimedia database navigation and face recognition primarily due to the complexity of the operations needed to support the transform from/to device dependent color spaces.

To overcome such problems and to serve the needs of network-centric applications and WWW-based color imaging systems, a new standardized color space based on a colorimetric RGB (sRGB) space has recently been proposed [56]. The aim of the new color space is to complement the current color space management strategies by providing a simple, yet efficient and cost effective method of handling color in the operating systems, device drivers and the Web using a simple and robust device independent color definition.

Since most computer monitors are similar in their key color characteristics and the RGB space is the most suitable color space for the devices forming a modern computer-based imaging systems, the colorimetric RGB space seems to be the best candidate for such a standardized color space.

In defining a colorimetric color space, two factors are of paramount importance:

- the viewing environment parameters with its dependencies on the Human Visual System
- the standard device space colorimetric definitions and transformations [56]

The viewing environment descriptions contain all the necessary transforms needed to support conversions between standard and target viewing environments. On the other hand, the colorimetric definitions provide the transforms necessary to convert between the new sRGB and the CIE-XYZ color space.

The reference viewing environment parameters can be found in [56] with the sRGB tristimulus values calculated from the CIE-XYZ values according to the following transform:

$$\begin{bmatrix} R_{sRGB} \\ G_{sRGB} \\ B_{sRGB} \end{bmatrix} = \begin{bmatrix} 3.2410 & -1.5374 & -0.4986 \\ -0.9692 & 1.8760 & 0.0416 \\ 0.0556 & -0.2040 & 1.0570 \end{bmatrix} \begin{bmatrix} X \\ Y \\ Z \end{bmatrix} \tag{1.122}$$

In practical image processing systems negative sRGB tristimulus values and sRGB values greater than 1 are not retained and typically removed by utilizing some form of clipping. In the sequence, the linear tristimulus values are transformed to nonlinear sR′G′B′ as follows:

1. If $R_{sRGB}, G_{sRGB}, B_{sRGB} \leq 0.0034$ then

$$sR' = 12.92 R_{sRGB} \tag{1.123}$$

$$sG' = 12.92 G_{sRGB} \tag{1.124}$$

$$sB' = 12.92 B_{sRGB} \tag{1.125}$$

2. else if $R_{sRGB}, G_{sRGB}, B_{sRGB} > 0.0034$ then

$$sR' = 1.055 R_{sRGB}{}^{\frac{1.0}{2.4}} - 0.055 \tag{1.126}$$

$$sG' = 1.055 G_{sRGB}{}^{\frac{1.0}{2.4}} - 0.055 \tag{1.127}$$

$$sB' = 1.055 B_{sRGB}{}^{\frac{1.0}{2.4}} - 0.055 \tag{1.128}$$

The effect of the above transformation is to closely fit a straightforward γ value of 2.2 with a slight offset to allow for invertibility in integer mathematics. The nonlinear R′G′B′ values are then converted to digital values with a black digital count of 0 and a white digital count of 255 for 24-bit coding as follows:

$$sR_d = 255.0 sR' \tag{1.129}$$

$$sG_d = 255.0 sG' \tag{1.130}$$

$$sB_d = 255.0 sB' \tag{1.131}$$

The backwards transform is defined as follows:

$$sR' = sR_d + 255.0 \tag{1.132}$$

$$sG' = sG_d + 255.0 \tag{1.133}$$

$$sB' = sB_d + 255.0 \tag{1.134}$$

and

1. if $R_{sRGB}, G_{sRGB}, B_{sRGB} \leq 0.03928$ then

$$R_{sRGB} = sR' + 12.92 \tag{1.135}$$

$$G_{sRGB} = sG' + 12.92 \tag{1.136}$$

$$B_{sRGB} = sB' + 12.92 \tag{1.137}$$

2. else if $R_{sRGB}, G_{sRGB}, B_{sRGB} > 0.03928$ then

$$R_{sRGB} = \left(\frac{sR' + 0.055}{1.055}\right)^{2.4} \tag{1.138}$$

$$G_{sRGB} = \left(\frac{sG' + 0.055}{1.055}\right)^{2.4} \tag{1.139}$$

$$B_{sRGB} = \left(\frac{sB' + 0.055}{1.055}\right)^{2.4} \tag{1.140}$$

with

$$\begin{bmatrix} X \\ Y \\ Z \end{bmatrix} = \begin{bmatrix} 0.4124 & 0.3576 & 0.1805 \\ 0.2126 & 0.7152 & 0.0722 \\ 0.0193 & 0.1192 & 0.9505 \end{bmatrix} \begin{bmatrix} R_{sRGB} \\ G_{sRGB} \\ B_{sRGB} \end{bmatrix} \tag{1.141}$$

The addition of a new standardized color space which supports Web-based imaging systems, device drivers, printers and monitors complementing the existing color management support can benefit producers and users alike by presenting a clear path towards an improved color management system.

1.13 Color Images

Color imaging systems are used to capture and reproduce the scenes that humans see. Imaging systems can be built using a variety of optical, electronic or chemical components. However, all of them perform three basic operations, namely: (i) image capture, (ii) signal processing, and (iii) image formation. Color-imaging devices exploit the trichromatic theory of color to regulate how much light from the three primary colors is absorbed or reflected to produce a desired color.

There are a number of ways to acquiring and reproducing color images, including but not limited to:

- **Photographic film**. The film which is used by conventional cameras contains three emulation layers, which are sensitive to red and blue light, which enters through the camera lens.
- **Digital cameras**. Digital cameras use a CCD to capture image information. Color information is captured by placing red, green and blue filters before the CCD and storing the response to each channel.
- **Cathode-Ray tubes**. CRTs are the display device used in televisions and computer monitors. They utilize a extremely fine array of phosphors that emit red, green and blue light at intensities governed by an electron gun, in accordance to an image signal. Due to the close proximity of the phosphors and the spatial filtering characteristics of the human eye, the emitted primary colors are mixed together producing an overall color.
- **Image scanners**. The most common method of scanning color images is the utilization of three CCD's each with a filter to capture red, green and blue light reflectance. These three images are then merged to create a copy of the scanned image.
- **Color printers**. Color printers are the most common method of attaining a printed copy of a captured color image. Although the trichromatic theory is still implemented, color in this domain is subtractive. The primaries which are used are usually cyan, magenta and yellow. The amount of the three primaries which appear on the printed media govern how much light is reflected.

1.14 Summary

In this chapter the phenomenon of color was discussed. The basic color sensing properties of the human visual system and the CIE standard color specification system XYZ were described in detail. The existence of three types of spectral absorption cones in the human eyes serves as the basis of the trichromatic theory of color, according to which all visible colors can be created by combining three . Thus, any color can be uniquely represented by a three dimensional vector in a color model defined by the three primary colors.

Table 1.3. Color Model

Color System	Transform (from RGB)	Component correlation
RGB	-	highly correlated
$R'G'B'$	non linear	
XYZ	linear	correlated
YIQ	linear	uncorrelated
YCC	linear	uncorrelated
I1I2I3	linear	correlated
HSV	non linear	correlated
HSI	non linear	correlated
HLS	non linear	correlated
$L^*u^*v^*$	non linear	correlated
$L^*a^*b^*$	non linear	correlated
Munsell	non linear	correlated

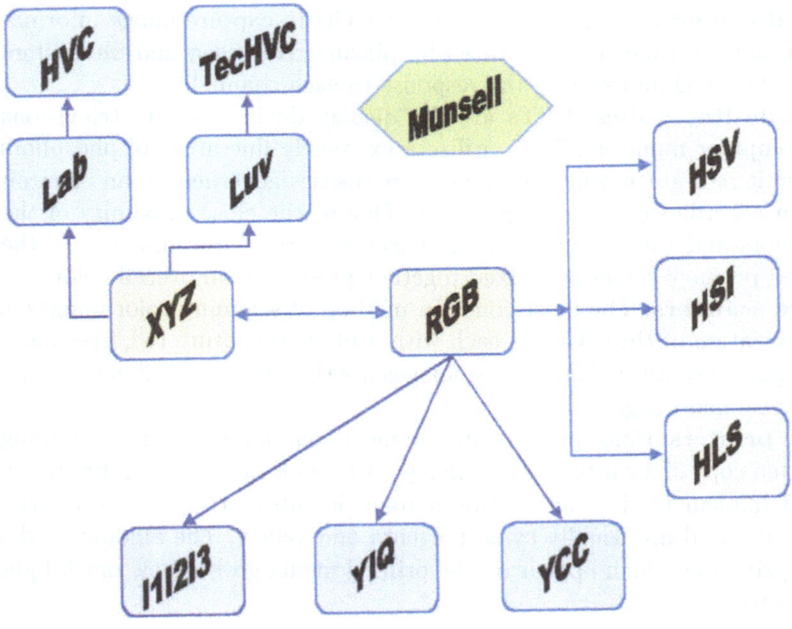

Fig. 1.16. A taxonomy of color models

Color specification models are of paramount importance in applications where efficient manipulation and communication of images and video frames are required. A number of color specification models are in use today. Examples include color spaces, such as the RGB, R′G′B′, YIQ, HSI, HSV, HLS,$L^*u^*v^*$, and $L^*a^*b^*$. The color model is a mathematical representation of spectral colors in a finite dimensional vector space. In each one of them the actual color is reconstructed by combining the basis elements of the vector

Color Spaces		
Models		**Applications**
Colorimetric	XYZ	colorimetric calculations
Device-oriented	- non-uniform spaces RGB, YIQ, YCC	storage, processing, analysis coding, color TV, storage (CD-ROM)
	- uniform spaces $L^*a^*b^*$, $L^*u^*v^*$	color difference evaluation analysis, color management systems
User-oriented	HSI, HSV, HLS, I1I2I3	human color perception multimedia, computer graphics
Munsell		human visual system

spaces, the so called primary colors. By defining different primary colors for the representation of the system different color models can be devised. One important aspect is the color transformation, the change of coordinates from one color system to another (see Table 1.3). Such a transformation associates to each color in one system a color in the other model. Each color model comes into existence for a specific application in color image processing. Unfortunately, there is no technique for determining the optimum coordinate model for all image processing applications. For a specific application the choice of a color model depends on the properties of the model and the design characteristics of the application. Table 1.14 summarizes the most popular color systems and some of their applications.

References

1. Gonzalez, R., Woods, R.E. (1992): Digital Image Processing. Addisson Wesley, Reading MA.
2. Robertson, P., Schonhut, J. (1999): Color in computer graphics. IEEE Computer Graphics and Applications, **19**(4), 18-19.
3. MacDonald, L.W. (1999): Using color effectively in computer graphics. IEEE Computer Graphics and Applications, **19**(4), 20-35.
4. Poynton, C.A. (1996): A Technical Introduction to Digital Video. Prentice Hall, Toronto, also available at http://www.inforamp.net/~poynton/Poynton-Digital-Video.html .
5. Wyszecki, G., Stiles, W.S. (1982): Color Science, Concepts and Methods, Quantitative Data and Formulas. John Wiley, N.Y. , 2nd Edition.
6. Hall, R.A. (1981): Illumination and Color in Computer Generated Imagery. Springer Verlag, New York, N.Y.
7. Hurlbert, A. (1989): The Computation of Color. Ph.D Dissertation, Massachusetts Institute of Technology.
8. Hurvich, Leo M. (1981): Color Vision. Sinauer Associates, Sunderland MA.
9. Boynton, R.M. (1990): Human Color Vision. Halt, Rinehart and Winston.
10. Gomes, J., Velho, L. (1997): Image Processing for Computer Graphics. Springer Verlag, New York, N.Y., also available at http://www.springer-ny.com/catalog/np/mar97np/DATA/0-387-94854-6.html .

11. Fairchild, M.D. (1998): Color Appearance Models. Addison-Wesley, Readings, MA.
12. Sharma, G., Yrzel, M.J., Trussel, H.J. (1998): Color imaging for multimedia. Proceedings of the IEEE, **86**(6): 1088–1108.
13. Sharma, G., Trussel, H.J. (1997): Digital color processing. IEEE Trans. on Image Processing, **6**(7): 901-932.
14. Lammens, J.M.G. (1994): A Computational Model for Color Perception and Color Naming. Ph.D Dissertation, State University of New York at Buffalo, Buffalo, New York.
15. Johnson, G.M., Fairchild, M.D. (1999): Full spectral color calculations in realistic image synthesis. IEEE Computer Graphics and Applications, **19**(4), 47-53.
16. Lu, Guoyun (1996): Communication and Computing for Distributed Multimedia Systems. Artech House Publishers, Boston, MA.
17. Kubinger, W., Vincze, M., Ayromlou, M. (1998): The role of gamma correction in colour image processing. in Proceedings of the European Signal Processing Conference, **2**: 1041–1044.
18. Luong, Q.T. (1993): Color in computer vision. in Handbook of Pattern Recognition and Computer Vision, Word Scientific Publishing Company): 311–368.
19. Young, T. (1802): On the theory of light and colors. Philosophical Transactions of the Royal Society of London, **92**: 20–71.
20. Maxwell, J.C. (1890): On the theory of three primary colors. Science Papers 1, Cambridge University Press: 445–450.
21. Padgham, C.A., Saunders, J.E. (1975): The Perception of Light and Color. Academic Press, New York, N.Y.
22. Judd, D.B., Wyszecki, G. (1975): Color in Business, Science and Industry. John Wiley, New York, N.Y.
23. Foley, J.D., vanDam, A., Feiner, S.K., Hughes, J.F. (1990): Fundamentals of Interactive Computer Graphics. Addison Wesley, Reading, MA.
24. CCIR (1990): CCIR Recommendation 709. Basic parameter values for the HDTV standard for studio and for international program exchange. Geneva, Switcherland.
25. CIE (1995): CIE Publication 116. Industrial color-difference evaluation. Vienna, Austria.
26. Poynton, C.A. (1993): Gamma and its disguises. The nonlinear mappings of intensity in perception, CRTs, film and video. SMPTE Journal: 1099–1108.
27. Kasson M.J., Ploaffe, W. (1992): An analysis of selected computer interchange color spaces. ACM Transaction of Graphics, **11**(4): 373-405.
28. Shih, Tian-Yuan (1995): The reversibility of six geometric color spaces. Photogrammetric Engineering and Remote Sensing, **61**(10): 1223–1232.
29. Levkowitz H., Herman, G.T. (1993): GLHS: a generalized lightness, hue and saturation color model. Graphical Models and Image Processing, **CVGIP-55**(4): 271–285.
30. McLaren, K. (1976): The development of the CIE $L^*a^*b^*$ uniform color space. J. Soc. Dyers Colour, 338–341.
31. Hill, B., Roer, T., Vorhayen, F.W. (1997): Comparative analysis of the quantization of color spaces on the basis of the CIE-Lab color difference formula. ACM Transaction of Graphics, **16**(1): 110–154.
32. Hall, R. (1999): Comparing spectral color computation methods. IEEE Computer Graphics and Applications, **19**(4), 36-44.
33. Hague, G.E., Weeks, A.R., Myler, H.R. (1995): Histogram equalization of 24 bit color images in the color difference color space. Journal of Electronic Imaging, **4**(1), 15-23.

34. Weeks, A.R. (1996): Fundamentals of Electronic Image Processing. SPIE Press, Piscataway, New Jersey.
35. Benson, K. B. (1992): Television Engineering Handbook. McGraw-Hill, London, U.K.
36. Smith, A.R. (1978): Color gamut transform pairs. Computer Graphics (SIGGRAPH'78 Proceedings), **12**(3): 12–19.
37. Healey, C.G., Enns, J.T. (1995): A perceptual color segmentation algorithm. Technical Report, Department of Computer Science, University of British Columbia, Vancouver.
38. Luo, M. R. (1998): Color science. in Sangwine, S.J., Horne, R.E.N. (eds.), The Colour Image Processing Handbook, 26–52, Chapman & Hall, Cambridge, Great Britain.
39. Celenk, M. (1988): A recursive clustering technique for color picture segmentation. Proceedings of the Int. Conf. on Computer Vision and Pattern Recognition, **1**: 437–444.
40. Celenk, M. (1990): A color clustering technique for image segmentation. Computer Vision, Graphics, and Image Processing, **52**: 145–170.
41. Cong, Y. (1998): Intelligent Image Databases. Kluwer Academic Publishers, Boston, Ma.
42. Ikeda, M. (1980): Fundamentals of Color Technology. Asakura Publishing, Tokyo, Japan.
43. Rhodes, P. A. (1998): Colour management for the textile industry. in Sangwine, S.J., Horne, R.E.N. (eds.), The Colour Image Processing Handbook, 307-328, Chapman & Hall, Cambridge, Great Britain.
44. Palus, H. (1998): Colour spaces. in Sangwine, S.J., Horne, R.E.N. (eds.), The Colour Image Processing Handbook, 67–89, Chapman & Hall, Cambridge, Great Britain.
45. Tektronix (1990): TekColor Color Management System: System Implementers Manual. Tektronix Inc.
46. Birren, F. (1969): Munsell: A Grammar of Color. Van Nostrand Reinhold, New York, N.Y.
47. Miyahara, M., Yoshida, Y. (1988): Mathematical transforms of (R,G,B) colour data to Munsell (H,V,C) colour data. Visual Communications and Image Processing, **1001**, 650–657.
48. Hering, E. (1978): Zur Lehe vom Lichtsinne. C. Gerond's Sohn, Vienna, Austria.
49. Jameson, D., Hurvich, L.M. (1968): Opponent-response functions related to measured cone photo pigments. Journal of the Optical Society of America, **58**: 429–430.
50. de Valois, R.L., De Valois, K.K. (1975): Neural coding of color. in Carterette, E.C., Friedman, M.P. (eds.), Handbook of Perception. Volume 5, Chapter 5, 117–166, Academic Press, New York, N.Y.
51. de Valois, R.L., De Valois, K.K. (1993): A multistage color model. Vision Research **33**(8): 1053–1065.
52. Holla, K. (1982): Opponent colors as a 2-dimensional feature within a model of the first stages of the human visual system. Proceedings of the 6th Int. Conf. on Pattern Recognition, **1**: 161–163.
53. Ohta, Y., Kanade, T., Sakai, T. (1980): Color information for region segmentation. Computer Graphics and Image Processing, **13**: 222–241.
54. von Stein, H.D., Reimers, W. (1983): Segmentation of color pictures with the aid of color information and spatial neighborhoods. Signal Processing II: Theories and Applications, **1**: 271–273.
55. Tominaga S. (1986): Color image segmentation using three perceptual attributes. Proceedings of CVPR'86, **1**: 628-630.

56. Stockes, M., Anderson, M., Chandrasekar, Sri., Motta, Ricardo (1997): A standard default color space for the Internet sRGB. International Color Consortium (ICC), contributed document electronic reprint (http://www.color.org).

2. Color Image Filtering

2.1 Introduction

The function of a filter is to transform a signal into another more suitable for a given purpose [1]. As such, filters find applications in image processing, computer vision, telecommunications, geophysical signal processing and biomedicine. However, the most popular application of filtering is the process of detecting and removing unwanted noise from a signal of interest. Noise affects the perceptual quality of the image decreasing not only the appreciation of the image but also the performance of the task for which the image was intended. Therefore, filtering is an essential part of any image processing system whether the final product is used for human inspection, such as visual inspection, or for an automatic analysis.

Noise introduces random variations into sensor readings, making them different from the ideal values, and thus introducing errors and undesirable side effects in subsequent stages of the image processing process. Noise may result from sensor malfunction, imperfect optics, electronic interference, or flaws in the data transmission procedure. In considering the signal-to-noise ratio over practical communication media, such as microwave or satellite links, there would be a degradation in quality due to low received signal power. Degradation of the image quality can also be a result of processing techniques, such as aperture correction, which amplifies both high frequency signals and noise [2], [3], [4].

In many cases, the noise characteristics vary within the same application. Such cases are the channel noise in image transmission as well as atmospheric noise corrupting multichannel satellite images. The noise encountered in digital image processing applications cannot always be described in terms of the commonly assumed Gaussian model. It can however, be characterized in terms of impulsive sequences which occur in the form of short duration, high energy spikes attaining large amplitudes with probability higher than that predicted by a Gaussian density model [5], [6], [7]. Thus, it is desirable for image filters to be robust to impulsive or generally heavy-tailed, non-Gaussian noise [1], [8]. In addition, when processing color images to remove noise, care must be taken to retain the chromatic information. The different filters applied to color images are required to preserve chromaticity, edges and fine image details. The preservation and the possible enhancement of these features is of paramount importance during processing.

Before the different filtering techniques developed over the last ten years to suppress noise are examined, the different kinds of noise corrupting color images should be defined. It is shown how they can be quantified and used in the context of digital color image processing. Statistical tools and techniques consistent with the color representation models which form the basis for most of the color image filters discussed in the second part of this chapter are also considered.

2.2 Color Noise

Based on the trichromatic theory of color, color images are encoded as scalar values in the three color channels, namely, red, green and blue. Color sensors, as any other sensor, can be affected by noise due to malfunction, interference or design flaw. As a result, instead of recording the ideal color value, a random fluctuation around this value is registered by each color channel. Although it is relatively easy to treat noise in the three chromatic channels separately and apply existing gray scale filtering techniques to reduce the scalar noise magnitudes, a different treatment of noise in the context of color images is needed. Color noise can be viewed as color fluctuation given to a certain color signal. As such the color noise signal should be considered as a 3-channel perturbation vector in the RGB color space, affecting the spread of the actual color vectors in the space [2].

Image sensors can be divided into two categories, photochemical and photoelectronic sensors [1]. The positive and negative photographic films are typical photochemical sensors. Although they have the advantage that they can detect and record the image at the same time, the image that they produce cannot by easily digitized. In photochemical sensors, such as films, the noise is mainly due to the silver grains that precipitate during the film exposure. They behave randomly during the film exposure and development and experimental studies have shown that this noise, often called film grain noise, can be modeled in its limit as a Poisson process or Gaussian process [9]. This type of noise is particularly dominant in images acquired with high speed film due to the film's large silver halide grain size. In addition to the film grain noise, photographic noise is due to dust that collects on the optics and the negatives during the film developing process [10].

Photoelectronic sensors have the advantage over the film that they can be used to drive an image digitizer directly. Among the several photoelectronic sensors, such as standard vidicon tubes, Charge Injection Devices (CID), Charge Coupled Devices (CCD), and silicon vidicon tubes, CCDs are the most extensively used [11]. CCD cameras consist of a two-dimensional array of solid state light sensing elements, the so-called cells. The incident light induces electric charges in each cell. These charges are shifted to the right from cell to cell by using a two-phase clock and they come to the read-out register. The rows of cells are scanned sequentially during a vertical scan and thus the image is recorded and sampled simultaneously. In photoelectronic sensors two kinds of noise appear, namely: (i) thermal noise, due to the various electronic circuits, which is usually modeled as additive *white*, zero-mean, Gaussian noise and (ii) photoelectronic noise, which is produced by the random fluctuation of the number of photons on the light sensitive surface of the sensor. Assuming a low level of fluctuation, it has a Bose-Einstein statistic and is modeled by a Poisson like distribution. On the other hand, when its level is high, the noise can be modeled as Gaussian process with standard deviation equal to the square root of the mean.

In the particular case of CCD cameras, transfer loss noise is also present. In CCD technology, charges are transfered from one cell to the other. However, in practice, this process is not complete. A fraction of the charges is not transferred and it represents the transfer noise. The noise occurs along the rows of cells and therefore, has strong horizontal correlation. It usually appears as a white smear located on one side of a bright image spot. Other types of noise due to capacitance coupling of clock lines and output lines or due to noisy cell re-charging are also present to the CCD camera [1].

2.3 Modeling Sensor Noise

This section focuses on a thermal type of noise which for analysis purposes it is assumed that the scalar (gray scale) sensor noise, is *white* Gaussian in nature, having the following probability distribution function:

$$p(x_n) = N(0,\sigma) = \frac{1}{(2\pi\sigma)^{\frac{1}{2}}} \exp\left(\frac{-x^2}{2\sigma^2}\right) \tag{2.1}$$

It can be reasonably assumed that all three color sensors have the same zero average noise magnitude with constant noise variance σ^2 over the entire image plane. To further simplify the analysis, it is assumed that the noise signals corrupting the three color channels are uncorrelated. Let the noise perturbation vector in the RGB color space be denoted as $p = (r^2 + g^2 + b^2)^{\frac{1}{2}}$, where r, g, b are the scalar perturbation quantities (magnitudes) in the red, green and blue chromatic channels respectively. Based on the assumption of identical noise distributions of variance σ^2, for the noise corrupting signal in the three sensors, it can be expected that the noise perturbation vector has a spatial probability density function, which depends only on the value of the perturbation magnitude p as follows:

$$p_r(p) = \int_{-p}^{p} \int_{-(p^2-r^2)^{\frac{1}{2}}}^{(p^2-r^2)^{\frac{1}{2}}} p_{r_r} p_{r_g} [p_{r_b} p_\alpha + p_{r_b}(-p_\alpha)] \, dg \, dr \tag{2.2}$$

$$p_r(p) = (\frac{1}{(2\pi\sigma)})^3 \exp(\frac{-p^2}{2\sigma^2}) \tag{2.3}$$

with $p_{r_r} = p_{r_g} = \frac{1}{(2\pi\sigma)^{\frac{1}{2}}} \exp(\frac{-x^2}{2\sigma^2})$ and $p_\alpha = (p^2 - r^2 - g^2)^{\frac{1}{2}}$.

The probability distribution function has its peak values at $p = (2\sigma)^{\frac{1}{2}}$, unlike the scalar zero-mean noise functions assumed at the beginning. In practical terms, that suggests that if a non-zero scalar noise distribution exists in an individual channel of a color sensor, then the RGB reading will be corrupted by noise, and the registered values will be different than the original ones [2].

Short-tailed, thermal noise modeled as Gaussian distribution is not the only type of noise corrupting color images. In some cases, filtering schemes under a different noise scenario need to be evaluated. One such possible scenario is the presence of noise modeled after a long tailed distribution, such as exponential or Cauchy distribution [1]. In gray scale image processing, the bi-exponential distribution is used for this purpose. The distribution has the form of $p(x) = \frac{\lambda}{2}\exp(-\lambda|x|)$, with $\lambda \geq 0$. For the case of color images, with the three channels, the multivariate analogous with the Euclidean distance is used instead of the absolute value used in the single channel case [4], [12]. That gives a spherically symmetric exponential distribution of:

$$p(\mathbf{x}) = K\exp(-\lambda(r^2 + g^2 + b^2)^{\frac{1}{2}}) \tag{2.4}$$

For this to be a valid probability distribution, K must be selected, such that

$$\int_{-\infty}^{\infty}\int_{-\infty}^{\infty}\int_{-\infty}^{\infty} p(\mathbf{x})\, dr\, dg\, db = 1 \tag{2.5}$$

Combining the above two equations and transforming to spherical coordinates the following is obtained:

$$p(\mathbf{x}) = \int_0^{2\pi}\int_0^{\pi}\int_0^{\infty} K\exp(-\lambda r_d) {r_d}^2 sin(\theta)\, dr_d\, d\theta\, d\phi = 1 \tag{2.6}$$

$$p(\mathbf{x}) = \frac{\lambda^3}{8\pi}\exp(-\lambda(r^2 + g^2 + b^2)^{\frac{1}{2}}) \tag{2.7}$$

where r_d is the length of the color vector in spherical coordinates. Evaluating the first and second moments of the distribution as: $n_i = E[X_i] = 0$, $i = 1,2,3$, $\Sigma_{ii} = E[x_i^2] = \frac{4}{\lambda^2}$, $i = 1,2,3$, and $\Sigma_{ij} = E[x_i x_j] = 0$, $i \neq j$, $i,j = 1,2,3$ and re-writing $\sigma = \frac{2}{\lambda}$ the distribution takes the following form:

$$p_r(p) = \frac{1}{\pi\sigma^3}\exp(\frac{-2}{\sigma}(r^2 + g^2 + b^2)^{\frac{1}{2}}) \tag{2.8}$$

2.4 Modeling Transmission Noise

Recording noise is not the only kind of noise encountered during the process. Image transmission noise is also present and there are various sources that can generate this type of noise. Among others, there are man made phenomena, such as car ignition systems, industrial machines in the vicinity of the receiver, switching transients in power lines and various unprotected switches. In addition, natural causes, such as lightning in the atmosphere and ice cracking in the antarctic region, can also affect the transmission process. The transmission noise also known in the case of gray scale imaging as salt-pepper noise, is modeled after an impulsive distribution. However, a problem in the study of the effect of the noise in the image processing process is the lack of model of multivariate impulsive noise. A number of simplified models have been introduced recently to assist in the performance evaluation of the different color image filters.

The three-variate impulsive noise model considered here is as follows [13], [14]:

$$n(\mathbf{x}) = \begin{cases} \mathbf{s} & \text{with probability } (1-p) \\ (d, s_2, s_3)^T & \text{with probability } p_1 p \\ (s_1, d, s_3)^T & \text{with probability } p_2 p \\ (s_1, s_2, d)^T & \text{with probability } p_3 p \\ (d, d, d)^T & \text{with probability } p_\Sigma p \end{cases} \tag{2.9}$$

where $n(\mathbf{x})$ is the noisy signal, $\mathbf{s} = (s_1, s_2, s_3)^T$ is the noise free color vector, d is the impulse value and

$$p_\Sigma = 1 - p_1 - p_2 - p_3 \tag{2.10}$$

where $\sum_{i=1}^{3} p_i \leq 1$ is the impulsive noise degree of contamination. Impulse d can have either positive or negative values. It is further assumed that $d \gg s_1 \gg s_2 \gg s_3$ and that the delta functions are situated at $(+255, -255)$. Thus, when an impulse is added or subtracted forcing the pixel value outside the $[0, 255]$ range, clipping is applied to force the corrupted noise value into the integer range specified by the 8-bit arithmetic.

In many practical situations an image is often corrupted by both additive Gaussian noise due to faulty sensors and transmission noise introduced by environmental interference or faulty communication. Thus, an image can be thought as corrupted by mixed noise according to the following model:

$$y(\mathbf{x}) = \begin{cases} s(\mathbf{x}) + n(\mathbf{x}) & \text{with probability } (1-p_I) \\ n_I(\mathbf{x}) & \text{otherwise} \end{cases} \tag{2.11}$$

where $s(\mathbf{x})$ is the noise-free $3 - \textit{variate}$ color signal with the additive noise $n(\mathbf{x})$ modeled as zero mean *white* Gaussian noise and $n_I(\mathbf{x})$ transmission noise modeled as multivariate impulsive noise with p_I the impulsive noise degree of contamination [14], [15]. From the discussion above, it can be concluded that the simplest model in color image processing, and the most commonly used, is the additive noise model. According to this model, it is assumed that variations in image colors are gradual. Thus, pixels which are significantly different from their neighbors can be attributed to noise. Therefore, most image filtering techniques attempt to replace those atypical readings, usually called outliers, with values derived from nearby pixels. Based on this principle, several filtering techniques have been proposed over the years. Each different filter discussed in this chapter considers color images as discrete two-dimensional sequences of vectors $[\mathbf{y}(N_1, N_2); N_1, N_2 \in \mathcal{Z}]$. In general, a color pixel $\mathbf{y}$ is a p-variate vector signal, with $p = 3$, when a color model such as RGB is considered. The index $\mathcal{Z}$ is the set of all integers $\mathcal{Z} = (..., -1, 0, 1, ...)$. For simplicity, let $k = (N_1, N_2)$, where $k \in \mathcal{Z}^2$. Each multivariate, image pixel $\mathbf{y}_k = [y_1(k), y_2(k), ..., y_p(k)]^T$, belongs to a pth dimensional vector space $\mathbf{R}^p$. Let the set of image vectors spanned by a $n = (2N+1)\times(2N+1)$ window centered at k be defined as $W(n)$. The color image filters will operate on the window's center sample $\mathbf{y}_k$ and this window will be moved across the set of vectors in $W(n)$ in the image plane in a raster scan fashion [25] with $W^*(n)$ denoting the set of vectors in $W(n)$ without the center pixel $\mathbf{y}_k$.

At a given image location the set of vectors $\mathbf{y}_i$, $i = 1, 2, ..., n$ which is the result of a constant vector-valued signal $\mathbf{x} = [x_1, x_2, ..., x_p]^T$ corrupted by additive zero-mean, p-channel noise $\mathbf{n}_k = [n_1, n_2, ..., n_p]^T$ are accounted for by [16], [4], [3]:

$$\mathbf{y}_k = \mathbf{x} + \mathbf{n}_k \tag{2.12}$$

The noise vectors are distributed according to some joint distribution function $f(\mathbf{n})$. Furthermore, the noise vectors at different instants are assumed to be independently and identically distributed (i.i.d) and uncorrelated to the constant signal.

As it was explained before, some of the observed color signal values have been altered due to the noise. The objective of the different filtering structures is to eliminate these outlying observations or reduce their influence without disturbing those color vectors which have not been significantly corrupted by noise. Several filtering techniques have been proposed over the years. Among them, there are the linear processing techniques, whose mathematical simplicity and existence of a unifying theory make their design and implementation easy. Their simplicity, in addition to their satisfactory performance in a variety of practical applications, has made them methods of choice for many years. However, most of these techniques operate under the assumption that the signal is represented by a stationary model, and thus try to optimize the parameters of a system suitable for such a model. However, many signal processing problems cannot be solved efficiently by using linear techniques. Unfortunately, linear processing techniques fail in image processing, since they cannot cope with the nonlinearities of the image formation model and cannot take into account the nonlinear nature of the human visual system [1]. Image signals are composed of flat regional parts and abruptly changing areas, such as edges , which carry important information for visual perception. Filters having good edge and image detail preservation properties are highly suitable for image filtering and enhancement. Unfortunately, most of the linear signal processing techniques tend to blur edges and to degrade lines, edges and other fine image details [1].

The need to deal with increasingly complex nonlinear systems coupled with the availability of increasing computing power has led to a reevaluation of the conventional filtering methodologies. New algorithms and techniques which can take advantage of the increase in computing power and which can handle more realistic assumptions are needed. To this end, nonlinear signal processing techniques have been introduced more recently. Nonlinear techniques, theoretically, are able to suppress non-Gaussian noise, to preserve important signal elements, such as edges and fine details, and eliminate degradations occurring during signal formation or transmission through nonlinear channels. In spite of an impressive growth in the past two decades, coupled with new theoretical results, the new tools and emerging applications, nonlinear filtering techniques still lack a unifying theory that can encompass existing nonlinear processing techniques. Instead, each class of nonlinear operators possesses its own mathematical tools which can provide a reasonably good analysis of its performance. As a consequence, a multitude of non-linear signal processing techniques have appeared in the literature. At present the following classes of nonlinear processing techniques can be identified:

- polynomial based techniques [17], [18]
- homomorphic techniques [1], [19].
- techniques based on mathematical morphology [20], [21], [22], [23]
- order statistic based techniques [24], [1], [25]

Polynomial filters, especially second order Volterra filters (quadratic filters), have been used for color image filtering, nonlinear channel modeling in telecommunications as well as in multichannel geophysical signal processing. Homomorphic filters and their extensions are one of the first classes of nonlinear filters and have been used extensively in digital image and signal processing. This filter class has been used in various practical applications, such as multiplicative and signal dependent noise removal, color image processing, multichannel satellite image processing and identification of fingerprints. Their basic characteristic is that they use nonlinearities (mainly the logarithm) to transform nonlinearly related signals to additive signals and then to process them by linear filters. The output of the linear filter is then transformed afterwards by the inverse nonlinearity. Morphological filters utilize geometric rather than analytical features of signals. Mathematical morphology can be described geometrically in terms of the actions of the operators on binary, monochrome or color images. The geometric description depends on small synthetic images called structuring elements. This form of mathematical morphology, often called structural morphology, is highly useful in the analysis and processing of images. Morphological filters are found in image processing and analysis applications. Specifically, areas of applications include image filtering, image enhancement and edge detection. However, the most popular family of nonlinear filters is that of the order statistics filters. The theoretical basis of order statistics filters is the theory of robust statistics [26], [27]. There exist several filters which are members of this class. The vector median filter (VMF) is the best known member of this family [24], [28].

The rationale of the approach is that unrepresentative or outlying observations in sets of color vectors can be seen as contaminating the data and thus hampering the methods of signal restoration. Therefore, the different order statistics based filters provide the means of interpreting or categorizing outliers and methods for handling them, either by rejecting them or by adopting methods of reducing their impact. In most cases, the filter employs some method of inference to minimize the influence of any outlier rather than rejecting or including it into our working data set. Outliers can be defined in scalar, univariate data samples although outliers exist in multivariate data, such as color image vectors [29]. The fundamental notion of an outlier as an observation which is statistically unexpected in terms of some basic model can also be extended to multivariate data and to color signals in particular. However, the expression of this notion and the determination of the appropriate procedures to identify and accommodate outliers is by no means as straightforward when more than one dimension is operated in, mainly due to the fact that a multivariate outlier no longer has a simple manifestation of an observation which *deviates the most* from the rest of the samples [30].

In univariate data analysis there is a natural ordering of data, which enables extreme values to be identified and the distance of these outlying values from the center to be computed easily. As such, the problem of identifying and isolating any individual values which are atypical of those in the rest of the data set is a simple one. For this reason, a plethora of filtering techniques based on the concept of univariate ordering have been introduced.

The popularity and the wide spread use of scalar order statistic filters lead to the introduction of similar techniques for the analysis of multivariate, multichannel signals, such as color vectors. However, in order for such filters to be devised the problem of ordering multivariate data should be solved. In this chapter techniques and methodologies for ordering multivariate signals with particular emphasis on color image signals are introduced, examined and analyzed. The proposed ordering schemes will then be used to define a number of nonlinear, multichannel digital filters suitable for color images.

2.5 Multivariate Data Ordering Schemes

A multivariate signal is a signal where each sample has multiple components. It is also called a vector valued, multichannel or multispectral signal. Color images are typical examples of multivariate signals. A color image represented by the three primaries in the RGB coordinate system is a two-dimensional three-variate (three-channel) signal [12], [14], [35], [36]. Let $\mathbf{X}$ denote a p-dimensional random variable, e.g. a p-dimensional vector of random variables $\mathbf{X} = [X_1, X_2, ..., X_p]^{\tau}$. The probability density function (pdf) and the cumulative density function (cdf) of this p-dimensional random variable will be denoted by $f(\mathbf{X})$ and $F(\mathbf{X})$, respectively. Now let $\mathbf{x}_1, \mathbf{x}_2, ..., \mathbf{x}_n$ be n random samples from the multivariate $\mathbf{X}$. Each one of the $\mathbf{x}_i$ are p-dimensional vectors of observations $\mathbf{x}_i = [x_{i1}, x_{i2}, ..., x_{ip}]^{\tau}$. The goal is to arrange the n values $(\mathbf{x}_1, \mathbf{x}_2, ..., \mathbf{x}_n)$ in some sort of order. The notion of data ordering, which is natural in the one dimensional case, does not extend in a straightforward way to multivariate data, since there is no unambiguous, universally acceptable way to order n multivariable samples. Although no such unambiguous form of ordering exists, there are several ways to order the data, the so called *sub-ordering* principles. The role of sub-ordering principles in multivariate data analysis was given in [34], [29].

Since, in effect, ranking procedures isolate outliers by properly weighting each ranked multivariate sample, these outliers can be discorded. The sub ordering principles are useful in detecting outliers in a multivariate sample set. Univariate data analysis is sufficient to detect any outliers in the data in terms of their extreme value relative to an assumed basic model and then employ a robust accommodation method of inference. For multivariate data however, an additional step in the process is required, namely the adaption of the appropriate sub-ordering principle as the basis for expressing extremeness of observations.

The sub-ordering principles are categorized in four types:

1. marginal ordering or M-ordering [34], [37], [38], [16], [39]
2. conditional ordering or C-ordering [34], [39], [40]
3. partial ordering or P-ordering [34], [41]
4. reduced (aggregated) ordering or R-ordering [34], [4], [16], [39]

2.5.1 Marginal Ordering

In the marginal ordering (M-ordering) scheme, the multivariate samples are ordered along each of the $p - dimensions$ independently yielding:

$$\begin{array}{l} \mathbf{x}_{1(1)} \leq \mathbf{x}_{1(2)} \leq \cdots \leq \mathbf{x}_{1(n)} \\ \mathbf{x}_{2(1)} \leq \mathbf{x}_{2(2)} \leq \cdots \leq \mathbf{x}_{2(n)} \\ \quad \cdots \qquad \cdots \\ \mathbf{x}_{p(1)} \leq \mathbf{x}_{p(2)} \leq \cdots \leq \mathbf{x}_{p(n)} \end{array} \tag{2.13}$$

According to the M-ordering principle, ordering is performed in each channel of the multichannel signal independently. The vector $\mathbf{x}_1 = [x_{1(1)}, x_{2(1)}, ..., x_{p(1)}]^{\tau}$

consists of the minimal elements in each dimension and the vector $\mathbf{x}_n = [x_{1(n)}, x_{2(n)}, ..., x_{p(n)}]^T$ consists of the maximal elements in each dimension. The marginal median is defined as $\mathbf{x}_{\nu+1} = [x_{1(\nu)}, x_{2(\nu)}, ..., x_{p(\nu)}]^T$ for $n = \nu + 1$, which may not correspond to any of the original multivariable samples. In contrast, in the scalar case there is a one-to-one correspondence between the original samples $\mathbf{x}_i$ and the order statistics $\mathbf{x}_{(i)}$.

The probability distribution of p-variate marginal order statistics can be used to assist in the design and analysis of color image processing algorithms. Thus, the cumulative distribution function (cdf) and the probability distribution function (pdf) of marginal order statistics is described. In particular, the analysis is focused in the derivation of three-variate (three-dimensional) marginal order statistics, which is of interest since three-dimensional vectors are used to describe the color signals in the different color systems, such as the RGB.

The three-dimensional space is divided into eight subspaces by a point (x_1, x_2, x_3). The requested cdf is given as:

$$F_{r_1,r_2,r_3}(x_1, x_2, x_3) =$$

$$\sum_{i_1=r_1}^{n} \sum_{i_2=r_2}^{n} \sum_{i_3=r_3}^{n} P[i_1 of X_{1i} \leq x_1, i_2 of X_{2i} \leq x_2, i_3 of X_{3i} \leq x_3] \tag{2.14}$$

of the marginal order statistic $X_{1(r_1)}, X_{2(r_2)}, X_{3(r_3)}$ when n three-variate samples are available [38].

Let n_i, $i = 0, 1, ..., 7$ denote the number of data points belonging to each of the eight subspaces. In this case:

$$P[i_1; X_{1i} \leq x_1, i_2; X_{2i} \leq x_2, i_3; X_{3i} \leq x_3] =$$

$$\sum_{n_0} \cdots \sum_{n_7} \frac{n!}{\prod_{i=0}^{7} n_i!} \prod_{i=0}^{7} F_i^{n_i}(x_1, x_2, x_3) \tag{2.15}$$

Given that the total number of points is $\sum_{i=0}^{7} n_i = n$, the following conditions hold for the number of data points lying in the different subspaces:

$$n_0 + n_2 + n_4 + n_6 = i_1$$

$$n_0 + n_1 + n_4 + n_5 = i_2$$

$$n_0 + n_1 + n_2 + n_3 = i_3 \tag{2.16}$$

Thus, combining (2.14) and (2.15) the cdf for the three-variate case is given by [38]:

$$F_{r_1,r_2,r_3}(x_1, x_2, x_3) =$$

$$\sum_{i_1=r_1}^{n} \sum_{i_2=r_2}^{n} \sum_{i_3=r_3}^{n} \sum_{n_0} \cdots \sum_{n_{2^3-1}} \frac{n!}{\prod_{i=0}^{2^3-1} n_i!} \prod_{i=0}^{2^3-1} F_i^{n_i}(x_1, x_2, x_3) \tag{2.17}$$

which is subject to the constraints of (2.16). The probability density function is given by:

$$f_{(r_1,r_2,r_3)}(x_1,x_2,x_3) = \frac{\partial^3 F_{r_1,r_2,r_3}(x_1,x_2,x_3)}{\partial x_1 \partial x_2 \partial x_{x3}} \quad (2.18)$$

The joint cdf for the three-variate case can be calculated as follows [38]:

$$F_{r_1,r_2,r_3 s_1,s_2,s_3}(x_1,x_2,x_3,t_1,t_2,t_3) = \sum_{j_1=s_1}^{n} \sum_{i_1=r_1}^{j_1} \cdots \sum_{j_3=s_3}^{n} \sum_{i_3=r_3}^{j_3} \phi(r) \quad (2.19)$$

with

$$\phi(r) = P[i_1 of X_{1i} \leq x_1, j_1 of X_{1i} \leq t_1, i_2 of X_{2i} \leq x_2, j_2 of X_{2i} \leq t_2, i_3 of X_{3i} \leq x_3, j_3 of X_{3i} \leq t_3] \quad (2.20)$$

for $X - i < t_i$ and $r_i < s_i$, $i = 1,2,3$. The two points (x_1,x_2,x_3) and (t_1,t_2,t_3) divide the three-dimensional space into 3^3 subspaces. If n_i, F_i, $i = 0,1,...,(3^3-1)$ denote the number of data points and the probability masses in each subspace then it can be prove that [38], [16]:

$$\phi(r) = \sum_{n_0} \cdots \sum_{(n_3^3)-1} \frac{n!}{\prod_{i=0}^{(n_3^3)-1} n_i!} \prod_{i=0}^{(n_3^3)-1} F_i^{n_i}(x_1,x_2,x_3) \quad (2.21)$$

under the constraints:

$$\sum_{i=0}^{3^3-1} n_i = n \quad (2.22)$$

$$\sum_{I_0=0} n_i = i_1, \quad \sum_{I_1=0} n_i = i_2$$

$$\sum_{I_2=0} n_i = i_3 \quad (2.23)$$

$$\sum_{I_0=0,1} n_i = j_1, \quad \sum_{I_1=0,1} n_i = j_2$$

$$\sum_{I_2=0,1} n_i = j_3 \quad (2.24)$$

where $i = (I_2, I_1, I_0)$ is an arithmetic representation of number i with base 3. Through (2.19)-(2.24) a numerically tractable way to calculate the joint cdf for the three-variate order statistics is possible.

2.5.2 Conditional Ordering

In conditional ordering (C-ordering) the multivariate samples are ordered conditional on one of the marginal sets of observations. Thus, one of the marginal components is ranked and the other components of each vector are listed according to the position of their ranked component. Assuming that the first dimension is ranked, the ordered samples would be represented as follows:

$$
\begin{aligned}
&\mathbf{x}_{1(1)} \leq \mathbf{x}_{1(2)} \leq \cdots \leq \mathbf{x}_{1(n)} \\
&\mathbf{x}_{2[1]} \leq \mathbf{x}_{2[2]} \leq \cdots \leq \mathbf{x}_{2[n]} \\
&\quad \cdots \qquad \cdots \\
&\mathbf{x}_{p[1]} \leq \mathbf{x}_{p[2]} \leq \cdots \leq \mathbf{x}_{p[n]}
\end{aligned}
\tag{2.25}
$$

where $\mathbf{x}_{1(i)}$, $i = 1, 2, ..., n$ are the marginal order statistics of the first dimension, and $\mathbf{x}_{j[i]}$, $j = 2, 3, ..., p$, $i = 1, 2, ..., n$ are the quasi-ordered samples in dimensions $j = 2, 3, ..., p$, conditional on the marginal ordering of the first dimension. These components are not ordered, they are simply listed according to the ranked components. In the two dimensional case ($p = 2$) the statistics $\mathbf{x}_{2(i)}$, $i = 1, 2, ..., n$ are called *concomitants* of the order statistics of $\mathbf{x}_1$. The advantage of this ordering scheme is its simplicity since only one scalar ordering is required to define the order statistics of the vector sample. The disadvantage of the C-ordering principle is that since only information in one channel is used for ordering, it is assumed that all or at least most of the important ordering information is associated with that dimension. Needless to say that if this assumption were not to hold, considerable loss of useful information may occur. As an example, the problem of ranking color signals in the YIQ color system may be considered. A conditional ordering scheme based on the luminance channel (Y) means that chrominace information stored in the I and Q channels would be ignored in ordering. Any advantages that could be gained in identifying outliers or extreme values based on color information would therefore be lost.

2.5.3 Partial Ordering

In partial (P-ordering), subsets of data are grouped together forming minimum convex hulls. The first convex hull is formed such that the perimeter contains a minimum number of points and the resulting hull contains all other points in the given set. The points along this perimeter are denoted *c-order group 1*. These points form the most extreme group. The perimeter points are then discarded and the process repeats. The new perimeter points are denoted *c-order group 2* and then removed in order for the process to be continued. Although convex hull or elliptical peeling can be used for outlier isolation, this method provides no ordering within the groups and thus it is not easily expressed in analytical terms. In addition, the determination of the convex hull is conceptually and computationally difficult, especially with higher-dimensional data. Thus, although trimming in terms of ellipsoids of minimum content [41] rather than convex hull has been proposed, P-ordering is rather infeasible for implementation in color image processing.

2.5.4 Reduced Ordering

In reduced (aggregating) or R-ordering , each multivariate observation $\mathbf{x}_i$ is reduced to single, scalar value by means of some combination of the component sample values. The resulting scalar values are then amenable to univariate ordering. Thus, the set $\mathbf{x}_1, \mathbf{x}_2, ..., \mathbf{x}_n$ can be ordered in terms of the values $R_i = R(\mathbf{x}_i)$, $i = 1, 2, ..., n$. The vector $\mathbf{x}_i$ which yields the maximum value $R_{(n)}$ can be considered as an outlier, provided that its extremeness is obvious comparing to the assumed basic model.

In contrast to M-ordering, the aim of R-ordering is to effect some sort of overall ordering on the original multivariate samples, and by ordering in this way, the multivariate ranking is reduced to a simple ranking operation of a set of transformed values. The type of ordering cannot be interpreted in the same manner as the conventional scalar ordering as there are no absolute minimum or maximum vector samples. Given that multivariate ordering is based on a reduction function $R(.)$, points which diverge from the 'center' in opposite directions may be in the same order ranks. Furthermore, by utilizing a reduction function as the mean to accomplish multivariate ordering, useful information may be lost. Since distance measures have a natural mechanism for identification of outliers, the reduction function most frequently employed in R-ordering is the generalized (Mahalanobis) distance [29], [30]:

$$R(\mathbf{x}, \bar{\mathbf{x}}, \Gamma) = (\mathbf{x} - \bar{\mathbf{x}})^{\tau} \Gamma^{1} (\mathbf{x} - \bar{\mathbf{x}}) \tag{2.26}$$

where $\bar{\mathbf{x}}$ is a location parameter for the data set, or underlying distribution, in consideration and Γ is a dispersion parameter with Γ^{-1} used to apply a differential weighting to the components of the multivariate observation inversely related to the population variability. The parameters of the reduction function can be given arbitrary values, such as $\bar{\mathbf{x}} = 0$ and $\Gamma = I$, or they can be assigned the true mean μ and dispersion Σ settings. Depending on the state of knowledge about these values, their standard estimates:

$$\bar{\mathbf{x}} = \frac{1}{n} \sum_{i=1}^{n} \mathbf{x}_i \tag{2.27}$$

and

$$S = \frac{1}{n-1} \sum_{i=1}^{n} (\mathbf{x} - \bar{\mathbf{x}})(\mathbf{x} - \bar{\mathbf{x}})^{\tau} \tag{2.28}$$

can be used instead. Within the framework of the generalized distance, different reduction functions can be utilized in order to identify the contribution of an individual multivariate sample. A list of such functions include, among others, the following [42], [43]:

$$q_i^2 = (\mathbf{x} - \bar{\mathbf{x}})^T (\mathbf{x} - \bar{\mathbf{x}}) \tag{2.29}$$

$$t_i^2 = (\mathbf{x} - \bar{\mathbf{x}})^T S (\mathbf{x} - \bar{\mathbf{x}}) \tag{2.30}$$

$$u_i^2 = \frac{(\mathbf{x} - \bar{\mathbf{x}})^T S (\mathbf{x} - \bar{\mathbf{x}})}{(\mathbf{x} - \bar{\mathbf{x}})^T (\mathbf{x} - \bar{\mathbf{x}})} \tag{2.31}$$

$$v_i^2 = \frac{(\mathbf{x} - \bar{\mathbf{x}})^T S^{-1} (\mathbf{x} - \bar{\mathbf{x}})}{(\mathbf{x} - \bar{\mathbf{x}})^T (\mathbf{x} - \bar{\mathbf{x}})} \tag{2.32}$$

$$d_i^2 = (\mathbf{x} - \bar{\mathbf{x}})^T S^{-1} (\mathbf{x} - \bar{\mathbf{x}}) \tag{2.33}$$

$$d_k^2 = (\mathbf{x} - \mathbf{x}_k)^T S^{-1} (\mathbf{x} - \mathbf{x}_k) \tag{2.34}$$

with $i < k = 1, 2, ...n$. Each one of the these functions identifies the contribution of the individual multivariate sample to specific effects as follows [43]:

1. q_i^2 isolates data which excessively inflate the overall scale.
2. t_i^2 determines which data has the greatest influence on the orientation and scale of the first few principal components [44], [45].
3. u_i^2 emphasizes more the orientation and less the scale of the principal components.
4. v_i^2 measures the relative contribution on the orientation of the last few principal components.
5. d_i^2 uncovers the data points which lie far away from the general scatter of points.
6. d_k^2 has the same objective as d_i^2 but provides far more detail of inter-object separation.

The following comments should be made regarding the reduction functions discussed in this section:

1. If outliers are present in the data then $\bar{\mathbf{x}}$ and Σ are not the best estimates of the location and dispersion for the data, since they will be affected by the outliers. In the face of outliers, robust estimators of both the mean value and the covariance matrix should be utilized. A robust estimation of the matrix S is important because outliers inflate the sample covariance and thus may mask each other making outlier detection even in the presence of only a few outliers. Various design options can be considered. Among them the utilization of the marginal median (median evaluated using M-ordering) as a robust estimate of the location. However, care must be taken since the marginal median of n multivariate samples is

not necessarily one of the input samples. Depending on the estimator of the location used in the ordering procedure the following schemes can be distinguished [15].

a) *R-ordering about the mean* (Mean R-ordering)
Given a set of n multivariate samples $\mathbf{x}_i$, $i = 1, 2, ...n$ in a processing window and $\bar{\mathbf{x}}$ the mean of the multivariates, the *mean R-ordering* is defined as:

$$(\mathbf{x}_{(1)}, \mathbf{x}_{(2)}, ..., \mathbf{x}_{(n)} : \bar{\mathbf{x}}) \tag{2.35}$$

where $(\mathbf{x}_{(1)}, \mathbf{x}_{(2)}, ..., \mathbf{x}_{(n)})$ is the ordering defined by:
$d_i^2 = (\mathbf{x} - \bar{\mathbf{x}})^T (\mathbf{x} - \bar{\mathbf{x}})$ and $(d_{(1)}^2 \leq d_{(2)}^2 \leq ... \leq d_{(n)}^2)$.

b) *R-ordering about the marginal median* (Median R-ordering)
Given a set of n multivariate samples $\mathbf{x}_i$, $i = 1, 2, ...n$ in a processing window and $\mathbf{x}_m$ the marginal median of the multivariates, the *median R-ordering* is defined as:

$$(\mathbf{x}_{(1)}, \mathbf{x}_{(2)}, ..., \mathbf{x}_{(n)} : \mathbf{x}_m) \tag{2.36}$$

where $(\mathbf{x}_{(1)}, \mathbf{x}_{(2)}, ..., \mathbf{x}_{(n)})$ is the ordering defined by:
$d_i^2 = (\mathbf{x} - \mathbf{x}_m)^T (\mathbf{x} - \mathbf{x}_m)$ and $(d_{(1)}^2 \leq d_{(2)}^2 \leq ... \leq d_{(n)}^2)$.

c) *R-ordering about the center sample* (Center R-ordering)
Given a set of n multivariate samples $\mathbf{x}_i$, $i = 1, 2, ...n$ in a processing window and $\mathbf{x}_{\bar{n}}$ the sample at the window center $\bar{n}$, the *center R-ordering* is defined as:

$$(\mathbf{x}_{(1)}, \mathbf{x}_{(2)}, ..., \mathbf{x}_{(n)} : \mathbf{x}_{\bar{n}}) \tag{2.37}$$

where $(\mathbf{x}_{(1)}, \mathbf{x}_{(2)}, ..., \mathbf{x}_{(n)})$ is the ordering defined by:
$d_i^2 = (\mathbf{x} - \mathbf{x}_{\bar{n}})^T (\mathbf{x} - \mathbf{x}_{\bar{n}})$ and $(d_{(1)}^2 \leq d_{(2)}^2 \leq ... \leq d_{(n)}^2)$. Thus, $\mathbf{x}_{(1)} = \mathbf{x}_{\bar{n}}$.

2. Statistic measures, such as d_i^2 and d_k^2 are invariant under non singular transformation of the data.
3. Statistics which measure the influence on the first few principal components, such as t_i^2, u_i^2, d_i^2 and d_k^2 are useful in detecting those outliers which inflate the variance, covariance or correlation in the data. Statistic measures, such as v_i^2 will detect those outliers that add insignificant dimensions and/or singularities to the data.

Statistical descriptions of the descriptive measures listed above can be used to assist in the design and analysis of color image processing algorithms. As an example, the statistical description of the d_i^2 descriptor will be presented. Given the multivariate data set $(\mathbf{x}_1, \mathbf{x}_2, ..., \mathbf{x}_n)$ and the population mean $\bar{\mathbf{x}}$, interest lies in determining the distribution for the distances d_i^2 or equivalently for $D_i = d_i^{2\frac{1}{2}}$. Let the probability density function (pdf) of D for the input be denoted as f_D and the pdf for the ith ranked distance be $f_{D_{(i)}}$. If the multivariate data samples are independent and identically distributed then D will be also independent and identically distributed (i.i.d). Based on this assumption $f_{D_{(i)}}$ can be evaluated in terms of f_D as follows [1], [39].

$$f_{D_{(i)}}(x) = \frac{n!}{(i-1)!(n-i)!} F_D^{i-1}(x)[1-F_D(x)]^{n-i} f_D(x) \tag{2.38}$$

with $F_D(x)$ the cumulative distribution (cdf) for the distance D.

As an example, assume that the multivariate samples $\mathbf{x}$ belong to a multivariate elliptical distribution with parameters μ_x, Σ_x and of the form:

$$f(\mathbf{x}) = K_p |\Sigma_x|^{\frac{-1}{2}} h((\mathbf{x}-\mu_x)^T \Sigma^{-1} (\mathbf{x}-\mu_x)) \tag{2.39}$$

for some function $h(.)$, where K_p is a normalizing constant and Σ_x is positive definite. This class of distributions includes the multivariate Gaussian distribution and all other densities whose contours of equal probability have an elliptical shape. If a distribution such as the multivariate Gaussian belonging to this class exists, then all its marginal distributions and its conditional distributions also belong to this class.

For the special case of the simple *Euclidean* distance $d_i = (\mathbf{x}-\bar{\mathbf{x}})^T (\mathbf{x}-\bar{\mathbf{x}})^{\frac{1}{2}}$ $f_{D_{(x)}}$ has the general form of:

$$f_{D_{(x)}} = \frac{2K_p \pi^{\frac{p}{2}}}{\Gamma(\frac{p}{2})} x^{p-1} h(x^2) \tag{2.40}$$

where $\Gamma(.)$ is the gamma function and $x \geq 0$. If the elliptical distribution assumed initially for the multivariate $\mathbf{x}_i$ samples is considered to be multivariate Gaussian with mean value μ_x and covariance $\Sigma_x = \sigma^2 I_p$, then the normalizing constant is $K_p = (2\pi\sigma^2)^{\frac{1}{2}}$ and the $h(x^2) = \exp(\frac{-x^2}{2\sigma^2})$, and thus $f_{D_{(x)}}$ takes the form of the Rayleigh distribution:

$$f_{D_{(x)}} = \frac{x^{p-1}}{\sigma^p 2^{\frac{p-2}{2}} \Gamma(\frac{p}{2})} \exp(\frac{-x^2}{2\sigma^2}) \tag{2.41}$$

Based on this distribution the kth moment of D is given as:

$$E[D^k] = (2\sigma)^{\frac{k}{2}} \frac{\Gamma(\frac{p+k}{2})}{\Gamma(\frac{p}{2})} \tag{2.42}$$

with $k \geq 0$. It can easily be seen from the above equation that the expected value of the distance D will increase monotonically as a function of the parameter σ in the assumed multivariate Gaussian distribution.

To complete the analysis, the cumulative distribution function F_D is needed. Although there is no closed form expression for the cdf of a Rayleigh random variable, for the special case where p is an even number, the requested cdf can be expressed as:

$$F_D(x) = 1 - \exp(\frac{-x^2}{2\sigma^2}) \sum_{k=0}^{(\frac{p}{2})-1} (\frac{1}{k!})(\frac{x^2}{2\sigma^2})^k \tag{2.43}$$

Using this expression the following pdf for the distance $D_{(i)}$ can be obtained:

$$f_{D_{(i)}}(x) = Cx^{p-1}\exp\left(\frac{-x^2}{2\sigma^2}\right)F_D(x)^{(i-1)}(1 - F_D(x))^{n-i} \tag{2.44}$$

where $C = \frac{(n!)\sigma^p \Gamma(\frac{p}{2})}{(i-1)!(n-i)!2^{\frac{p-2}{2}}}$ is a normalization constant.

In summary, R-ordering is particularly useful in the task of multivariate outlier detection, since the reduction function can reliably identify outliers in multivariate data samples. Also, unlike M-ordering, it treats the data as vectors rather than breaking them up into scalar components. Furthermore, it gives all the components equal weight of importance, unlike C-ordering. Finally, R-ordering is superior to P-ordering in its simplicity and its ease of implementation, making it the sub ordering principle of choice for multivariate data analysis.

2.6 A Practical Example

To better illustrate the effect of the different ordering schemes discussed here, the order statistics for a sample set of data will be provided. For simplicity, two dimensional data vectors will be considered. In the example, seven vectors will be used. The data points are:

$$D_o : \begin{cases} \mathbf{x}_1 = (1,1) \\ \mathbf{x}_2 = (5,3) \\ \mathbf{x}_3 = (7,2) \\ \mathbf{x}_4 = (3,3) \\ \mathbf{x}_5 = (5,4) \\ \mathbf{x}_6 = (6,5) \\ \mathbf{x}_7 = (6,8) \end{cases} \tag{2.45}$$

(I) Marginal ordering. For the case of M-ordering the first and the second components are ordered independently as follows:

$$[1,5,7,3,5,6,6] \Rightarrow [1,3,5,5,6,6,7] \tag{2.46}$$

and

$$[1,3,2,3,4,5,8] \Rightarrow [1,2,3,3,4,5,8] \tag{2.47}$$

and thus, the ordered vectors are:

$$D_M : \begin{cases} \mathbf{x}_{(1)} = (1,1) \\ \mathbf{x}_{(2)} = (3,2) \\ \mathbf{x}_{(3)} = (5,3) \\ \mathbf{x}_{(4)} = (5,3) \\ \mathbf{x}_{(5)} = (6,4) \\ \mathbf{x}_{(6)} = (6,5) \\ \mathbf{x}_{(7)} = (7,8) \end{cases} \tag{2.48}$$

with the median vector $(5,3)$ and the minimum /maximum vectors $(1,1)$ and $(6,8)$ respectively.

(II) Conditional ordering. For the case of C-ordering the second channel will be used for ordering, with the second components ordered as follows:

$$[1,3,2,3,4,5,8] \Rightarrow [1,2,3,3,4,5,8] \tag{2.49}$$

and thus, the corresponding vectors ordered as:

$$D_C : \begin{cases} \mathbf{x}_{(1)} = (1,1) \\ \mathbf{x}_{(2)} = (7,2) \\ \mathbf{x}_{(3)} = (5,3) \\ \mathbf{x}_{(4)} = (3,3) \\ \mathbf{x}_{(5)} = (5,4) \\ \mathbf{x}_{(6)} = (6,5) \\ \mathbf{x}_{(7)} = (6,8) \end{cases} \tag{2.50}$$

where the median vector is $(3,3)$ and the minimum / maximum defined as $(1,1)$ and $(6,8)$ respectively.

(III) Partial ordering.
For the case of P-ordering the ordered sub groups for the data set examined here are:

$$D_P : \begin{cases} C_1 = [(1,1),(6,8),(7,2)] \\ C_2 = [(6,5),(5,3),(3,3)] \\ C_3 = [(5,4)] \end{cases} \tag{2.51}$$

As it can be seen, there is no ordering within the groups and thus no way to distinguish a median or most central vector. The only information received is that C_3 is the most central group with C_1 the most extreme group.

(IV) Reduced ordering.
For the case of R-ordering, the following reduction function is used:

$$q_i = ((\mathbf{x} - \bar{\mathbf{x}})^T (\mathbf{x} - \bar{\mathbf{x}}))^{\frac{1}{2}} \tag{2.52}$$

where $\bar{\mathbf{x}} = \frac{1}{7}\sum_{i=1}^{7} \mathbf{x}_i = (4.7, 3.7)$. Allowing the q_i's to be calculated as:

$$q_i : \begin{cases} q_1 = 4.58 \text{ for } \mathbf{x}_1 = (1,1) \\ q_2 = 0.76 \text{ for } \mathbf{x}_2 = (5,3) \\ q_3 = 2.86 \text{ for } \mathbf{x}_3 = (7,2) \\ q_4 = 1.85 \text{ for } \mathbf{x}_4 = (3,3) \\ q_5 = 0.42 \text{ for } \mathbf{x}_5 = (5,4) \\ q_6 = 1.82 \text{ for } \mathbf{x}_6 = (6,5) \\ q_7 = 4.49 \text{ for } \mathbf{x}_7 = (6,8) \end{cases} \tag{2.53}$$

and thus, the ordered data set is as follows:

$$D_R : \begin{cases} \mathbf{x}_{(1)} = (5,4) \\ \mathbf{x}_{(2)} = (5,3) \\ \mathbf{x}_{(3)} = (6,5) \\ \mathbf{x}_{(4)} = (3,3) \\ \mathbf{x}_{(5)} = (7,2) \\ \mathbf{x}_{(6)} = (6,8) \\ \mathbf{x}_{(7)} = (1,1) \end{cases} \tag{2.54}$$

with $\mathbf{x}_{(1)} = (5,4)$ the most centrally located point and $\mathbf{x}_{(7)} = (1,1)$ the most outlying data sample.

2.7 Vector Ordering

The sub-ordering principles discussed here can be used to rank any kind of multivariate data. However, to define an ordering scheme which is attractive for color image processing, this should be geared towards the ordering of color image vectors. Such an ordering scheme should satisfy the following criteria:

1. The proposed ordering scheme should be useful from a robust estimation perspective, allowing for the extension of the operations of scalar order statistic filters to the color, multivariate domain.
2. The proposed ordering scheme should preserve the notion of varying levels of *extremeness* that was present in the scalar ordering case.
3. The proposed ordering scheme should take into consideration the type of multivariate data being used. Therefore, since the RGB coordinate system will be used throughout this work for color image filtering, the ordering scheme should give equal importance to the three primary color channels and should consider all the information contained in each of the three channels.

Based on these three principles, the ordering scheme that will be utilized is a variation of the R-ordering scheme that employs a dissimilarity (or alternatively similar) measure to the set of $\mathbf{x}_i$. That is to say that the aggregate measure of point $\mathbf{x}_i$ from all other points:

$$R_a(\mathbf{x}_i) = \sum_{j=1}^{n} R(\mathbf{x}_i, \mathbf{x}_j) \tag{2.55}$$

is used for ranking purposes. The scalar quantities $R_{ai} = R_a(\mathbf{x}_i)$ are then ranked in order of magnitude and the associated vectors will be correspondingly ordered:

$$R_{a1} \leq R_{a2} \leq \ldots \leq R_{an} \tag{2.56}$$

$$\mathbf{x}_{(1)} \leq \mathbf{x}_{(1)} \leq \ldots \leq \mathbf{x}_{(n)} \tag{2.57}$$

Using the ordering scheme proposed here, the ordered $\mathbf{x}_{(i)}$ have a one-to-one relationship with the original samples $\mathbf{x}_i$, unlike marginal ordering and furthermore all the components are given equal weight or importance unlike conditional ordering.

The proposed ordering scheme focuses on inter relationships between the multivariate samples, since it computes similarity or distance between all pairs of data points in the sample set. The output of the ranking procedure depends critically on the type of data from which the computation is to be made, and the function $R(\mathbf{x}_i, \mathbf{x}_j)$ selected to evaluate the similarity $s(i,j)$ or distance $d(i,j)$ between the two vectors $\mathbf{x}_i$ and $\mathbf{x}_j$. In the rest of the chapter measures suitable for the task will be introduced and discussed.

2.8 The Distance Measures

The most commonly used measure to the quantify distance between two p-D signals is the generalized Minkowski metric (L_p norm). It is defined for two vectors $\mathbf{x}_i$ and $\mathbf{x}_j$ as follows [44]:

$$d_M(i,j) = \left(\sum_{k=1}^{p} |(x_i^k - x_j^k)|^p\right)^{\frac{1}{p}} \tag{2.58}$$

where p is the dimension of the vector $\mathbf{x}_i$ and x_i^k is the kth element of $\mathbf{x}_i$. Three special cases of the L_M metric are of particular interest. Namely:

1. The City-Block distance (L_1 norm) corresponding to $M = 1$. In this case, the distance between the two p-D vectors is considered to be the summation of the absolute values between their components:

$$d_1(i,j) = \sum_{k=1}^{p} ||x_i^k - x_j^k|| \tag{2.59}$$

2. The Euclidean distance (L_2 norm) corresponding to $M = 2$. In this model, the distance between the two p-D signals is set to be the square root of the summation of the square distances among their components:

$$d_2(i,j) = \left(\sum_{k=1}^{p} (x_i^k - x_j^k)^2\right)^{\frac{1}{2}} \tag{2.60}$$

3. The Chess-board distance (L_∞ norm) corresponding to $p = \infty$. In this case, the distance between the two p-D vectors is considered equal to the maximum distance among their components:

$$d_\infty(i,j) = \max_k ||x_i^k - x_j^k|| \tag{2.61}$$

The Euclidean distance is relatively expensive, since it involves the evaluation of the squares of the componentwise distances and requires the calculation of the square root. To accommodate such operations, floating point arithmetic is required for the evaluation of the distance. On the other hand both the L_1 and L_∞ norms can be evaluated using integer arithmetic resulting in computationally attractive distance evaluation algorithms.

In addition, to alleviate the problem, fast approximations to the Euclidean distance recently have been proposed. These approximate distances use a linear combination of the absolute componentwise distances to approximate the L_2 norm.

The general form of the approximate Euclidean distance (L_{2a} norm) is as follows [46], [47]:

$$d_2(i,j) = \sum_{k=1}^{p} a_k |x_i^k - x_j^k| \tag{2.62}$$

with $a_k = (k)^{\frac{1}{2}} - (k-1)^{\frac{1}{2}}$, $k = 1, 2, ..., p$ the weights in the approximation formula.

For multichannel signals, with relatively small dimensions ($p < 5$), the computations are sped up further by rounding up to negative powers of 2, such that the weights can be determined as $a_k = \frac{1}{2^{p-1}}$, so that the multiplications between the weights and the vector components can be implemented by bit shifting, which proves to be a very fast operation.

The Minkowski metric discussed above is only one of many possible methods [44], [43]. Other measures can be devised in order to quantify distances among multichannel signals. Such a measure is the Canberra distance defined as follows [43]:

$$d_c(i,j) = \sum_{i=1}^{p} \frac{|x_i^k - x_j^k|}{|x_i^k + x_j^k|} \tag{2.63}$$

where p is the dimension of the vector $\mathbf{x}_i$ and x_i^k is the kth element of $\mathbf{x}_i$. The Canberra metric applies only to non-negative multivariate data which is the case when color vectors described in the RGB reference system are considered. Another distance measure applicable only to vectors with non-negative components, such as color signals, is the Czekanowski coefficient defined as follows [43]:

$$d_z(i,j) = 1 - \frac{2\sum_{k=1}^{p} \min(x_{ik}, x_{jk})}{\sum_{k=1}^{p} (x_{ik}, x_{jk})} \tag{2.64}$$

If the variables under study are on very different scales or of different quantities, then it would make sense to standardize the data prior to applying any of these distance measures in order to ensure that no single variable will dominate the results.

Of course, there are many other measures by which a distance function can be constructed. Depending on the nature of the problem and the constraints imposed by the design, one method may be more appropriate than the other. Furthermore, measures other than distance can be used to measure similarity between multivariate vector signals, as the next section will attest.

2.9 The Similarity Measures

Distance metrics are not the only approach to the problem of defining similarity between two multidimensional signals. Any non-parametric function $s(\mathbf{x}_i, \mathbf{x}_j)$ can be used to compare the two multichannel signals $\mathbf{x}_i$ and $\mathbf{x}_j$. This can be done by utilizing a symmetric function, whose value is large when $\mathbf{x}_i$ and $\mathbf{x}_j$ are similar. An example of such a function is the normalized inner product defined as [44]:

$$s_1(\mathbf{x}_i, \mathbf{x}_j) = \frac{\mathbf{x}_i \mathbf{x}_j^t}{|\mathbf{x}_i||\mathbf{x}_j|} \tag{2.65}$$

which corresponds to the cosine of the angle between the two vectors $\mathbf{x}_i$ and $\mathbf{x}_j$. Therefore, the angle between the two vectors can be considered as a measure of their similarity.

The cosine of the angle (or the magnitude of the angle) discussed here is used to quantify their similarity in orientation. Therefore, in applications where the orientation difference between two vector signals is of importance, the normalized inner product or equivalently the angular distance ,

$$\theta = cos^{-1}\left(\frac{\mathbf{x}_i \mathbf{x}_j^t}{|\mathbf{x}_i||\mathbf{x}_j|}\right) \tag{2.66}$$

can be used instead of the L_M metric functions to quantify the dissimilarity between the two vectors. As an example, color images where the color signals appear as three-variate vectors in the RGB color space are considered. It was argued in [12] that similar colors have almost parallel orientations. On the other hand, significantly different colors point in different overall directions in the three-variate color space. Thus, the angular distance, which quantifies the orientation difference between two color signals, is a meaningful measure of their similarity.

It is obvious that a generalized similarity measure model which can effectively quantify differences among multichannel signals should take into consideration both the magnitude and the orientation of each vector signal. The distance or similarity measures discussed thus far, utilize only part of the information carried by the vector signal. It is anticipated that a generalized measure based on both the magnitude and the orientation of the vectors will provide a robust solution to the problem of similarity between two vectors.

To this end, a new similarity measure was introduced [48]. The proposed measure defines similarity between two vectors $\mathbf{x}_i$ and $\mathbf{x}_j$ as follows:

$$s_2(\mathbf{x}_i, \mathbf{x}_j) = (\frac{\mathbf{x}_i \mathbf{x}_j^t}{|\mathbf{x}_i||\mathbf{x}_j|})(1 - \frac{||\mathbf{x}_i| - |\mathbf{x}_j||}{\max(|\mathbf{x}_i|, |\mathbf{x}_j|)}) \tag{2.67}$$

As can be seen, this similarity measure takes into consideration both the direction and the magnitude of the vector inputs. The first part of the measure is equivalent to the angular distance defined previously and the second part is related to the normalized difference in magnitude. Thus, if the two vectors under consideration have the same length, the second part of (2.67) becomes unity and only the directional information is used. On the other hand, if the vectors under consideration have the same direction in the vector space (collinear vectors) the first part (orientation) is unity and the similarity measure of (2.67) is based only on the magnitude difference.

The proposed measure can be considered a member of the generalized 'content model' family of measures, which can be used to define similarity between multidimensional signals [49]-[51]. The main idea behind the 'content model' family of similarity measures is that similarity between two vectors is regarded as the degree of common content in relation to the total content of the two vectors [52]-[58]. Therefore, given the common quantity, commonality C_{ij}, and the total quantity, totality T_{ij}, the similarity between $\mathbf{x}_i$ and $\mathbf{x}_j$ is defined as:

$$s(\mathbf{x}_i, \mathbf{x}_j) = \frac{C_{ij}}{T_{ij}} \tag{2.68}$$

Based on the general framework of (2.68), different similarity measures can be obtained by utilizing different commonality and totality concepts.

Given two input signals $\mathbf{x}_i$ and $\mathbf{x}_j$, assume that the angle between them is θ and their magnitudes are $|\mathbf{x_i}|$ and $|\mathbf{x_j}|$ respectively. As before, the magnitudes of the vectors represent the intensity and the angle between the vectors quantifies the orientation difference between them. Based on these elements, commonality can be defined as the sum of the projections of one vector over the other and totality as the sum of their magnitudes. Therefore, their similarity model can be written as:

$$s_3(\mathbf{x}_i, \mathbf{x}_j) = \frac{h_i + h_j}{|\mathbf{x_i}| + |\mathbf{x_j}|} = \frac{|\mathbf{x_i}|cos(\theta) + |\mathbf{x_i}|cos(\theta)}{|\mathbf{x_i}| + |\mathbf{x_j}|} = cos(\theta) \tag{2.69}$$

where $h_i = |\mathbf{x_i}|cos(\theta)$. Although, content model in [55], [56] is equivalent to the normalized inner product (cosine of the angle) similarity model of (2.65), different similarity measures can be devised if commonality is defined and/or totality between the two vectors differently. Experimental studies have revealed that there is a systematic deviation between empirically measured similarity values and those obtained through the utilization of the model in [52], especially in applications where the magnitudes of the vectors are of importance. To compensate for the discrepancy, the totality T_{ij} was redefined as the vector sum of the two vectors under consideration. In such a case similarity was defined as:

$$s_4(\mathbf{x}_i, \mathbf{x}_j) = \frac{h_i + h_j}{(|\mathbf{x_i}|^2 + |\mathbf{x_j}|^2 + 2|\mathbf{x_i}||\mathbf{x_j}|cos(\theta))^{\frac{1}{2}}} \tag{2.70}$$

In the special case of vectors with equal magnitudes, the similarity measure is solely based on the orientation differences between the two vectors and it can be written as:

$$s_4(\mathbf{x}_i, \mathbf{x}_j) = \frac{cos(\theta)}{cos(\frac{\theta}{2})} \tag{2.71}$$

These are not the only similarity measure, which can be devised based on the content-model approach. For example, it is also possible to define commonality between two vectors as a vector algebraic sum, instead of a simple sum, of their projections. That gives a mathematical value of commonality lower than the one used in the models reported earlier. Using the two totality measures two new similarity measures can be compromised as:

$$s_5(\mathbf{x}_i, \mathbf{x}_j) = \frac{(|h_i|^2 + |h_j|^2 + 2|h_i||h_j|cos(\theta))^{\frac{1}{2}}}{|\mathbf{x}_i| + |\mathbf{x}_j|} \tag{2.72}$$

or

$$s_5(\mathbf{x}_i, \mathbf{x}_j) = \frac{cos(\theta)(|\mathbf{x}_i|^2 + |\mathbf{x}_j|^2 + 2|\mathbf{x}_i||\mathbf{x}_j|cos(\theta))^{\frac{1}{2}}}{|\mathbf{x}_i| + |\mathbf{x}_j|} \tag{2.73}$$

If only the orientation similarity between the two vectors is of interest, assuming that $|\mathbf{x_i}| = |\mathbf{x_j}|$, the above similarity measure can be rewritten as:

$$s_5(\mathbf{x}_i, \mathbf{x}_j) = cos(\theta)cos(\frac{\theta}{2}) \tag{2.74}$$

If, on the other hand, the totality T_{ij} is defined as the algebraic sum of the original vectors and define commonality C_{ij} as the algebraic sum of the corresponding projections, the resulting similarity measure can be expressed as:

$$s_6(\mathbf{x}_i, \mathbf{x}_j) = \frac{cos(\theta)p_{(i,j)}}{p_{(i,j)}} = cos(\theta) \tag{2.75}$$

with

$$p_{(i,j)} = (|\mathbf{x}_i|^2 + |\mathbf{x}_j|^2 + 2|\mathbf{x}_i||\mathbf{x}_j|cos(\theta))^{\frac{1}{2}} \tag{2.76}$$

which is the same expression obtained through the utilization of the inner product in (2.65).

Other members of the content based family of similarity measures can be obtained by modifying either the commonality or the totality or both of them. The formula of (2.68) can be seen as a guideline for the construction of specific models where the common part and the total part are specified.

As a general observation, it can be claimed that when totality and commonality were derived according to the same principle, e.g. sum of vectors, the cosine of the angle between the two vectors can be used to quantify similarity. On the other hand, when commonality and totality were derived according to different principles, similarity was defined as a function of both the angle between the vectors and their corresponding magnitudes.

Content-based measures can also be used to define dissimilarity among vector signals. This is the approach taken in [57] where the emphasis is on what is uncommon to the two vectors instead of on what is common. In his dissimilarity model, the uncommon part to the vectors divided by the total part was assumed to be the measure of their dissimilarity. It was suggested in [57] that the part not in common is specified as the distance between the two vector termini with the totality defined as the vector sum of the two vectors under consideration. Further, assuming that similarity and distance are complimentary, the following similarity measure were proposed:

$$s_7(\mathbf{x}_i, \mathbf{x}_j) = 1 - \frac{\left(|\mathbf{x}_i|^2 + |\mathbf{x}_j|^2 - 2|\mathbf{x}_i||\mathbf{x}_j|cos(\theta)\right)^{\frac{1}{2}}}{\left(|\mathbf{x}_i|^2 + |\mathbf{x}_j|^2 + 2|\mathbf{x}_i||\mathbf{x}_j|cos(\theta)\right)^{\frac{1}{2}}} \tag{2.77}$$

where the numerator of the ratio represents the distance between the two vector termini, e.g. vector difference, and the denominator is an indication of the totality.

The different non-metric similarity measures described here can be used instead of the Minkowski type distance measures to quantify distance among a vector under consideration and the ideal prototype in our membership function mechanism, as discussed earlier.

Although in the $s_7(\mathbf{x}_i, \mathbf{x}_j)$ model it was assumed that distance and similarity are complimentary, judgments of differences may be related to similarity in various ways [56], [59], [60]. The most commonly used approach is that suggested in [58] and used in [48], where difference judgments are correlated negatively with similarity judgments. In most applications difference judgments are often the inverse of similarity judgments and the choice between the two rests on practical considerations.

It should be emphasized at this point, that a satisfactory approximation of the similarity (or difference) mechanism with a static model, such as those considered here, can be obtained only when the comparison of vector signals is concentrated to a relatively small part of the p-variate space. That is to say that relatively high homogeneity is required [57].

Other forms of similarity can also be used to to rank multivariate, vector-like signals. Assuming that two vector signals $\mathbf{x}_i, \mathbf{x}_j$ are available, their degree of similarity can be obtained by any of the following methods [61]:

1. Correlation coefficient method.
   Defining $\bar{x}_i = \frac{1}{p}\sum_{k=1}^{p} x_{ik}$ and $\bar{x}_j = \frac{1}{p}\sum_{k=1}^{p} x_{jk}$ the correlation coefficient between the two vectors is given as follows:

$$s_{ij} = \frac{\sum_{k=1}^{p} |x_{ik} - \bar{x}_i||x_{jk} - \bar{x}_j|}{(\sum_{k=1}^{p} (x_{ik} - \bar{x}_i)^2)^{\frac{1}{2}} (\sum_{k=1}^{p} (x_{ik} - \bar{x}_i)^2)^{\frac{1}{2}}} \tag{2.78}$$

2. Exponential similarity method.

$$s_{ij} = \frac{1}{p}\sum_{k=1}^{p} (\exp{(\frac{-3}{4})}\frac{(x_{ik} - x_{jk})^2}{\beta_k^2}) \tag{2.79}$$

   with the parameter $\beta_k > 0$ a design parameter, the value of which is data determined.
3. The absolute-value exponent method.

$$s_{ij} = \exp{(-\beta)}(\sum_{k=1}^{p} |x_{ik} - x_{jk}|) \tag{2.80}$$

   as before the parameter $\beta_k > 0$ a design parameter used to regulate the rate of similarity with its value determined by the designer.
4. The absolute-value reciprocal method.

$$s_{ij} = \begin{cases} 1 & \text{if } i = j \\ 1 - \frac{\beta}{\sum_{k=1}^{p} |x_{ik} - x_{jk}|} & \text{if } i \neq j \end{cases} \tag{2.81}$$

   where β is selected so that $0 \leq s_{ij} \leq 1$.
5. Maximum-minimum method.

$$s_{ij} = \frac{\sum_{k=1}^{p} \min{(x_{ik}, x_{jk})}}{\sum_{k=1}^{p} \max{(x_{ik}, x_{jk})}} \tag{2.82}$$

6. Arithmetic-mean minimum method.

$$s_{ij} = \frac{\sum_{k=1}^{p} \min{(x_{ik}, x_{jk})}}{\frac{1}{2}\sum_{k=1}^{p} (x_{ik} + x_{jk})} \tag{2.83}$$

7. Geometric-mean minimum method.

$$s_{ij} = \frac{\sum_{k=1}^{p} \min{(x_{ik}, x_{jk})}}{\sum_{k=1}^{p} (x_{ik}x_{jk})^{\frac{1}{2}}} \tag{2.84}$$

Of course there are many other methods by which a similarity or distance value between two vector signals can be constructed. Depending on the nature and the objective of the problem on hand, one method may be more appropriate than the other. The fundamental idea, however, is that through the reduction function, a multivariate space is mapped into a scalar space. Techniques other than distance or similarity measures can be utilized to assist with the mapping. One such technique is the space filling curves. Space filling curves can be defined as a set of discrete curves that make it possible to cover all the points of a p-dimensional multivariate space. In particular, a space filling curve must pass through all the points of the space only once, and make it possible to realize a mapping of the p-dimensional space into a scalar interval, thus it allows for ranking multivariate data. That is to say, it is possible to associate with each point in the p-dimensional space a scalar value which is directly proportional to the length of the curve necessary to reach the point itself starting from the origin of the coordinates. Then, as for all vector ordering schemes, vector ranking can be based on sorting the scalar values associated with each vector.

Through the utilization of the space filling curves it is possible to reduce the dimensionality of the space. A bi-dimensional space is considered here for demonstration purposes. A generic curve γ allows an association of a scalar value with a p-variate vector as follows:

$$\gamma(t_k) = (x_{1k}(t_k), x_{2k}(t_k)) \tag{2.85}$$

with γ: $\mathcal{Z} \to K$, $K \subset \mathcal{Z}^2$.

A filling curve makes it possible to cover, as the parameter t_k varies, all the points of the discrete space K, so that each point is crossed only once, $\mathbf{x}_k \in K$ then exists $\gamma(t_k) = \mathbf{x}(t_k)$ and if $t_k, t_l \in \mathcal{Z}$ then $t_k \neq t_l \to \gamma(t_k) \neq \gamma(t_k)$. In accordance with the above definitions, a filling curve substantially makes a scanning operation of the K space and generates a list of vectors in which there is no repetition of the same elements of $\mathbf{x}_k$. The filling curve itself is invertible thus, if $\gamma(t_k) = \mathbf{x}_k$ then:

$$\exists \gamma^{-1} : K \to \mathcal{Z} : \gamma^{-1}(\mathbf{x}_k) = t_k \tag{2.86}$$

An important observation which derives from (2.86) is that, by means of parameter t_k, it is possible to make a scalar indexing operation for each bi-dimensional vector and then to reduce the bi-dimensional space and use the set of transformed values for scalar ordering. To design a space filling curve able to be used for color image processing, it is necessary to extend the notion of space filling to the three channel RGB color space. The three-variate filling curve can be imagined as an expansion of successive increasing layers, ordered according to the maximum value of each three dimensional color vector. A possible implementation strategy is to impose that the three variate filling curve crosses all points at the same maximum value in a continuous way, e.g. by covering in an ordered way the three sides of a cube in the RGB color space [62], [63].

2.10 Filters Based on Marginal Ordering

The use of marginal ordering (M-ordering) is the most straightforward multivariate approach to color image filtering based on data ordering. The three color image channels, in the RGB color space, are ordered independently. Several multichannel nonlinear filters that are based on marginal ordering can be proposed. The marginal median filter (MAMF) is the running marginal median operator $\mathbf{y}_{(\nu+1)}$ for $n = 2\nu + 1$. The marginal rank order filter is the running order statistic $\mathbf{y}_{(i)}$ [38]. Based on similar concepts defined for univariate (one-dimensional) order statistics, a number of nonlinear filters, such as the median, the α-trimmed mean and the L-filter have been devised for color images by using marginal ordering.

Theoretical analysis and experimental results had led to the conclusion that the marginal median filter is robust in the sense that it discards (filters out) impulsive noise while preserving important signal features, such as edges. However, its performance in the suppression of additive white Gaussian noise, which is frequently encountered in image processing, is inferior to that of the moving average or other linear filters. Therefore, a good compromise between the marginal median and the moving average or mean filter is required. Such a filter is the α-trimmed mean filter, which is the robust estimator for the normal (Gaussian) distribution. In gray scale images the α-trimmed mean filter is implemented as a local area operation, where after ordering the univariate pixel values in the local window, the top α% and the bottom α% are rejected and the mean of the remaining pixels is taken as the output of the filter, thus achieving a compromise between the median and mean filters.

Now, using the marginal ordering scheme as defined previously, the α-trimmed mean filter for p-dimensional vector images has the following form [4], [65]:

$$\mathbf{y}_{\alpha_n} = \begin{bmatrix} \frac{1}{n(1-2\alpha_1)} \sum_{i=\alpha_1 n+1}^{n-\alpha_1 n} y_{1(i)} \\ \cdots \\ \frac{1}{n(1-2\alpha_p)} \sum_{i=\alpha_p n+1}^{n-\alpha_p n} y_{p(i)} \end{bmatrix} \tag{2.87}$$

The α-trimmed mean filter, as defined is 2.87, will reject 2α% of the outlying multivariate samples while still using $(1-2\alpha)$ of the pixels to take the average. The trimming operation should cause the filter to have good performance in the presence of long tailed or impulsive noise and should help to preserve sharp edges, while averaging or mean operation should cause the filter to also perform well in the presence of short tailed noise, such as Gaussian.

Trimming can also be obtained by rejecting data that lie far away from their marginal median value. The remaining data can be averaged to form the modified trimmed mean filter as follows:

$$\mathbf{y}_{\tilde{\alpha}_n} = \frac{\sum_W \alpha_r \mathbf{y}_{i+r}}{\sum_W \alpha_r} \tag{2.88}$$

with

$$\alpha_r = \begin{cases} 1 & (\mathbf{y}_i - \mathbf{y}_{(\nu+1)})^T \Gamma^{-1} (\mathbf{y}_i - \mathbf{y}_{(\nu+1)}) \leq d \\ 0 & \text{otherwise} \end{cases} \tag{2.89}$$

where W is the filter window and Γ is a matrix related to data dispersion.

The α-trimmed filter is a member of the family of marginal order statistic filters, also called L-filters [66], whose output is defined as a linear combination of the order statistics of the input signal sequence. The design of an optimal L-filter for estimating a constant signal corrupted by additive white noise have been proposed in [66] and has been extended to the design of L-filters for multivariate signals based on marginal ordering (M-ordering).

The following estimator will be called the p-variate marginal L-filter:

$$\mathcal{T}(\mathbf{y}_{\mathcal{L}}) = \sum_{i_1=1}^{n} \cdots \sum_{i_p=1}^{n} \mathbf{A}_{(i_1,i_2,\ldots,i_p)} \mathbf{y}_{(i_1,i_2,\ldots,i_p)} \tag{2.90}$$

where $\mathbf{y}_{(i_1,i_2,\ldots,i_p)} = [x_{1(i_1}, \ldots, x_{p(i_p)}]^{\mathcal{T}}$ are the marginal order statistics and $\mathbf{A}_{(i_1,i_2,\ldots,i_p)}$ are $p{\times}p$ matrices. The performance of the marginal L-filter depends on the choice of the matrices $\mathbf{A}_{(i_1,i_2,\ldots,i_p)}$. The L-filter of (2.90) coincides with the p-variate marginal median for the following choice of matrices $\mathbf{A}_{(i_1,i_2,\ldots,i_p)}$:

$$\mathbf{A}_{(i_1,i_2,\ldots,i_p)} = \mathbf{0} \quad i_j \neq \nu + 1$$

$$\mathbf{A}_{nu+1,\ldots,nu+1} = \mathbf{I}_{p \times p} \tag{2.91}$$

Similarly, the marginal maximum $\mathbf{y}_{(n)}$, the marginal minimum $\mathbf{y}_{(1)}$ and the moving average (mean) as well as the α-trimmed mean filter are special cases of (2.90).

The robustness of the L-filters in the presence of multivariate outliers can be found by using the p-variate influence function [38], [37]. The influence function is a tool used in robust estimation for qualitatively characterizing the behavior of a filter in the presence of outliers. It relates to the asymptotic bias caused by the contamination of the observations. As the name implies, the function measures the influence of an outlier on the filter's output. To evaluate the influence function in the p-variate case it is assumed that the vector filter is expressible as a functional $\mathcal{T}$ of the empirical distribution F of the data samples. When the sample size n is sufficiently large $\mathcal{T}(F_n)$ converges in probability to an asymptotic functional $\mathcal{T}(F)$ of the underlying distribution F.

Then the influence function $IF(\mathbf{y}, \mathcal{T}, F)$ which measures the change of T caused by an additional observation at point $\mathbf{y}$ is calculated as follows [26], [27]:

$$IF(\mathbf{y}, \mathcal{T}, F) = \lim_{t \to 0} \frac{\mathcal{T}[(1-t)F - t\Delta_{\mathbf{y}}] - \mathcal{T}[F]}{t} \tag{2.92}$$

where Δ_y is $\Delta x_1 \Delta x_2 \ldots \Delta x_p$, a product of unit step functions at $x_1, x_2, \ldots, x_p$ respectively. Each component of the influence function indicates the standardized change that occurs in the corresponding component of the filter when the assumed underlying distribution F is perturbed due to the presence of t

outliers. If the change is bounded, the filter has good robustness properties and an outlier cannot destroy its performance. Therefore, the robustness of the filter can be measured in terms of its gross error sensitivity [38]:

$$\gamma^*(\mathcal{T},F) = \sup_{\mathbf{x}} ||IF(\mathbf{y},\mathcal{T},F)||_2 \tag{2.93}$$

where $||.||_2$ denotes the Euclidean norm. It can be proved, under certain conditions, that the L-filter is asymptotically normal and its covariance matrix is given by:

$$V(\mathcal{T},F) = \int IF(\mathbf{y},\mathcal{T},F)IF(\mathbf{y},\mathcal{T},F)^{\tau}\, dF(\mathbf{y}) \tag{2.94}$$

In cases, such as the one considered here, where the actual signal $\mathbf{x}$ is approximately constant in the filter's window, the performance of the filter is measured by the dispersion matrix of the output:

$$D(\mathcal{T}) = E[(\mathcal{T}(\mathbf{y}_{\mathcal{L}}) - \mu_T)(\mathcal{T}(\mathbf{y}_{\mathcal{L}}) - \mu_T)^{\tau}] \tag{2.95}$$

where $\mu_T = E[\mathcal{T}(\mathbf{y}_{\mathcal{L}})]$. The smaller the elements of the output dispersion matrix, the better the performance of the filter. The dispersion matrix is related asymptotically to the covariance matrix $V(\mathcal{T},F)$ as follows [38]:

$$D(\mathcal{T}) = \frac{1}{n}V(\mathcal{T},F) \tag{2.96}$$

The coefficients of the L-filter can be optimized for a specific noise distribution with respect to the mean squared error between the filter output and the desired, noise-free color signal, provided that the latter is available to the designer and constant within the filter window. The structural constraints of unbiasness and location invariance can also be incorporated in the filter design. To this end, the mean square error (MSE) is used between the filter output $\hat{\mathbf{y}} = \mathcal{T}(\mathbf{y}_{\mathcal{L}})$ and the constant, noise-free, multivariate signal $\mathbf{x}$ expressed in the following way:

$$\begin{aligned} \epsilon &= E[(\hat{\mathbf{y}} - \mathbf{x})^{\tau}(\hat{\mathbf{y}} - \mathbf{x})] \\ &= E[\sum_{i=1}^{n}\sum_{j=1}^{n} \mathbf{y}_{(i)} A_i^{\tau} A_j \mathbf{y}_{(j)}] - 2\mathbf{x}^{\tau} E[A_i \mathbf{y}_{(i)}] + \mathbf{x}^{\tau}\mathbf{x} \end{aligned} \tag{2.97}$$

After some manipulation, (2.97) becomes:

$$\epsilon = \sum_{i=1}^{n}\sum_{j=1}^{n} tr[A_i R_{ij} A_j^{\tau}] - 2\mathbf{x}^{\tau}\sum_{i=1}^{n} A_i \mu_i + \mathbf{x}^{\tau}\mathbf{x} \tag{2.98}$$

where R_{ij} is the $(p \times p)$ correlation matrix of the jth and i^{th} order statistics $R_{ij} = E[\mathbf{y}_i \mathbf{y}_j]$, $i,j = 1,2,...,n$ and μ_i, $i = 1,2,...,n$ denotes the $(p \times 1)$ mean vector of the ith order statistic $\mu_i = E[\mathbf{y}_{(i)}]$.

Let $\mathbf{a}_i$ denote the $(np\times1)$ vector that is made up by the ith row of matrices $A_1,\ldots,A_n$. Also, the $(np\times1)$ vector $\bar{\mu}_p$ is defined in the following way:

$$\bar{\mu}_p = [\mu_1^T, \mu_1^T, \ldots, \mu_p^T]^T$$

where μ_j denote the mean vector of the order statistics in channel j, as well as the $(np\times np)$ matrix $\bar{R}_p$

$$\bar{R}_p = \begin{bmatrix} R_{11} & R_{12} & \cdots & R_{1p} \\ R_{12}^T & R_{22} & \cdots & R_{2p} \\ \cdots & \cdots & \cdots & \cdots \\ R_{1p}^T & R_{2p}^T & \cdots & R_{pp} \end{bmatrix} \tag{2.99}$$

Using the previous notation, after some manipulation the MSE is given by:

$$\epsilon = \sum_{i=1}^{p} \mathbf{a}_{(i)}^T \bar{R}_p \mathbf{a}_{(i)} - 2\mathbf{x}^T [\mathbf{a}_{(1)}^T, \mathbf{a}_{(2)}^T, \ldots, \mathbf{a}_{(p)}^T]^T + \mathbf{x}^T\mathbf{x} \tag{2.100}$$

The minimization of the MSE in (2.100) results in the following p sets of equations:

$$\bar{R}_p \mathbf{a}_{(m)} = \mathbf{x}_m \bar{\mu}_p \tag{2.101}$$

with $m = 1, 2, \ldots p$, which yields the optimal p-variate L-filter coefficients:

$$\mathbf{a}_{(1)}^* = \mathbf{x}_1 \bar{R}_p^{-1} \bar{\mu}_p$$

$$\mathbf{a}_{(m)}^* = \frac{\mathbf{x}_m}{\mathbf{x}_1} \mathbf{a}_{(1)}^* \tag{2.102}$$

where $m = 2, \ldots p$.

That completes the derivation of the multivariate L-filters based on the marginal sub-ordering principle and the MSE fidelity criterion. In addition, the constrained minimization subject to the constraints of the unbiased and location-invariant estimation can be found in [66]. Simulation results reported in [33], [66] suggest that multivariate filters based on marginal data ordering are superior to simple moving average, marginal median and single channel L filters when applied to color images.

2.11 Filters Based on Reduced Ordering

Reduced ordering (R-ordering) is another sub-ordering principle which has been extensively used in the development of multivariate color image filters. R-ordering orders p-variate, vector valued signals according to their distance from some reference vector. As a consequence, multivariate ordering is reduced to scalar ordering. Reduce ordering is rather easy to implement, it can provide cues about outliers and is the sub-ordering principle that is the most natural for vector valued observations, such as color image signals. It is obvious that the choice of an appropriate reference vector is crucial for the reduced ordering scheme. Depending on the reference vectors, different ranking schemes, such as the median R-ordering, the center R-ordering and the mean R-ordering, the marginal median, the center value or the window average have been used as the reference vector respectively. The choice of the appropriate reference vector depends on the design characteristics and is application dependent.

Assuming that a suitable reference vector and an appropriate reduction function are available, the set of vectors $W(n)$ can be ordered. It can be expected that any outliers be located at the upper extreme ranks of the sorted sequence. Therefore, an order statistics $\mathbf{y}_{(j)}$, $j = 1, 2, ..., m$ with $m \leq n$ can be selected where it can be safely assumed that the color vectors are not outliers.

For analysis purposes, the Euclidean distance will be used as a reduction function and that mean R-ordering, that is ordering around the mean value $\bar{\mathbf{y}}$ of the samples in the processing window, is utilized. Then, let $d_{(j)}$ define the radius of a hyper-sphere centered around the sample mean value. The hyper-sphere defines a region of confidence. If the sample $\mathbf{y}_k$ lies within the hyper-sphere it can be assumed that this color vector is not an outlier and thus, it should not be altered by the filter operation. Otherwise, if $\mathbf{y}_k$ is beyond the specific volume, that is if $L_2(\mathbf{y}_k, \bar{\mathbf{y}}) = ||\mathbf{y}_k - \bar{\mathbf{y}}||_2 = (\mathbf{y}_k - \bar{\mathbf{y}}))^T(\mathbf{y}_k - \bar{\mathbf{y}})$ is greater than $d_{(j)}$, then the window center value is replaced with the nearest vector signal contained in the set $W^*(n) = [(\mathbf{y}_{(1)}, \mathbf{y}_{(2)}, ..., \mathbf{y}_{(j)}]$. Therefore, the resulting reduced ordering R_E filter can be defined as follows [25]:

$$\hat{\mathbf{y}}_k = \begin{cases} \mathbf{y}_k & \text{if } L_2(\mathbf{y}_k, \bar{\mathbf{y}}) \leq d_j \\ [\mathbf{y}_j \in [\mathbf{y}_{(1)}, \mathbf{y}_{(2)}, ..., \mathbf{y}_{(m)}]; & \\ \qquad \min_j ||\mathbf{y}_j - \mathbf{y}_k||_2] & \text{otherwise} \end{cases} \tag{2.103}$$

Based on the above definition, although the filter threshold is $d_{(j)}$, the output of the filter when the replacement occurs is not necessarily $\mathbf{y}_j$ since there may exist another sample which is closer to $\mathbf{y}_k$.

The threshold order statistic $\mathbf{y}_j$ is a design parameter which defines the volume of the hyper-sphere around the reference point, in this case the mean value. Thus, it defines the likelihood of an input vector to be modified by the filtering process. The filter's replacement probability can be used as an indication of the extent of smoothing being performed by the R_E estimator. In (2.103) a vector replacement occurs if the center sample $\mathbf{y}_k$ has a distance, from the mean, d_k greater than that of the jth ranked distance $d_{(j)}$ in the set $W^*(n)$. The probability of a filter replacement p_f can then be expressed as:

$$p_f = \int_{r=0}^{\infty} Pr[d_k > r | d_{(j)} = r] Pr[d_{(j)} = r] \, dr \tag{2.104}$$

By excluding the center sample from the ranked set $W^*(n)$ the probability of $[d_k > r]$ is independent of the event $[d_{(j)} = r]$. Therefore, the conditional probability in (2.104) can be reduced. In addition, since the samples in the observation set are assumed independent and identically distributed (i.i.d.) the filter replacement probability is given as:

$$p_f = \int_{x=0}^{\infty} [1 - F_d(x)] f_{d(j)}(x)\, dx \tag{2.105}$$

where F_d is the cumulative distribution function (cdf) and $f_{d(j)}$ the probability distribution function for the jth ranked vector distance.

If the value of j is large enough towards the upper rank order statistics, fewer replacements will be attempted by the filter. This design parameter can be used to balance the need for noise suppression through vector replacements and detail preservation and it can be tuned to achieve the desired objective.

However, the ranked order statistics threshold is not the only design parameter in the filter. The kind of reduction function used also affects the performance of the filter. The Euclidean distance (L_2 norm), which is usually employed, fixes the confidence interval as a hyper-sphere of constant volume. However, in some applications the performance of the R_E filter can be improved by modifying the region of confidence to match certain source distributions. This can be obtained by using the generalized distance (Mahalanobis distance) which takes into account the dispersion matrix of the data. If needed, other reduction functions, such as the (q_i^2) or (u_i^2) measures can also be used. Different reduction measures define different confidence volumes. If a-priori information about the noise characteristics is available then the confidence volume can be related to the covariance matrix of the noise distribution [67], [68].

It was mentioned previously that when outliers are present, the estimates of the location and dispersion will be affected by them. Therefore, robust estimates of both the mean value and the sample covariance should be utilized. Various design options are available. The most commonly used robust estimates are the multivariate running M-estimates of the location $\bar{\mathbf{y}}_M$ and the covariance S_M defined as follows:

$$\bar{\mathbf{y}}_M = \frac{\sum_{i=1}^{n} w_i \mathbf{y}_i}{\sum_{i=1}^{n} w_i} \tag{2.106}$$

$$S_M = \frac{\sum_{i=1}^{n} w_i^2 (\mathbf{y}_i - \bar{\mathbf{y}}_M)(\mathbf{y}_i - \bar{\mathbf{y}}_M)^T}{\sum_{i=1}^{n} (w_i^2 - 1)} \tag{2.107}$$

The denominator in (2.107) can be different, with some authors preferring to use $(\sum_{i=1}^{n} w_i^2)$ or (n) instead of $(\sum_{i=1}^{n} (w_i^2 - 1))$ proposed in [71].

The weights w_i are calculated iteratively using the Mahalanobis distance from the previously calculated $\bar{\mathbf{y}}_M$ until stability in the weights is achieved. To further reduce the influence of possible outliers on the M-filter's weights a weighting function $\phi(.)$ is used. During the successive iterations, weights are calculated according to the following formula:

$$w_i = \frac{\phi(d_i)}{d_i} \tag{2.108}$$

A re-descending function which limits the influence of observations resulting in large distances is used in (2.108). A different number of weighting function can be used to achieve this task [69], [43]. For example, a simple, yet effective, function is given by :

$$w_i = \frac{1+p}{1+d_i^2} \tag{2.109}$$

where the weight of each vector depends directly on its $d_i = (\mathbf{y}_i - \bar{\mathbf{y}})(\mathbf{y}_i - \bar{\mathbf{y}})^T$ value. Other functions can be used instead. For example, the designer may wish to give full weight to data that have relatively small distances and down-weight those observations that occupy the extreme ranks of the ordered sequence. In such cases, the weighting function can be defined as follows [69], [70]:

$$w_i = \begin{cases} 1 & \text{if } d_i \leq d_0 \\ \frac{d_0}{d_i} & \text{otherwise} \end{cases} \tag{2.110}$$

Another, more complex, was used in [71] resulting in weights defined as:

$$w_i = \begin{cases} d_i = (\mathbf{y}_i - \bar{\mathbf{y}}_M)S_M^{-1}(\mathbf{y}_i - \bar{\mathbf{y}}_M)^T & \text{if } d_i \leq d_0 \\ d_0 \exp\left(\frac{-1}{2}\frac{(d_i - d_0)^2}{c_2^2}\right) & \text{otherwise} \end{cases} \tag{2.111}$$

where d_0, b_2 are tuning parameters used to control the range of the weights. The parameter values can be either data dependent, or can be fixed. In the latter case, experimental analysis, suggests that values, such as $d_0 = \sqrt{p} + \frac{c_1}{\sqrt{2}}$ with:

1. $c_1 = \infty$, c_2 immaterial

2. $c_1 = 2$, $c_2 = \infty$

3. $c_1 = 2$, $c_2 = 1.25$

should provide acceptable results in most applications [71]. Comparing the functions suggested above the first set of parameters yields a value of 1 for all samples, the second set gives a step function with two possible values, and the third option leads to a full descending set of weights over the whole data set [43].

Given the robust estimates of the location and the dispersion matrix, the robust version of the Mahalanobis distance can be calculated as:

$$d_j^M = ||\mathbf{y}_i - \bar{\mathbf{y}}_M||_M = (\mathbf{y}_i - \bar{\mathbf{y}}_M)S_M^{-1}(\mathbf{y}_i - \bar{\mathbf{y}}_M)^T$$

and thus a new confidence volume around the M-estimate of the location (robust mean) can be formed. Therefore, similar to the R_E filter, a new filter called the R_M filter can be defined based on the R-ordering principle and the robust Mahalanobis distance [25]:

$$\hat{\mathbf{y}}_k = \begin{cases} \mathbf{y}_k & \text{if } ||\mathbf{y}_j - \mathbf{y}_k||_M \leq d_j^M \\ \mathbf{y}_j \in [\mathbf{y}_{(1)}, \mathbf{y}_{(2)}, ..., \mathbf{y}_{(m)}] & \\ \quad ; \min_j ||\mathbf{y}_j - \mathbf{y}_k||_M & \text{otherwise} \end{cases} \tag{2.112}$$

where $||\mathbf{y}_j - \mathbf{y}_k||_M = (\bar{\mathbf{y}}_M - \mathbf{y}_k) S_M^{-1} (\bar{\mathbf{y}}_M - \mathbf{y}_k)^{\tau}$ is the generalized Mahalanobis distance between vectors $\mathbf{y}_j$ and $\mathbf{y}_k$.

The performance of the R_M filter depends primarily on the robust estimate of the dispersion matrix. In non-smooth parts of the image or in image regions where strong edges, line or other structures are spanned by the estimator's window, special consideration is required since sample estimate S_M may no longer be appropriate [25].

Both the R_E and the R_M filters discussed above are based on the mean R-ordering principle, where the mean value of the samples, or a robust estimator of the location, is used as the reference vector. However, in areas of the image which are rich in details, the mean filter, or the marginal median, tend to smooth out details, such as lines and small structures. However, due to the masking properties of the human eye, detail retention is more important than noise reduction. Thus, in such a case the center sample of $W(n)$ should be used as reference vector instead of the mean or the marginal median (center R-ordering). In non-stationary color images, the flat image regions often follow areas filled with details. Thus, a filter based on R-ordering where its reference vector moves adaptively towards the mean, the marginal median, or the center sample according to the structure of the noisy image, seems appealing.

To this end, a new R-ordering multivariate filter was introduced in [14], [15]. The filter output has the minimum sum of (Euclidean) distances to the mean, marginal median and the center sample. Furthermore, the filter output is one of the multivariate samples included in the $W(n)$ set. Thus, given the set $W(n) = [\mathbf{y}_1, \mathbf{y}_2, ..., \mathbf{y}_n]$ the output of the proposed R_C filter is defined as:

$$\hat{\mathbf{y}}_k = \mathbf{y}_j; \quad \min_j [||\mathbf{y}_j - \bar{\mathbf{y}}||_2 + ||\mathbf{y}_j - \mathbf{y}_m||_2 + ||\mathbf{y}_j - \mathbf{y}_k||_2] \tag{2.113}$$

where $||.||_2$ is the Euclidean distance (L_2) norm, $\mathbf{y}_m$ is the marginal median with $\bar{\mathbf{y}}$ the mean of the $W(n)$ set.

Other multivariate filters based on R-ordering can be defined by modifying distance criteria used in (2.113). For example, it was suggested in [15] that the sum of distances can be replaced with that of squared distances. In that case, the multivariate filter leads to a single computation:

$$\hat{\mathbf{y}}_k = \mathbf{y}_j; \quad \min_j [||\mathbf{y}_j - \bar{\mathbf{y}}||_2{}^2 + ||\mathbf{y}_j - \mathbf{y}_m||_2{}^2 + ||\mathbf{y}_j - \mathbf{y}_k||_2{}^2] \tag{2.114}$$

From (2.114) it is clear that the R_C filter can be reduced to the following form:

$$\hat{\mathbf{y}}_k = \mathbf{y}_j; \quad \min_j ||\mathbf{y}_j - \mathbf{y}_{\mathcal{A}}||_2 \tag{2.115}$$

where

$$\mathbf{y}_{\mathcal{A}} = \frac{[\bar{\mathbf{y}} + \mathbf{y}_m + \mathbf{y}_k]}{3}$$

Simple inspection of the R_C variants reveals that the proposed equations (2.114) and (2.115) cannot achieve the same noise attenuation as the R_E or R_M filters due to the presence of the center sample. However, a number of properties can be associated with the R_C design. In particular, as a direct consequence of the properties of the Euclidean distance it can be proven that the R_C variants are invariant to scale and bias. Furthermore, if in the set $W(n)$ the center sample $\mathbf{y}_k$ is a convex combination of $\mathbf{y}_m$ and $\bar{\mathbf{y}}$ then the input signal is a root signal of the R_C filter [15]. A special case of this property is that if a multivariate input $\mathbf{y}_i$ is a root signal of the marginal median, it is also a root of the R_C filter . That is to say, that the filter possesses more root signals and thus, preserves more details than the marginal median having the same window size [15].

In the R_C variants presented above, the mean, the marginal median and the center sample have equal importance. However, both the center sample and the mean are sensitive to outliers. On the other hand, the marginal median or a robust estimate of the mean may result in excessive smoothing of image details. Therefore, the R_C filter cannot suppress impulsive noise so efficiently as the marginal median and cannot preserve the image details as well as the identity filter of the R_E filter. To overcome these drawbacks and to enhance the performance in noise suppression an adaptive version of the filter was proposed [15]. The output of the R_{aC} filter is defined as:

$$\begin{aligned}\hat{\mathbf{y}}_k = \mathbf{y}_j; \quad \min_j[(1-\alpha)\beta||\mathbf{y}_j - \bar{\mathbf{y}}||_2{}^2 + (1-\beta)||\mathbf{y}_j - \mathbf{y}_m||_2{}^2 \\ + \alpha\beta||\mathbf{y}_j - \mathbf{y}_k||_2{}^2]\end{aligned} \tag{2.116}$$

where $0 \leq \alpha \leq 1$, $0 \leq \beta \leq 1$ are the weights which control the output of the adaptive filter. In the case of $\alpha = \frac{1}{2}$ and $\beta = \frac{2}{3}$, the above expression reduces to the filter defined in (2.115). Similar to the R_C, its adaptive version can also be simplified to the:

$$\hat{\mathbf{y}}_k = \mathbf{y}_j; \quad \min_j ||\mathbf{y}_j - \mathbf{y}_{Co}|| \tag{2.117}$$

with

$$\begin{aligned}\mathbf{y}_{Co} = (1-\alpha)\beta||\mathbf{y}_j - \bar{\mathbf{y}}||_2{}^2 + (1-\beta)||\mathbf{y}_j - \mathbf{y}_m||_2{}^2 \\ + \alpha\beta||\mathbf{y}_j - \mathbf{y}_k||_2{}^2\end{aligned} \tag{2.118}$$

The output $\mathbf{y}_{Co}$ is itself an estimate of the noisy vector at the window center, since it constitutes a weighted sum of the mean, the marginal mean and the center sample. Actually, the calculation of the adaptive weights α and β can be performed either on the R_{Co} or R_C filter since the later is simply the sample closest to $\mathbf{y}_k$ in $W(n)$.

The weights in (2.118) are varied adaptively according to the local activity of the signal and the noise. The two parameters are determined separately. In the procedure described in [15] the minimization of the parameter α is attempted first assuming that $\beta = 1$. This implies that the image area that is processed is regarded as being free of outliers. Since only additive, white Gaussian noise is assumed present, the mean square error (MSE) is the criterion which the filter output seeks to minimize. Similarly to [1], the minimization of the MSE yields:

$$\alpha = \begin{cases} 1 - \frac{\sigma_n^2}{\sigma_y^2} & \text{if } \sigma_y^2 > \sigma_n^2 \\ 0 & \text{otherwise} \end{cases} \tag{2.119}$$

To complete the calculations the covariance σ_n^2 of the additive noise corrupting the image is assumed to be known a-priori. Furthermore, information regarding the actual signal characteristics is needed. In areas of the image where edges or other strong details exist the activity can be attributed to signal and not to the additive noise. These areas can then be used to calculate the characteristics of the image signal as follows:

$$\mathbf{y}_a = \frac{1}{N}\sum_{i=1}^{N} \mathbf{y}_i \tag{2.120}$$

$$\sigma_a^2 = \frac{1}{N}\sum_{i=1}^{N} \mathbf{y}_i - \mathbf{y}_a)(\mathbf{y}_i - \mathbf{y}_a)^{\tau} \tag{2.121}$$

where $N = (2K+1)\times(2K+1)$ is the window size. In the above expression, image samples near an edge tend to have larger dispersion and can be used to estimate the characteristics of the actual image data.

In the computation of α the characteristics of the image signal are estimated from samples. Thus, it is possible for outliers to be mistaken for image details. To suppress outlying observations without disturbing details, the pixel values are compared with a predetermined threshold. If a given pixel value is beyond the threshold, then that particular vector is declared an outlier and it is replaced with the marginal median within the window. In the R_{Co} filter of (2.118) the parameter β is used for this purpose. If the current pixel is considered an outlier, then $\beta \rightarrow 0$, otherwise, β should approach 1 in order to preserve the image detail. Based on these considerations a minmax operation inside the processing window is used to adaptively determine the value of the parameter β, assuming that the other parameter has been already evaluated by (2.119). After some manipulation, the parameter value is calculated as [15]:

$$\beta = \begin{cases} 0 & \text{if } \beta_n \leq 0 \\ \beta_n & \text{if } 0 < \beta_n < 1 \\ 1 & \text{otherwise} \end{cases} \tag{2.122}$$

with

$$\beta_n = \frac{(\alpha+1)\sigma_y^2 - \alpha\sigma_n^2}{\mathbf{y}_k^2} \tag{2.123}$$

The above equation completes the heuristic procedure introduced in [15] for the calculations of the weights. The R_{C_o} filter should be able to remove both impulsive as well as additive Gaussian noise while keeping most of the details of the image unchanged. However, the filter is computationally expensive and the computation of its weights is based on local statistics, where a number of assumptions were made based on experimental justifications.

The R-ordering principle can be used in the derivation of multivariate L filters for vector signals, such as color images. The L filters based on the R-ordering are similar to those discussed in the previous section and their coefficients can be optimized for a specific noise distribution with respect to the mean square error between the filter output and the desired, noise free signal, provided that the latter is constant within the filtering window. As in the case of marginal ordering, a p-variate multichannel filter based on R-ordering is defined by the following input-output relation:

$$\hat{\mathbf{y}}_k = \sum_{i=1}^{n} A_i \mathbf{y}_{(i)} \tag{2.124}$$

where A_i are the $(p \times p)$ coefficient matrices and $\mathbf{y}_{(i)}$ are the R-ordered input vectors of the $W(n)$ set. According to (2.124) each component of the output vector $\hat{\mathbf{y}}$ is a linear combination of all $y_{(i)}l$, $i = 1, 2, ..n$, $l = 1, 2, ..., p$. To calculate the coefficients in (2.124), an optimization procedure based on the MSE error between the filter output and the constant signal value $\mathbf{x}$ is utilized. Similar to the analysis used in the derivation of the marginal based L filter, the following is obtained [72]:

$$\epsilon = E[(\hat{\mathbf{y}} - \mathbf{x})^T (\hat{\mathbf{y}} - \mathbf{x})]$$

$$\epsilon = E[\sum_{i=1}^{n}\sum_{j=1}^{n} \mathbf{y}_{(i)} A_i^T A_j \mathbf{y}_{(j)}] - 2\mathbf{x}^T E[A_i \mathbf{y}_{(i)}] + \mathbf{x}^T \mathbf{x} \tag{2.125}$$

After some manipulation, the previous equation becomes:

$$\epsilon = \sum_{i=1}^{n}\sum_{j=1}^{n} tr[A_i R_{ij} A_j^T] - 2\mathbf{x}^T \sum_{i=1}^{n} A_i \mu_i + \mathbf{x}^T \mathbf{x} \tag{2.126}$$

where R_{ij} is the $(p \times p)$ correlation matrix of the jth and i^{th} order statistics $R_{ij} = E[\mathbf{y}_{(i)} \mathbf{y}_{(j)}]$, $i, j = 1, 2, ..., n$ and μ_i, $i = 1, 2, ..., n$ denotes the $(p \times 1)$ mean vector of the ith order statistic $\mu_i = E[\mathbf{y}_{(i)}]$.

By denoting with $\mathbf{a}_i$ the $(np\times1)$ vector that is made up by the ith row of matrices $A_1, ..., A_n$ and also defining the $(np\times np)$ matrix $\bar{R}_p$ and the $(np\times1)$ vector $\bar{\mu}_p$ in the following way:

$$\bar{\mu}_p = [\mu_1^T, \mu_1^T, ..., \mu_p^T]^T$$

where μ_j denote the mean vector of the order statistics in channel j, and

$$\bar{R}_p = \begin{bmatrix} R_{11} & R_{12} & \cdots & R_{1p} \\ R_{12}^T & R_{22} & \cdots & R_{2p} \\ \cdots & \cdots & \cdots\cdots & \\ R_{1p}^T & R_{2p}^T & \cdots & R_{pp} \end{bmatrix} \tag{2.127}$$

the following is obtained:

$$\epsilon = \sum_{i=1}^{p} \mathbf{a}_{(i)}^T \bar{R}_p \mathbf{a}_{(i)} - 2\mathbf{x}^T[\mathbf{a}_{(1)}^T, \mathbf{a}_{(2)}^T, ..., \mathbf{a}_{(p)}^T]^T + \mathbf{x}^T\mathbf{x} \tag{2.128}$$

The minimization of the MSE in (2.128) results in the following p sets of equations:

$$\bar{R}_p \mathbf{a}_{(m)} = \mathbf{x}_m \bar{\mu}_p$$

with $m = 1, 2, ...p$. By solving these equations for $\mathbf{a}$ the following expression for the optimal unconstrained filter coefficients is obtained:

$$\mathbf{a}_{(m)} = \mathbf{x}_m \bar{R}_p^{-1} \bar{\mu}_p \tag{2.129}$$

where $m = 1, 2, ...p$.

2.12 Filters Based on Vector Ordering

The R-ordering based filters discussed in the previous section rely on an appropriate reference vector, such as the marginal median or the mean vector. However, multivariate filters can be defined by utilizing the vector ordering scheme. If the aggregate distance of the sample $\mathbf{y}_i$ to the set of vectors $W(n) = [\mathbf{y}_1, \mathbf{y}_2, ..., \mathbf{y}_n]$ is employed as a reduction function then,

$$d_i = \sum_{j=1}^{n} d(\mathbf{y}_i, \mathbf{y}_j) \tag{2.130}$$

where $d(\mathbf{y}_i, \mathbf{y}_j)$ represents an appropriate distance (or similarity) measure. The arrangements of the d_i's in ascending order, associates the same ordering to the multivariate samples. Thus, an ordering

$$d_{(1)} \leq d_{(2)} \leq ... \leq d_{(n)} \tag{2.131}$$

implies the same ordering to the corresponding $\mathbf{y}_i$'s:

$$\mathbf{y}_{(1)} \leq \mathbf{y}_{(2)} \leq \cdots \leq \mathbf{y}_{(n)} \tag{2.132}$$

The vector median filter (VMF) can be defined as the vector $\mathbf{y}_{VM}$ contained in the given set whose distance to all other vectors is minimum [24], [28], [73]:

$$\sum_{j=1}^{n} d(\mathbf{y}_{VM}, \mathbf{y}_j) \leq \sum_{j=1}^{n} d(\mathbf{y}_i, \mathbf{y}_j) \quad \forall i \in [1, 2, ..., n] \tag{2.133}$$

In the ordered sequence of (2.132), $\mathbf{y}_{VM} = \mathbf{y}_{(1)}$. The distances between the vectors in (2.130) can be calculated in several different ways. The measures of similarity or dissimilarity discussed in this chapter can be used for this purpose. The most commonly used measure is the L_1 norm (City Block distance) . However, a more general class of vector median filters is obtained by using the L_p norm with the Euclidean distance (L_2 norm), the method of choice in many practical applications. The selected distance measure affects the noise reduction and detail preservation properties of the vector median filter. In order to optimize its performance the properties of the color space and the noise characteristics should be taken into consideration in the selection of the distance measure. The Euclidean distance has been proven to be better when the noise in signal components is correlated while the L_1 norm provides better results when non-correlated noise is present.

The VMF can also be defined as the maximum likelihood estimate of the location parameter of a bi-exponential distribution. This is in complete analogy to the scalar case where the scalar median is the maximum likelihood estimate of the exponential distribution [1]. For a p-dimensional bi-exponential distribution :

$$f(\mathbf{x}) = \lambda \quad \exp(-\alpha \; d(\beta, \mathbf{x})) \tag{2.134}$$

The maximum likelihood estimate $\hat{\beta}$ for the sample population $\mathbf{y}_i \quad i = 1, 2, ..., n$ can be defined by maximizing the expression:

$$L(\beta) = \prod_{i=1}^{n} \lambda \exp(-\alpha d(\beta, \mathbf{y}_i)) \tag{2.135}$$

which reduces to minimizing:

$$\sum_{i=1}^{n} d(\beta, \mathbf{y}_i) \tag{2.136}$$

The equation (2.136) has no closed-form solution. However, it is possible to obtain one if $\hat{\beta}$ is confined to one of the input vectors $\mathbf{y}_i \quad i = 1, 2, ..., n$. In this case, (2.133) can be derived again to define the vector median. As for all the multivariate filters discussed in this chapter, the VMF can be used as a color image processing filter by utilizing a window $W(n)$ which slides on the image plane. For each window position, the vector sample at the window center $\mathbf{y}_k$ is replaced by the vector median of the samples in $W(n)$. Several properties follow from the definition of the vector median [24]. In particular:

1. If the vector dimension is $p = 1$ then the VMF reduces to the scalar median.
2. The VMF is scale, translation and rotation invariant.
3. The output of the VMF is always one of the input vectors. That guarantees the existence of root signals since repeated filtering of the color vectors with the VMF will turn the signal to a root signal for this vector median filter. This property has been proven only for filter length 3, experimental analysis however shows that the property holds for other filter lengths.
4. A step edge is a root signal for the vector median filter.

The vector median operation can be combined with linear filtering for the case where the median is not adequate in filtering out the noise, such as in the case of additive Gaussian noise [74], [13]. The resulting filtering operator, named extended vector median filter (EVMF), selects the vector median $\mathbf{y}_{VM}$ or the average $\bar{\mathbf{y}}$ of the set $W(n)$ according to:

$$\mathbf{y}_{EVM} = \begin{cases} \bar{\mathbf{y}} & \text{if } \sum_{j=1}^{n} d(\bar{\mathbf{y}}, \mathbf{y}_j) \leq \sum_{j=1}^{n} d(\mathbf{y}_{VM}, \mathbf{y}_j) \\ \mathbf{y}_{VM} & \text{otherwise} \end{cases} \tag{2.137}$$

where $\bar{\mathbf{y}} = \sum_{j=1}^{n} \mathbf{y}_j$. The definition of the $\mathbf{y}_{EVM}$ can be derived by minimizing (2.135) with the additional constraint that β be one of the $\mathbf{y}_i$ or $\bar{\mathbf{y}}$. As in the case of VMF a number of distance or similarity measures can be used with the L_1 norm or the Euclidean distance, the most commonly used. The EVMF, in a sense, adapts to the input characteristics so that near edges or areas with high details behave like the VMF and thus, preserve edges and fine details, whereas, in the smooth parts of the image it more often chooses the mean vector to be the output value, resulting in improved noise attenuation.

The mean filter used in (2.137) is sensitive to outliers which can be mistaken for image edges and compromise the performance of the filter. In such situations, where low additive Gaussian noise and high percentage of impulsive noise (outliers) is assumed present, good noise attenuation can be achieved by utilizing an α-trimmed marginal median, or a robust estimator of the location, instead of the average mean.

The corresponding filter returns a vector value according to:

$$\mathbf{y}_{\alpha EVM} = \begin{cases} \mathbf{y}_{\alpha} & \text{if } \sum_{j=1}^{n} d(\mathbf{y}_{\alpha}, \mathbf{y}_j) \leq \sum_{j=1}^{n} d(\mathbf{y}_{VM}, \mathbf{y}_j) \\ \mathbf{y}_{VM} & \text{otherwise} \end{cases} \tag{2.138}$$

where $\mathbf{y}_{\alpha}$ is an α-trimmed marginal filter.

The VMF can also be combined with FIR filters which operate in subwindows of the filter window W. Assuming an input set $W(n)$, a multivariate FIR-median hybrid filter (VFMHF) was defined in [24] with an output $\mathbf{y}_{VFMH}$ given by:

$$\mathbf{y}_{VFMH} = VM[\frac{1}{k}\sum_{j=1}^{k-1} \mathbf{y}_j, \mathbf{y}_k, \frac{1}{k}\sum_{j=k+1}^{n} \mathbf{y}_j] \tag{2.139}$$

The main reason behind the popularity and widespread use of filters, such as the VMF, is their simplicity. The computations involved however in the evaluation of the aggregated distances during ordering are extensive. Given a square $(n \times n)$ processing window, $n^2(n^2 - 1)$ distances must be computed at each window location. It is evident that the computational complexity of the VMF depends heavily on the distance metric adopted to compute distances among the color samples. The use of the Euclidean distance results in an expensive algorithm, since it involves the computation of the squares and, possibly, the computation of the square root for each distance. The fast approximations to the Euclidean distance can be used to speed up the calculations [46]. The approximated distance measure is more computationally effective than the classical Euclidean distance and can considerably reduce the computational complexity of the VMF. Vector median filter implementations based on the L_1 norm are considerably faster, although their complexity is still high enough for many practical applications. To speed up the filtering procedures, appropriate fast algorithms, such as running median algorithms can also be used [1]. In such approaches, distances which have already been calculated are not recomputed at each step. Thus, the number of distances to be evaluated can be reduced to $n^2(n-1)+0.5n(n-1)$, resulting in a computational complexity of $O(n^3)$ which is significantly lower than the $O(n^4)$ of the original VMF implementation.

2.13 Directional-based Filters

The multivariate sub-ordering schemes and the associated filters discussed until now are based on the Cartesian representation of the color signals. However, in many cases, polar representation of multivariate data is of interest. Vector directional filters (VDF) are a class of multivariate filters which are based on vector ordering principles and polar coordinates [75], [83]. The novelty in this class of filters is that the ordering criterion is the angle between the color image vectors. As the name implies, VDF actually operates on the direction of the image vectors. As a result, the processing of multivariate data is separated into directional processing and magnitude processing. Although the term has been also used by other authors to denote processing in certain directions in the image plain, here it is used in the context of vector spaces, denoting the direction of the image vectors in the color space. Similarly, magnitude processing denotes the processing of image data where only the color vector magnitudes are taken into consideration. These terms will be used in the sequel with the meaning given here [12].

The VDF family operates on the direction of the image vectors, aiming at eliminating vectors with atypical directions in the vector space. To achieve its objective, VDF utilizes the angular distance of section 2.6 to order the input vectors inside a processing window. As a result of this process, a set of input vectors with approximately the same direction in the vector space is produced as the output set. Since the vectors in this set are approximately collinear, a magnitude processing operation can be applied in a second step to produce the requested filtered output.

The basic vector directional filter (BVDF) is a ranked-order, nonlinear filter which parallelizes the VMF operation. However, it employs the angle between the two color vectors as the distance criterion.

If the aggregate distance is employed as a reduction function of the sample $\mathbf{y}_i$ to the set of vectors $W(n) = [\mathbf{y}_1, \mathbf{y}_2, ..., \mathbf{y}_n]$ then,

$$d_i = \sum_{j=1}^{n} d(\mathbf{y}_i, \mathbf{y}_j) \tag{2.140}$$

where

$$d(\mathbf{y}_i, \mathbf{y}_j) = cos^{-1}(\frac{\mathbf{y}_i \mathbf{y}_j^T}{|\mathbf{x}_y||\mathbf{y}_j|}) \tag{2.141}$$

The arrangements of the d_i's in ascending order associates the same ordering to the multivariate samples. Thus, an ordering

$$d_{(1)} \leq d_{(2)} \leq ... \leq d_{(n)} \tag{2.142}$$

implies the same ordering to the corresponding $\mathbf{y}_i$'s:

$$\mathbf{y}_{(1)} \leq \mathbf{y}_{(2)} \leq ... \leq \mathbf{y}_{(n)} \tag{2.143}$$

The basic vector directional filters (BVDF) can be defined as the vector $\mathbf{y}_{BVDF}$ contained in the given set whose angular distance to all other vectors is minimum:

$$\mathbf{y}_{BVDF} = \mathbf{y}_{(1)} \tag{2.144}$$

In other words, the BVDF chooses the vector most centrally located without considering the magnitudes of the input vectors. It can be proven that the BVDF is the sample spherical median of the set of vectors $W(n)$ inside the processing window with the added constraint for the filter output to be one of the input vectors. Indeed, in analogy to the spatial median used in the derivation of the VMF the spherical (directional) median for a spherical distribution (Θ, Φ) can be defined [76], [77]. The spherical median is defined by utilizing the notion of the distance D on the surface of the sphere as the minimum arc length between two points. This leads to the notion of the spherical median (SM) direction as the value $(\bar{\theta}, \bar{\phi})$ which minimizes the following quantity:

$$E[d((\Theta, \Phi), (\bar{\theta}, \bar{\phi}))] = E[\Theta^*], \Theta^* = cos^{-1}(\check{\lambda}\lambda + \check{\mu}\mu + \check{\nu}\nu) \tag{2.145}$$

over all choices (θ, ϕ), where $(\check{\lambda}, \check{\mu}, \check{\nu})$ and (λ, μ, ν) are the directions of (Θ, Φ) and (θ, ϕ) respectively [29]. Thus, (2.145) minimizes the expected angular difference between two unit vectors on the sphere. Assuming that a random sample from a spherical distribution $(\theta_1, \phi_1), (\theta_2, \phi_2), ..., (\theta_n, \phi_n)$ is available and denoting $(\lambda_i, \mu_i, \nu_i) = (sin(\theta_i)cos(\phi_i), sin(\theta_i)cos(\phi_i), cos(\theta_i))$ for the direction cosines of the spherical samples, then the sample spherical median (SSM) is defined as the point from which the sum of the arc lengths to the data points is minimized [76], [78]. In other words, for a given point (λ, μ, ν), this sum is calculated as:

$$d((\lambda_i, \mu_i, \nu_i), (\lambda, \mu, \nu)) = \sum_{i=1}^{n} cos^{-1}(\lambda_i \lambda + \mu_i \mu + \mu_i \nu) \tag{2.146}$$

From the above definitions, it is obvious that the direction of the BVDF output is the sample spherical median with the constraint that the output vector be one of the input vectors in order to avoid iterative algorithms for finding the solution.

Simple inspection of the BVDF definition reveals that the BVDF filter is similar to the VMF. The former results from the spherical median and is constrained to one of the input vectors, whereas the latter results from the spatial median with the same constraint. From an ordering point of view, both filters result from the vector ordering principle using an aggregate distance criterion. The difference lies in the distance criterion utilized. BVDF utilize the angle between the color image vectors, whereas VMF employ Minkowski type of distances between the color vectors.

The BVDF enjoys many deterministic properties that make it appropriate for color image processing. Among them, the following four are the most important [75], [77]:

1. Preservation of step edges. A step edge for a vector value signal is a root of the BVDF regardless of the window size.
2. Invariance under scaling and rotation. Scaling by a scalar value and rotation of the coordinate system do not affect the angle between two vectors, therefore the BVDF is invariant under these operations. However, the BVDF is not invariant to bias since the addition of a constant vector changes the angle between vectors.
3. Existence and convergence to root signals. A step edge is a root signal of the BVDF, which proves the existence of root signals. Furthermore, repeated application of the BVDF will eventually produce a signal which is the root signal. For presentation purposes, a two-variate signal is considered. In such a case, a signal $\mathbf{y}_i$ is a root of the BVDF of length $n = 2m + 1$ if, $\forall n$ which satisfies:

$$\frac{\mathbf{y}_{(l)2}}{\mathbf{y}_{(l)1}} \geq \frac{\mathbf{y}_{(k)2}}{\mathbf{y}_{(k)1}} \geq \frac{\mathbf{y}_{(j)2}}{\mathbf{y}_{(j)1}} \tag{2.147}$$

where $(k - m) \leq l < k < j \leq (k + m)$, and $\mathbf{y}_{(i)j}$ denotes the jth component of the sample $\mathbf{y}_{(i)}$. This condition stems from the fact that in two dimensions, the BVDF is always the vector which lies in the middle of all the vectors.

In the case of color image processing, the spherical median, and thus BVDF, provides the least error estimation of the angle location. Consequently, BVDF performs well when the vector magnitudes are of no importance and the direction of the vectors is the dominant factor since this filter disregards the color vectors' magnitudes and treats them as purely directional data. However, in practice, color image data are not pure spherical data since the magnitudes of the image vectors vary at different locations. To improve the performance of the basic vector directional filters a generalized filter structure was proposed [12], [77]. The new filter, appropriately called the generalized vector directional filter (GVDF) generalizes BVDF in the sense that its output is a superset of the single BVDF output. Instead of a single output, the GVDF outputs the set of vectors whose angle from all other vectors is small as opposed to the BVDF which outputs the vector whose angle from all the other vectors is minimum. Thus, the GVDF's produced output initially consists of a set of l input vectors with approximately the same direction in the color space.

$$(\mathbf{y}_{(1)}, \mathbf{y}_{(2)}, ..., \mathbf{y}_{(l)}) = GVDF[(\mathbf{y}_{(1)}, \mathbf{y}_{(2)}, ..., \mathbf{y}_{(n)})] \tag{2.148}$$

where $1 \leq l \leq n$. Consequently, GVDF achieves, in a sense, to produce a single channel signal since the set of vectors produced contains color samples in the same direction. The function of the VDF can be demonstrated if color image processing from the perspective of the RGB color cube is considered. In the RGB color space, a particular color vector intersects the Maxwell triangle (the triangle drawn between the three primaries R,G,B) at a given point. That point indicates the hue and saturation, the chromaticity properties, of the color. Therefore, the operation of the VDF can be described in terms of color chromaticity. Since the BVDF results in the least error estimate of the angle location, directional filters render the color vector with the least chromaticity error. In the case of GVDF the set of colors with similar chromaticities is rendered. In other words, the VDF family operates on the chromaticity components of a color by filtering out color vectors with large chromaticity errors.

The parameter l, the number of input vectors included in the GVDF's output set, is a design parameter. There are two ways of choosing l, namely adaptive and non-adaptive. The case of adaptive selection of l is of interest since it may produce a better output set. When there is a high variation of the color in the input image, such as in edge areas, only vectors that are from the same part of the edge as the center vector should be included in the final output set. On the other hand, at a uniform region, a lot of vectors should be included in the output set to improve the noise suppression capability of the filter. In the non-adaptive case, a preselected value is utilized. Experimental analysis has revealed that a value $l = \lfloor \frac{n}{2} \rfloor + 1$, where $\lfloor$ defines integer part, provides reasonable results in most practical applications [77].

The GVDF needs to be combined with an appropriate gray scale (magnitude processing) filter in order to produce a single output vector at each pixel. Since the GVDF's output set consists of vectors with approximately the same direction in the color space, any gray scale filter can be used for the magnitude processing module. What exact filter will be utilized is again based on the problem on hand and the constraints imposed in the design. Smoothing filters, such as α-trimmed (scalar) mean filter, the (scalar) median and the arithmetic mean filters can be used in the magnitude module. If prior information about the noise corruption is available, the designer may select the most appropriate magnitude processing module to maximize the GVDF's performance. However, this is seldom the case in a realistic application scenario, where information on the actual noise corruption is not available. In such a case the applicability of the GVDF is questionable.

To overcome the deficiencies of the GVDF, a new directional filter known as the distance-direction filter (DDF) was proposed [79], [80]. The DDF retains the structure of the BVDF, but utilizes a new distance criterion to order the vectors inside the processing window. A new distance criterion was utilized by the designers of the DDF in hopes to derive a filter which combines the properties of both these filters. Specifically, in the case of the DDF the distance inside W is defined as:

$$\beta_i = \sum_{j=1}^{n} A(\mathbf{y}_i, \mathbf{y}_j) \sum_{j=1}^{n} ||\mathbf{y}_i, \mathbf{y}_j|| \tag{2.149}$$

where $A(\mathbf{x}_i, \mathbf{x}_j)$ is the directional (angular) distance defined in (2.141) with the second term in (2.149) to account for the differences in magnitude in terms of the L_2 or L_1 metric.

As for any other ranked-order, multichannel, nonlinear filter, it is assumed that an ordering of the β_i 'distance'

$$\beta_{(1)} \leq \beta_{(2)} \leq \dots \leq \beta_{(n)} \tag{2.150}$$

implies the same ordering to the corresponding input vectors $\mathbf{x}_i$'s:

$$\mathbf{y}_{(1)} \leq \mathbf{y}_{(2)} \leq \dots \leq \mathbf{y}_{(n)} \tag{2.151}$$

Thus, the DDF defines the minimum order vector as its output: $\mathbf{y}_{DDF} = \mathbf{y}_{(1)}$. The simultaneous minimization of the distance functions used in the designs of the VMF and the BVDF was attempted in the design of the DDF in order to obtain a filter that can smooth out long tailed noise and also preserve the chromaticity component of the image vectors. Although the concept is appealing, and the resulting vector processing structure is simple, fast and without the additional module of the GVDF, there are a number of problems. Most notably, contrary to the distance (similarity) measures discussed in this chapter, the distance measure defined in (2.149) is heuristic, window-dependent, and has no ties to the characteristics of the individual color vectors. Furthermore, there is no analysis of the relative importance of the two components in the suggested distance. Thus, although the DDF can provide, in some cases, better results than those obtained by the BVDF or the GVDF, it cannot be considered as an effective, general purpose nonlinear filter.

However, its introduction inspired a new set of heuristic vector processing filters which try to capitalize on the same appealing principle, namely the simultaneous minimization of the distance functions used in the VMF and the BVDF. Such a filter is the hybrid directional filter introduced in [81]. This filter operates on the direction and the magnitude of the color vectors independently and then combines them to produce a unique final output. This hybrid filter, which can be viewed as a nonlinear combination of the VMF and BVDF filters, produces an output according to the following rule:

$$\mathbf{y}_{HyF} = \begin{cases} \mathbf{y}_{VMF} & \text{if } \mathbf{y}_{VMF} = \mathbf{x}_{BVDF} \\ \left(\frac{|\mathbf{y}_{VMF}|}{|\mathbf{y}_{BVDF}|}\right)\mathbf{y}_{BVDF} & \text{otherwise} \end{cases} \tag{2.152}$$

where $\mathbf{y}_{BVDF}$ is the output of the BVDF filter, $\mathbf{y}_{VMF}$ is the output of the VMF and $|.|$ denotes the magnitude of the vector.

Another more complex hybrid filter, which involves the utilization of an arithmetic (linear) mean filter (AMF), has also been proposed [81]. The structure of this so-called adaptive hybrid filter is as follows:

$$\mathbf{y}_{aHyF} = \begin{cases} \mathbf{y}_{VMF} & \text{if } \quad \mathbf{y}_{VMF} = \mathbf{y}_{BVDF} \\ \mathbf{y}_{out1} & \text{if } \quad \sum_{i=1}^{n} |\mathbf{y}_i - \mathbf{y}_{out1}| < \sum_{i=1}^{n} |\mathbf{y}_i - \mathbf{y}_{out2}| \\ \mathbf{y}_{out2} & \text{otherwise} \end{cases} \tag{2.153}$$

where $\mathbf{y}_{out1} = \left(\frac{|\mathbf{y}_{VMF}|}{|\mathbf{y}_{BVDF}|}\right)\mathbf{y}_{BVDF}$, $\mathbf{y}_{out2} = \left(\frac{|\mathbf{y}_{AMF}|}{|\mathbf{y}_{BVDF}|}\right)\mathbf{y}_{BVDF}$, and $\mathbf{y}_{AMF}$ denotes the output of an arithmetic (linear) mean filter operating inside the same processing window. According to its designers, the magnitude of the output vector will be that of the mean vector in smooth regions and that of the median operator near edges.

Although these two hybrid filters attempt to parallelize the operation of the DDF, it is obvious from the definitions that the proposed filters constitute only heuristic approximations. As in the case of the DDF, the combination of magnitude and orientation information was done in an algorithmic level resulting in a heuristic structure which, contrary to the DDF, cannot be classified as a ranked-order nonlinear estimator. Furthermore, the hybrid filters defined in (2.152) and (2.153) are computationally demanding, since they require evaluation of both the VMF and VDF outputs.

In addition to the filters based on vector magnitude or vector direction polynomial and rational filters have been recently introduced as a successful tool for applications ranging from noise smoothing to image enhancement and interpretation of encoded signals [84]. Their design is usually based on a weighted combination of nonlinear filters having lowpass of highpass behavior. In most applications the choice of the filter components is performed heuristically, although it is possible to calculate weighting coefficients using linear adaptive algorithms [85]. Straight forward application of the rational filters to color image processing would be based on processing the three color channels separately. To utilize the inherent corellation between the channels that exist in the RGB color space, extensions of the basic structure have been introduced recently [86]. A rational filter of particular interest to color image processing is the vector median rational hybrid filter (VMRHF) the output of which is the result of a vector rational function taking into account three input sub-functions which form an input function set Φ_1, Φ_2, Φ_3:

$$\mathbf{y}_{VMRHF} = \mathbf{y}_{\Phi_2} + \frac{\sum_{j=1}^{3} a_j \mathbf{y}_{\Phi_j}}{h + k||\mathbf{y}_{\Phi_1} - \mathbf{y}_{Phi_2}||} \tag{2.154}$$

where $a^T = [a_1, a_2, a_3]$ is a vector coefficient determined a-priori and k, h are positive, user-defined constants that are used to control the amount of the nonlinear effect. In recent applications, the filter coefficients are selected so they satisfy the condition:

$$\sum_{j=1}^{3} a_j = 0$$

and the sub-filters Φ_1, Φ_3 are chosen so that an acceptable compromise between noise reduction and chromaticity preservation can be achieved. Due to its structure, and through its parameters, VMRHF operates as a linear low-pass filter between three nonlinear sub-filters reducing the smoothing effect and preserving details and edges in the image [86].

2.14 Computational Complexity

Apart from the numerical behavior of any proposed algorithm, its computational complexity is a relevant measure of its practicality and usefulness since it determines the required computing power and processing (execution) time. A general framework to evaluate the computational requirements of recursive algorithms is given in [87], [88]. The framework of that analysis is used here in order to evaluate the computational requirements of the algorithms. Two assumptions are introduced in order to have a meaningful comparison among the different algorithms. First, it's assumed that the filter window is symmetric $(n \times n)$ and that n^2 vector samples are contained in it. Each color vector is assumed to be a point in $\mathcal{R}^p$. Secondly, the fundamental operations involved in the algorithms are matrix and vector operations. A detailed analysis of the computations involved in such operations is provided in [87], [89]. The interested reader can refer to them for more information on the subject. In this context, the total time required to complete an operation (or a sequence of operations) is proportional to the normalized total number of equivalent scalar operations, defined as:

$$Time = k \times (4 \times (MULTS) + (ADDS) + 6 \times (DIVS) + 25 \times (SQRTS))$$

where *MULTS* is the number of scalar multiplications required, *ADDS* is the number of scalar additions required, *DIVS* is the number of scalar divisions required and *SQRTS* is the number of the scalar square roots. The weights used in the above formula do not pertain to any particular machine. Rather, they can be considered mean values of those coefficients commonly encountered. All the qualitative results presented in the sequence hold even if the weighting coefficients in the above formula are different for a specific computing platform [89]. The computational complexity of the filters discussed in this chapter depends mainly on the approach adopted to sort the multivariate samples and the distance (similarity) measure [90].

In marginal ordering, multivariate samples are ordered independently along each dimension. The complexity of the marginal median can be easily determined by noting that it corresponds to p applications of the scalar median filter. Since, in this case, the input values are integer numbers, a fast running algorithm based on the window histogram can be used to implement the scalar median filter. The complexity of such a running scalar median is $O(n)$ comparisons, resulting in a marginal median filter with complexity of $p \cdot O(n)$.

In the R-ordering scheme, color vectors are sorted on the basis of their distance to a given reference point, such as the sample mean or the sample at the window center. It is obvious that both the distance metric and the central point affect the filter complexity. The R_C filter discussed in [15] will be taken as an example. In (2.113) the filter output is defined as the sample in the window which minimizes the sum of distances to the sample mean, marginal median and window central point. The computational complexity of the R_C can be determined as follows:

1. Computation of the sample mean and the marginal median: $2np$ ADDS, p DIVS, $p \cdot O(n)$ comparisons.
2. Computation of the Euclidean distances to the sample mean, marginal median and the central sample: $3n^2$ Euclidean distances
3. Summation of the different distances: $2n^2$ ADDS
4. selection of the minimum sum point: $n^2 - 1$ comparisons

To compute the Euclidean distance 1 SQRT, p MULTS and $2p$ ADDS are needed. The overall complexity of the R_C filter, when (2.113) is used, is $O(n^2)$. However, the need for $3n^2$ SQRTS makes the algorithm less attractive from a computational point of view. Since the squared Euclidean distance can be calculated using only p MULTS and $2p$ ADDS we may select to implement the R_C filter using the (2.114) instead. Using the linear combination of (2.115) only n^2 distances must be computed instead of $3n^2$.

In vector ordering, straightforward application of (2.133) requires the computation of the distances between all vector couples resulting in a computational complexity of $O(n^4)$. Running algorithms can be introduced to reduce the computational burden. Improvements can be achieved by calculating only the distances relative to the vectors which, at each new location, enter the filter window. The computational requirements of such a running algorithms can be determined as follows [88]:

1. Computation of distances between new and old vectors: $n(n-2)$ distances
2. Computation of distances between new samples: $\frac{n(n-1)}{2}$ distances
3. Updating of sums of distances relative to old vectors: $2n^2(n-1)$ ADDS
4. Computation of sums of distances relative to new samples: $n(n^2-1)$ ADDS
5. Choice of minimum vector: n^2-1 comparisons

To complete the computational analysis of the different filters based on vector ordering, the distance metric or similarity measure used in the calculations must be taken into account. Given the fact that 1 SQRT, p MULTS and $2p$ ADDS are needed to implement the Euclidean distance, the Euclidean metric based VMF has the highest complexity. On the other hand since the City block distance (L_1 norm) can be evaluated using $2p$ ADDS and p comparisons, the L_1-based VMF exhibits the lowest complexity with only $O(n^3)$ ADDS and $O(n^3)$ comparisons. Table 2.1 summarizes the computational complexity of four filters.

Table 2.1. Computational Complexity

Elementary Operations	VMF_1	VMF_2	$BVDF$	R_c
ADDS	$O(n^3)$	$O(n^3)$	$O(n^3)$	$O(n^2)$
MULTS/DIVS	-	$O(n^3)$	$O(n^3)$	$O(n^2)$
SQRTS	-	$O(n^3)$	$O(n^3)$	-
ARC COS	-	-	$O(n^3)$	-
COMPARISONS	$O(n^3)$	$O(n^2)$	$O(n^2 logn)$	$O(n^2)$

VMF_1 and VMF_2 are vector median filters implemented using the City Block and Euclidean distance respectively, BVDF symbolizes the basic vector directional filter and R_c is the reduced based ordering filter of (2.115).

2.15 Conclusion

The performance of the different nonlinear filters based on order statistics depends heavily on the problem under consideration. Since most of them are designed to perform well in a specific application, their performance deteriorates rapidly under different operation scenarios. Thus, a color image filter, which performs equally well in a wide variety of applications, is of great importance. Such a filter should be computationally efficient, reliable and should deliver acceptable results without requiring specific a-priori information about the noise characteristics. The adaptive algorithms discussed in the next chapter are an attempt to design a fast and efficient structure aimed at improved efficiency for practical realization.

References

1. Pitas, I., Venetsanopoulos, a.N. (1990): Nonlinear Digital Filters: Principles and Applications. Kluwer Academic Publishers, Boston, MA.
2. Sung, Kah-Yay (1993): A Vector Signal Processing Approach to Color. M.S. Thesis, Department of Electrical Engineering and Computer Science, Massachusetts Institute of Technology.
3. Vinaygamoorthy, S., Plataniotis, K.N., Androutsos, D., Venetsanopoulos, A.N. (1996): A multichannel filter for TV signal processing. IEEE Trans. on Consumer Electronics, **42**(2), 199-206.
4. Sanwalka, Sunil (1992): Vector Order Statistic Filters for Color Image Processing. M.A.Sc. Thesis, Department of Electrical and Computer Engineering, University of Toronto.
5. van Hateren, J.H. (1993): Spatial, temporal and pre-processing for color vision. Journal of Royal Statistical Society B, **251**, 61-68.
6. Zheng, J., Valavanis, K.P., Cauch, J.M. (1993): Noise removal for color images. Journal of Intelligent and Robotic Systems, **7**, 257-285.
7. Kayargadde, V., Martens, J.B. (1996): An objective measure for perceived noise. Signal Processing, **49**, 187-206.
8. Plataniotis, K.N., Androutsos, D., Vinayagamoorthy, S., Venetsanopoulos, A.N. (1997): Color image processing using adaptive multichannel filters. IEEE Trans. on Image Processing, **6**(7), 933-950.
9. Mees, C.E.K. (1954): The Theory of Photographic Process. McMillan Publishing Company.
10. Weeks, Arthur R. Jr. (1996): Fundamentals of Electronic Image Processing, SPIE/IEEE Series on Imaging Science & Engineering.
11. Jenkins, T.E. (1987): Optimal Sensing Techniques and Signal Processing. Prentice Hall.
12. Trahanias, P.E., Pitas, I., Venetsanopoulos, A.N. (1994): Color Image Processing. (Advances In 2D and 3D Digital Processing: Techniques and Applications, edited by C.T. Leondes), Academic Press.
13. Viero, T., Oistamo, K., Neuvo, Y. (1994): Three-dimensional median-related filters for color image sequence filtering. IEEE Trans. on Circuits and Systems for Video Technology, **4**(2), 129-142.
14. Tang, K., Astola, J., Neuvo, Y. (1994): Multichannel edge enhancement in color image processing. IEEE Trans. on Circuits and Systems for Video Technology, **4**(5), 468-479.
15. Tang, K., Astola, J., Neuvo, Y. (1995): Nonlinear multivariate image filtering techniques. IEEE Trans. on Image Processing, **4**(6), 788-797.
16. Pitas, I., Tsakalides, P. (1991): Multivariate ordering in color image restoration. IEEE Trans. on Circuits and Systems for Video Technology, **1**(3), 247-260.
17. Cotropoulos, C., Pitas, I (1994): Adaptive nonlinear filter for digital signal/image processing. (Advances In 2D and 3D Digital Processing, Techniques and Applications, edited by C.T. Leondes), Academic Press, **67**, 263-317.
18. M. Schetzen, M. (1982): The Voltera and Wiener Theories of Nonlinear Filters. J. Wiley & Sons, New York, USA.
19. Oppenheim, A.V., Schafer, R.W., Stockham, T.G. (1968): Nonlinear filtering of multiplied and convolved signals. Proceedings of IEEE, **56**, 1264-1291.

20. Serra, J. (1982): Image Analysis and Mathematical Morphology. Academic Press, New York, N.Y., USA.
21. E.R. Dougherty, E.R., Astola, J. (1994): An introduction to Nonlinear Image Processing. SPIE Press, Bellingham, WA., USA.
22. Maragos, P., Schafer, R.W. (1990): Morphological systems for multidimensional signal processing. Proceedings of IEEE, vol. 78, no. 4, pp. 690-710, 1990.
23. P. Maragos, P. (1996): Differential morphology and image processing. IEEE Trans. on Image Processing, **5**(6), 922-937.
24. Astola, J., Haavisto, P., Neuvo, Y. (1990): Vector median filters. Proceedings of the IEEE, **78**, 678-689.
25. Hardie, R.C., Arce, G.R. (1991): Ranking in R^P and its use in multivariate image estimation. IEEE Trans. on Circuits and Systems for Video Technology, **1**(2), 197-208.
26. David, H.A. (1981): Order Statistics. John Wiley and Sons, New York, N.Y.
27. Huber, P.S. (1981): Robust Statistics. John Wiley and Sons, New York, N.Y.
28. Astola, J., Haavisto, P., Neuvo, Y. (1988): Median type filters for color signals. Proceedings of the 1988 IEEE Symposium on Circuits and Systems, **2**(3), 1753-1756.
29. Barnett, V., Lewis, T. (1994): Outliers in Statistical Data. Third Edition, John Wiley and Sons, New York, N.Y.
30. Gnanadesikan, R. (1977): Methods for Statistical Data Analysis of Multivariate Observations. John Wiley and Sons, New York, N.Y.
31. David, H.A. (1981): Order Statistics. John Wiley and Sons, New York, N.Y.
32. Huber, P.S. (1981): Robust Statistics. John Wiley and Sons, New York, N.Y.
33. Astola, J., Kuosmanen, P. (1997): Fundamentals of Nonlinear Digital Filtering. CRC Press, Boca Raton, FLA.
34. Barnett, V. (1976): The ordering of multivariate data. Journal of Royal Statistical Society A, **139**(2), 331-354.
35. Kurekin, A., Lukin, V., Zelensky, A., Koivisto, P., Astola, J. (1999): Comparison of component and vector filter performance with application to multichannel and color image processing. Proceedings of the 1999 IEEE Workshop on Nonlinear Signal and Image Processing, **I**, 119-123.
36. Huttunen, H., Tico, M., Rusu, C., Kuosmanen, P. (1999): Ordering methods for multivariate RCRS filters. Proceedings of the 1999 IEEE Workshop on Nonlinear Signal and Image Processing, **II**, 506-510.
37. Galambos, J. (1975): Order statistics of samples from multivariate distributions. Journal of American Statistical Association, **70**, 674-680.
38. Pitas, I. (1990): Marginal order statistics in color image filtering. Optical Engineering, **29**(95), 493-503.
39. Hardie, R.C., Arce, G.R. (1991): Ranking in R^P and its use in multivariate image estimation. IEEE Trans. on Circuits and Systems for Video Technology, **1**(2), 197-208.
40. Watterson, G.A. (1959): Linear estimation in censored samples from multivariate normal populations. Ann. Math. Statistics, **30**(8), 814-824.
41. Titterington, D.M. (1978): Estimation of correlation coefficients by ellipsoidal trimming. Applied Statistics, **27**, 227-234.
42. Granadesikan, R., Kettenring, O.R. (1972): Robust estimates, residuals and outlier detection with multi response data. Biometrics, **28**, 81-124.
43. Krzanowski, K., Marriott, F.H.C. (1994): Multivariate Analysis I: Distributions, ordination and inference. Halsted Press, New York, N.Y.
44. Duda, R.O., Hart, P.E. (1973): Pattern Classification and Scene Analysis. John Wiley and Sons, New York, N.Y.

45. Tou, J.T., Gonzalez, R.C. (1974): Pattern Recognition Principles. Addison-Wesley.
46. Barni, M., Cappellini, V., Mecocci A. (1994): Fast vector median filter based on Euclidean norm approximation. IEEE Signal Processing Letters, **1**(6) 92-94.
47. Chaudhuri, J., Murthy, C.A., Chaudhuri, B.B. (1992): A modified metric to compare distances. Pattern Recognition, **25**(5) 667-677.
48. Plataniotis, K.N., Androutsos, D., Vinayagamoorthy, S., Venetsanopoulos, A.N. (1996): An adaptive nearest neighbor multichannel filter. IEEE Trans. on Circuits and Systems for Video Technology, **6**(6), 699-703.
49. Plataniotis, K.N., Androutsos, D., Venetsanopoulos, A.N. (1997): Content based color image filters. Electronic Letters, **33**(3), 202-203.
50. Plataniotis, K.N., Venetsanopoulos, A.N. (1999): A taxonomy of similarity operators for color image filtering. Proceedings of the 1999 IEEE Workshop on Nonlinear Signal and Image Processing, **I**, 119-123.
51. Plataniotis, K.N., Androutsos, D., Venetsanopoulos, A.N. (1999): Adaptive fuzzy systems for multichannel signal processing. Proceedings of IEEE, **87**(9).
52. Ekman, G.A. (1963): A direct method for multidimensional ratio scaling. Psychometrica, **28**, 3-41.
53. Ekman, G.A., Sjoberg, L. (1965): Scaling. Annual Rev. Psychol., **16**, 451-474.
54. Ekehammar, B. (1972): A comparative study of some multidimensional vector models for subjective similarity. Scandinavian Journal of Psychology, **82**(2), 190-206.
55. Sjoberg, L. (1975): Models of similarity and intensity. Psychological Bulletin, **82**(2), 191-206.
56. Sjoberg, L. (1977): Similarity and multidimensional ratio estimation with simultaneous qualitative and quantitative variation. Scandinavian Journal of Psychology, **18**, 307-316.
57. Goude, G. (1972): A multidimensional scaling approach to the perception of art: I. Scandinavian Journal of Psychology, **13**, 258-271.
58. Borg, I., Lingoes, J. (1987): Multidimensional Similarity Structure Analysis. Springer Verlag.
59. Shepard, R.N. (1981): Toward a universal law of generalization for psychological science. Science, **237**, 1317-1323.
60. Tversky, A. (1977): Features of similarity. Psychological Review, **84**(4), 327-352.
61. Leung, Y. (1998): Spatial Analysis and Planning under Imprecision. North-Holland.
62. Plataniotis, K.N., Regazzoni, C.S., Teschioni, A., Venetsanopoulos, A.N. (1996): A new distance measure for vectorial rank order filters based on space filling curves. IEEE Conference on Image Processing, **ICIP-96**(I), 411-414.
63. Regazzoni, C.S., Teschioni, A. (1997): A new approach to vector median filtering based on space filling curves. IEEE Trans. on Image Processing, **6**(7), 1025-1037.
64. Zervakis, M.E., Venetsanopoulos, A.N. (1991): Linear and nonlinear image restoration under the presence of mixed noise. IEEE Trans. on Circuits and Systems, **38**(3), 258-271.
65. Pitas, I. (1996): Multichannel order Statistical Filtering. (Circuits and Systems Tutorial, Chris Toumazou Editor), IEEE press, Piscataway N.J., USA, 41-50.
66. Cotropoulos, C., Pitas, I (1994): Multichannel L filters based on marginal data ordering. IEEE Trans. on Image Processing, **42**(10), 2581-2595.
67. Koivunen, V. (1996): Nonlinear filtering of multivariate images under robust error criterion. IEEE Trans. on Image Processing, **5**(6), 1054-1060.

68. Koivunen, V., Himayat, N., Kassam, S.A. (1997): Nonlinear techniques for multivariate images. Design and robustness characterization. Signal processing, **57**, 81-91.
69. Maronna, R.A. (1976): Robust M-estimators of multivariate location and scatter. Annals of Statistics, **4**, 51-67.
70. Devlin, S.L., Gnanadesikan, R., Kettenring, J.R. (1981): Robust estimation of dispersion matrices and principal components. Journal of the American Statistical Association, **76**, 354-362.
71. Campbell, N.A. (1980): Robust procedures in multivariate analysis I: Robust covariance estimation. Applied Statistics, **29**(3), 678-689.
72. Nikolaidis, N., Pitas I. (1996): Multichannel L-filters based on reduced ordering. IEEE Trans. on Circuits and Systems for Video Technology, **6**(5), 570-582.
73. Rantanen, H., Karlsoon, M., Pohjala, P., Kalli, S. (1992): Color video signal processing with median filters. IEEE Trans. on Consumer Electronics, **38**(3), 157-161.
74. Heinonen, P., Neuvo, Y. (1988): Vector FIR-Median hybrid filters for multispectral signals. Electronic Letters, **24**(1), 6-7.
75. Trahanias, P.E., Venetsanopoulos, A.N. (1993): Vector directional filters. A new class of multichannel image processing filters. IEEE Trans. on Image Processing, **2**, 528-534.
76. Small, C. (1990): A survey of multidimensional medians. International Statistics Review, **58**(3), 263-277.
77. Trahanias, P.E., Karakos D., Venetsanopoulos, A.N. (1996): Directional processing of color images: theory and experimental results. IEEE Trans. on Image Processing, **5**(6), 868-880.
78. Ko, D., Chang, T. (1993): Robust M estimators on spheres. Journal of Multivariate Analysis, **45**, 104-136.
79. Karakos, D., Trahanias, P.E. (1995): Combining vector median and vector directional filters. The directional-distance filters. Proceedings of the IEEE Conf. on Image Processing, **ICIP-95**(I), 171-174.
80. Karakos, D., Trahanias, P.E. (1997): Generalized multichannel image filtering structures. IEEE Trans. on Image Processing, **6**(7), 1038-1045.
81. Gabbouj, M., Cheickh, F.A. (1996): Vector median-vector directional hybrid filter for color image restoration, Proceedings of the European Signal Processing Conference, **VIII**, 879-881.
82. Nikolaidis, N., Pitas, I. (1994): Directional statistics in nonlinear vector field filtering. Signal Processing, 299-316.
83. Nikolaidis, N., Pitas, I. (1998): Nonlinear processing and analysis of angular signals. IEEE Trans. on Signal Processing, **46**(12), 3181-3194.
84. Kroner, S., Ramponi, G. (1999): Design constraints for polynomial and rational filters. in Proceedings, IEEE Workshop on Nonlinear Signal and Image Processing, **III**, 501-505.
85. Leung, H., Haykin, S. (1994): Detection and estimation using an adaptive rational function filter. IEEE Transaction on Signal Processing, **42**(12), 3366-3376.
86. Khriji, L., Gabbouj, M. (1999): A new class of multichannel image processing filters. in Proceedings, IEEE Workshop on Nonlinear Signal and Image Processing, **II**, 516-519.
87. Katsikas, S.K., Likothanasis, S., Lainiotis, D.G. (1991): On the parallel implementation of linear Kalman and Lainiotis filters and their efficiency. Signal Processing, **25**, 289-306.
88. Barni, M., Cappellini, V. (1998): On the computational complexity of multivariate median filters. Signal Processing, **71**, 45-54.

89. Plataniotis, K.N. (1994): Distributed Parallel Processing State Estimation Algorithms. Ph.D Dissertation, Department of Electrical and Computer Engineering, Florida Institute of Technology, Melbourne, Fl.
90. Plataniotis, K.N., Venetsanopoulos, A.N. (1997): Vector processing. in Sangwine, S.J., Horne, R.W.E., (eds.), The Colour Image Processing Handbook, 188-209, Chapman & Hall, Cambridge, Great Britain.

3. Adaptive Image Filters

3.1 Introduction

The nonlinear filters described in the previous chapter are usually optimized for a specific type of noise. However, the noise statistics, e.g. the standard deviation, and even the noise probability density function vary from application to application. Sometimes the noise characteristics vary in the same application from image to image. Such cases include the channel noise in image transmission and the atmospheric noise in satellite images. In these environments non-adaptive filters cannot perform well because their characteristics depend on noise and signal characteristics which are unknown. In the area of color image filtering adaptive designs have been recently introduced to address the problem of varying noise characteristics and to guarantee acceptable filtering results even in the case of partially known signaling models [1].

Adaptive filters attempt to overcome difficulties associated with the uncertainty about the data by utilizing estimation procedures based on local statistics [2], [3]. The parameters of the adaptive filter are determined in a data-dependent way. The performance of such filters depends heavily on the accuracy of the estimation of certain signal statistics. A number of test statistics have to be used to estimate the local nature of data. The weights of the adaptive filter are then adjusted according to the values of the test statistics within each processing window. The main problem with a particular adaptive design is that exact statistical analysis is difficult to accomplish and, in general, is time consuming. Another popular adaptive filtering approach is based on the determination of the local nature of the data by appropriate tests applied to the image before the selection of the filter. Adaptive versions of L-filters have been considered recently [4]. It has been found that these adaptive filters have good performance in a variety of different noise characteristics. Another family of training-based filters used in image processing is that of neural-based filters. The attractive generalization properties of neural networks, their ability to perform complex mappings from a set of noise signals to the noise-free signal and their parallel implementation make them the method of choice in many digital signal processing applications [5], [6].

There are a number of problems associated with such designs [3]:

1. A-priori knowledge about the signal and the desired response is required. Then the coefficients of the adaptive filter can be optimized for a specific noise distribution with respect to a specific error criterion. However, such information is not available in realistic signal processing applications.
2. Least Mean Square (LMS) or other Wiener-like filters are based on the assumption that the input signal and the available desired response are stationary ergodic processes. This is not true for many practical applications. Similarly, adaptive schemes based on noise statistics estimation are often assumed ergodic in order to justify the use of the sample mean and sample noise covariance in the calculations although it is known that that assumption does not always hold.
3. Adaptive schemes based on training signals are iterative processes with heavy computational requirements. The real-time implementation of such algorithms is usually not feasible.

Recently, a number of adaptive techniques based on fuzzy logic principles have been proposed [7–10]. Fuzzy logic based techniques have mainly been used in the past for high level analysis of signals and images, computer vision applications, systems control, pattern recognition and decision modeling. Different approaches ranging from fuzzy clustering to fuzzy entropy and decision under fuzzy constraints have been used for scene detection, object recognition and decision directed image analysis. However more recently, fuzzy techniques have been used for low level signal and image processing tasks, such as non-Gaussian noise elimination, nonlinear/non-Gaussian stochastic estimation, image enhancement, video coding, signal sharpening and edge detection [12]-[23].

Most of the fuzzy techniques in use today adopt a window-based rule driven approach leading to data-dependent fuzzy filters, which are constructed by fuzzy rules in order to remove additive noise while preserving important signal characteristics, such as signal edges. Since the antecedents of fuzzy rules can be composed of several local characteristics, it is possible for the fuzzy filter to adapt to local data. Local correlation in the data is utilized by applying the fuzzy rules directly on the signal elements which lie within the operational window. Thus, the output of the fuzzy filter depends on the fuzzy rule and the defuzzification process, which combines the effects of the different rules into an output value.

Through the utilization of linguistic terms, a fuzzy rule based approach to signal processing allows for the incorporation of human knowledge and intuition into the design, which cannot be achieved via traditional mathematical modeling techniques. However, there is no optimal way to determine the number and type of fuzzy rules required for the fuzzy image operation. Usually, a large number of rules are necessary and the designer has to compromise between quality and number of rules, since even for a moderate processing window, a large number of rules are required [7], [12], [18]. To overcome these difficulties data dependent filters adopting fuzzy reasoning have been proposed. These designs combine fuzzy concepts, such as membership functions, fuzzy rules, and fuzzy aggregators with nonlinear filters, such as the α-trimmed mean filter and the weighted average mean filter in order to remove Gaussian and non-Gaussian noise while preserving useful signal characteristics, such as edges or image details and texture. In addition, based on the adoption of a fuzzy positive Boolean function, a new class of operators named fuzzy stack filters have recently been proposed [9]. These operators extend the smoothing capabilities of the classical stack filters and can provide efficient and cost effective solutions provided that an adequate set of training signals is available. Recently, neuro-fuzzy filters and genetic optimization techniques have been combined in hopes to derive a nonlinear filter which can cancel noise and preserve signal details at the same time [10]. As is the case of nonlinear techniques in general, the fuzzy signal processing techniques available today lack a unifying theory. Cross-fertilization among the different fuzzy techniques as well as with other nonlinear techniques has shown to be promising. For example, mathematical morphology and fuzzy concepts have been blended together in the case of fuzzy stack operators, and fuzzy designs and order statistic filters have been efficiently integrated into one class even though they come from completely different origins [9], [15], [19], [20].

3.2 The Adaptive Fuzzy System

Consider window $W(n)$ which moves across the noise corrupted input image. The filtered estimate $\hat{\mathbf{y}}_k$ at the window center $W(n)$ is obtained by processing the noisy image vectors set $W(n) = (\mathbf{y}_1, \mathbf{y}_2, ..., \mathbf{y}_n)$.

Since the most commonly used methodology to decrease the level of random noise present in the signal is smoothing, a fuzzy averaging operation in order to replace the noisy vector at the window center with a suitable representative vector. Thus, the general form of the adaptive fuzzy system proposed here is given as a fuzzy weighted average of the input vectors inside the window W, estimating the uncorrupted multivariate image signal by determining the centroid of the input set as follows [22]:

$$\hat{\mathbf{y}} = \sum_{j=1}^{n} w_j \mathbf{y}_j = \sum_{j=1}^{n} \frac{\mu_j^{\lambda}}{\sum_{j=1}^{n} \mu_j^{\lambda}} \mathbf{y}_j \tag{3.1}$$

where λ is a parameter such that $\lambda \in [0, \infty)$.

For the special case of $\lambda = 1$ the following filtered output can be obtained [22]:

$$\hat{\mathbf{y}} = \sum_{j=1}^{n} w_j \mathbf{y}_j = \sum_{j=1}^{n} \frac{\mu_j}{\sum_{j=1}^{n} \mu_j} \mathbf{y}_j \tag{3.2}$$

In this adaptive design the weights provide the degree to which an input vector contributes to the output of the filter. The relationship between the image vector at the window center (vector under consideration) and each vector within the window should be reflected in the decision for the weights of the filter. Through the normalization procedure two constraints necessary to ensure that the output is an unbiased estimator are satisfied. Namely:

1. Each weight is a positive number, $w_j \geq 0$,
2. The summation of all the weights is equal to unity $\sum_{j=1}^{n} w_j = 1$.

The weights of the filter are determined adaptively using transformations of a distance criterion between the input vectors. These weighting coefficients are transformations of the distance between the center of the window (the vector under consideration) and all other vector samples inside the filter window. The transformation can be considered to be a membership function with respect to the specific window component. The adaptive algorithm evaluates a membership function based on a given vector signal and then uses the membership values to calculate the final filtered output.

This design qualifies as an adaptive, fuzzy system since it utilizes sample input data and inference procedures, in the form of transformed distance metrics, to define a fuzzy system at each time instant. Through the adaptation mechanism, it utilizes the system structure changes over time resulting in a time-varying mapping between input values and filtered output. This time-varying mapping defines an adaptive fuzzy system that changes over time. As it was argued in [21], adaptation, or learning, is essentially parameter changing. Thus, by changing the fuzzy weights in (3.1), an adaptive fuzzy system capable of learning new associations between input patterns and new functional dependencies has been developed. In the framework described here, this can be accomplished without the use of linguistic fuzzy rules or local statistics estimation. Features extracted from local data, such as distances among neighboring input vectors, are used to define the fuzzy weights.

The noise smoothing problem described above can also be seen as a problem of prototype estimation when given a set of signal inputs. In this sense, filtering is the process of replacing the noise-corrupted multichannel signal at the window center by a prototype signal, such that the differences between this prototype and all its neighbors inside the window are minimized in some sense. This operation is essentially a defuzzification procedure. It determines the most appropriate signal value, a vector signal in the case of multichannel inputs, to represent a collection of elements whose membership functions have been constructed over a universe of discourse.

Although a number of different defuzzification strategies exist, the centroid defuzzification approach, known as the Center of Gravity (CoG), is often utilized in practice. The CoG method generates a defuzzified value which is at the center of the values of a fuzzy set. Its defuzzified output actually corresponds to the membership graded weighted mean of the square (Euclidean) distance.

To clarify this, consider a fuzzy set A that is defuzzified as [21]:

$$\mathbf{A} = (\frac{\mu_1}{\mathbf{y}_1}, \frac{\mu_2}{\mathbf{y}_2}, ..., \frac{\mu_n}{\mathbf{y}_n}) \tag{3.3}$$

where μ_n is the membership function associated with the input value $\mathbf{y}_n$. If a quadratic cost function is considered:

$$\mathbf{K}(\bar{\mathbf{y}}) = \sum_{i=1}^{n} |\mathbf{y}_i - \bar{\mathbf{y}}|^{\tau} \mu_i |\mathbf{y}_i - \bar{\mathbf{y}}| \tag{3.4}$$

the Center of Gravity (CoG) defuzzified value is obtained when $\mathbf{K}(\bar{\mathbf{y}})$ is minimized by differentiation [21], [24]:

$$\bar{\mathbf{y}} = \frac{\sum_{i=1}^{n} \mathbf{y}_i \mu_i}{\sum_{i=1}^{n} \mu_i} \tag{3.5}$$

Simple inspection of the CoG defuzzified value obtained reveals the similarity with the adaptive filtered output of (3.1). Therefore the output of the adaptive filter can be considered as the output of the CoG defuzzification strategy with the noisy multichannel signals as members of a fuzzy set and the membership functions μ_i, $i - 1, 2, ..., n$ defined over them.

In such a design, the overall performance of the processing system is determined by the defuzzification procedure selected. The quadratic cost function discussed above can be generalized to include any arbitrary function of μ. Under such a scenario, it is assumed that the cost function associated with the selection of the defuzzified value to represent the fuzzy set A is :

$$\mathbf{K}(\bar{\mathbf{y}}) = \sum_{i=1}^{n} |\mathbf{y}_i - \bar{\mathbf{y}}|^{\tau} f(\mu_i) |\mathbf{y}_i - \bar{\mathbf{y}}| \tag{3.6}$$

where $f(\mu_i)$ is a function of the associated membership function. By minimizing the above quadratic form, a defuzzified (crisp) value can be obtained as:

$$\bar{\mathbf{y}} = \frac{\sum_{i=1}^{n} \mathbf{y}_i f(\mu_i)}{\sum_{i=1}^{n} f(\mu_i)} \tag{3.7}$$

A variant of particular importance is the power function $f(\mu_i) = \mu_i^{\lambda}$ with $\lambda \in [0, \infty)$. In this case the defuzzified value can be obtained through the following equation:

$$\bar{\mathbf{y}} = \frac{\sum_{i=1}^{n} \mathbf{y}_i \mu_i^{\lambda}}{\sum_{i=1}^{n} \mu_i^{\lambda}} \tag{3.8}$$

which is identical to the form used to generate the filtered output in the adaptive design of (3.1). It can easily be seen that, in the generalized defuzzification rule of (3.8), if $\lambda = 1$ the widely used CoG strategy can be obtained.

The defuzzified vector valued signal obtained through the CoG strategy is a vector valued signal which was not part of the original set of input vectors. However, if the output of the adaptive fuzzy system is required to be a member of the original input set, a different defuzzification strategy should be used. By defining $\mu_{(max)}$ to be the largest membership value, the adaptive weights in (3.1) can be re-written as follows:

$$w_j = \frac{\mu_j^\lambda}{\sum_{j=1}^{n} \mu_j^\lambda} = \frac{(\frac{\mu_j}{\mu_{(max)}})^\lambda}{\sum_{j=1}^{n} (\frac{\mu_j}{\mu_{(max)}})^\lambda} \tag{3.9}$$

Given that $\mu_j < \mu_{(max)}$, as $\lambda \rightarrow \infty$ then

$$w_j = \begin{cases} 1 \text{ if } \mu_j = \mu_{(max)} \\ 0 \text{ if } \mu_j \neq \mu_{(max)} \end{cases} \tag{3.10}$$

Equation (3.10) represents the 'Maximum Defuzzifier' strategy [24]. If the maximum value occurs at a single point only, the maximum defuzzifier strategy coincides with the 'Mean of Maxima' (MOM) defuzzification process. Through the Maximum Defuzzifier the output of an adaptive fuzzy system is defined as [22]:

$$\hat{\mathbf{y}} = \mathbf{y}_j; \quad \mu_j = \mu_{(max)} \tag{3.11}$$

3.2.1 Determining the Parameters

The most crucial step in the design of the adaptive fuzzy system is the determination of the membership function to be used in the construction of its weights. The problem of defining the appropriate membership function is one of paramount importance in the design and implementation of fuzzy systems in general. The difficulties associated with the meaning and measurement of the membership function hinder the applicability of fuzzy techniques in many practical applications. From an application point of view, it is important to clarify where the membership function arises, how is it used and measured, and how it can be manipulated in order to provide meaningful results. Since there are different interpretations of fuzziness, the meaning of the membership function changes depending on the application or methodology adopted. In general, apart from the formal definition, a membership function can be seen as a 'graded membership' in a set. Depending on the interpretation of fuzziness, various solutions to the problem of membership definition and 'graded membership' can be obtained. Viewing membership values as similarity indicators is often used in prototype theory where membership is a notion of being similar to a representative of a category [25]. Thus, a membership function value can be used to quantify the degree of similarity of an element to the set in question. The assumption behind this approach is that there exists a perfect (ideal) example of the set, which belongs to the set to the full degree. The valuation of membership for the rest of the elements in the set, can be regarded as the comparison of a given input $\mathbf{y}_i$ with the ideal input $\mathbf{y}_r$, which results in a distance $d(\mathbf{y}_i, \mathbf{y}_r)$.

The membership function must be scaled accordingly to adopt a view of similarity . If the input under consideration has all the features of the ideal prototype then the distance should be zero, and this object should belong to the set to the full degree. On the other hand, if no similarity between the ideal prototype and the input exists, the distance should be infinite and this notion should be reflected in the membership value.

Assuming that a certain degree of membership function is assigned to each element in the set this membership function can be defined as [26], [27]:

$$\mu_i = \frac{1}{1 + f(d(\mathbf{y}_i, \mathbf{y}_r))} \tag{3.12}$$

Based on the definition above:

1. $\mu_i = 0$ if $d(\mathbf{y}_i, \mathbf{y}_r) \rightarrow \infty$
2. $\mu_i = 1$ if $d(\mathbf{y}_i, \mathbf{y}_r) = 0$

Equation (3.12) is only a transformation rule from one numerical representation into another. To complete the process, the exact form of the distance function has to be specified. Depending on the specific distance measure that is applied to the input data, a different fuzzy membership function can be devised.

However, the definition of a distance (or similarity) measure requires an appropriate metric space on which the different distance (similarity) measures will be defined and evaluated. Although the notion of distance is very natural in the case of scalar signals (univariate signals), it cannot be extended in a straightforward way for the case of vector signals. However, as discussed in Chap. 2, different measures can be used to quantify similarity or dissimilarity among multivariate inputs. It should be emphasized in this point that all the different distance measures, as well as the sub-ordering principles discussed there, can be used in conjunction with the proposed adaptive techniques.

3.2.2 The Membership Function

Having discussed the different measures to quantify distance (or similarity) between two vector inputs, attention is turned to the problem of membership function specification. The generic form of the function was given in [26], [28] as:

$$\mu_i = \frac{1}{1 + f(d_i)} \tag{3.13}$$

where $f(.)$ is a function of the distance between the multivariate vector $\mathbf{y}_i$ and the ideal prototype $\mathbf{y}_r$. Membership functions are either monotonically increasing functions from 0 to 1, monotonically decreasing functions from 1 to 0 or can be divided into monotonically increasing or decreasing parts. Each increasing or decreasing part is specified by a cross-over or dispersion point. The particular function $f(d_i)$ used in (3.13) will determine the actual shape of the membership function [27]-[30]. The approach of [26] suggests that since the relationship between distances measured in physical units and perception is generally exponential, an exponential type of function should be used in the generic membership function [26]. The resulting type of an S-shaped function deduced from this proposition can be defined as:

$$\mu_i(\mathbf{y}_i) = \frac{1}{1 + exp(\frac{-c(s(\mathbf{y}_i - \mathbf{y}_r) - a_1)}{\sigma_1})} \tag{3.14}$$

with $\lim_{s \to \infty} \mu_i = 1$ and $\lim_{s \to 0} \mu_i = 0$ for a monotonically increasing function, and:

$$\mu_i(\mathbf{y}_i) = \frac{1}{1 + exp(\frac{-c(d(\mathbf{y}_i - \mathbf{y}_r) - a_2)}{\sigma_2})} \tag{3.15}$$

with $\lim_{d \to \infty} \mu_i = 0$ and $\lim_{d \to 0} \mu_i = 1$ for a monotonically decreasing function.

The resulting membership function is in the *S-shape* format [31]. Due to the lack of break points, S-shaped functions are best suited to represent natural, continuous behavior. The cross-over point, the point assigned a membership value of 0.5, is also the inflection point between the convex and the concave part in the S-shaped function defined in (3.14)-(3.15). By construction, the function never reaches absolute truth (or falsehood) values due to the asymptotic behavior of both values. If this constitutes a problem for a particular application domain it can be resolved by introducing the appropriate break-points at $l = a_1 + \sigma_1$ ($l = a_1 - \sigma_1$) and $l = a_2 + \sigma_2$ ($l = a_2 - \sigma_2$).

In addition, the membership function needs a parameter c, which controls its dispersion characteristics. Dispersion is defined as the range between the cross-over points and the nearest entry which receives the maximum value of 1. The dispersion value regulates the fuzzification process and is a design parameter which can be tuned to modify the fuzziness of the membership function. The parameters for cross-over points and dispersion are the minimum requirements for determining a fuzzy membership function.

In the case in which a distance measure is used to quantify dissimilarity between the vector under consideration and the ideal prototype, the decreasing form of the function is utilized. If a similarity measure is used instead, the monotonically increasing version of the membership function is considered.

Although not supported by the general form of the membership function, introduced in [26], the exponential (Gaussian) kernel was used by the authors as an approximation to the membership function, especially when Minkowski metrics are used, to calculate distance between multichannel signals.

In this case the proposed membership function can be defined as:

$$\mu_i(\mathbf{y}_i) = exp(-\frac{-d_p(\mathbf{y}_i - \mathbf{y}_r)^{\lambda}}{\beta}) \tag{3.16}$$

where λ is a positive constant and β is a distance threshold. The two parameters are design parameters. That is, their actual values vary with the application. The above parameters correspond to the denominational and exponential fuzzy generators controlling the amount of fuzziness. The membership model proposed, although empirically defined, complies with certain psychometrical experiments [32], according to which quantification of similarity of an unknown stimulus to known stimuli belonging to the category can be expressed as a simple exponential decay or Gaussian function of a normalized distance in a psychological space.

3.2.3 The Generalized Membership Function

Although the model of (3.14)-(3.16) satisfies the requirement imposed by the adaptive fuzzy framework, it is computationally expensive since it involves the evaluation of the exponential function. In addition, its parameters cannot be evaluated easily. Therefore, other functions which can retain the same characteristics and are easier to implement are needed. Such a membership form was proposed in [33]. This function is continuously increasing (decreasing) and satisfies the same boundaries conditions and complies with the generic membership form of (3.12). It also retains the properties of the S-shaped membership function. However, unlike the function in (3.14)-(3.15) it can be written as a rational function of two polynomials. It can be proven that for any input value z, the membership function can be completely characterized by only four parameters. Namely, the interval $[a, b]$ of the input parameter z, the sharpness λ of the membership function and the inflection point ν of the S-shaped function. Based on these parameters a membership function can be defined as:

$$\mu(z) = \frac{(1-\nu)^{\lambda-1}(z-a)^{\lambda}}{(1-\nu)^{\lambda-1}(z-a)^{\lambda} + \mu^{\lambda-1}(b-z)^{\lambda}} \tag{3.17}$$

for a monotonically increasing function, and

$$\mu(z) = \frac{(1-\nu)^{\lambda-1}(b-z)^{\lambda}}{(1-\nu)^{\lambda-1}(b-z)^{\lambda} + \mu^{\lambda-1}(z-a)^{\lambda}} \tag{3.18}$$

for a monotonically decreasing function, with the inflection point ν defined via $\mu(z_\nu) = \nu$ and $z_\nu = (b-a)\nu + a$ or $z_\nu = (a-b)\nu + b$ for the case of monotonically increasing or decreasing functions respectively.

The sharpness of the function (an indicator of increasing/decreasing membership) can be defined as:

$$\lambda = f'(z-\nu)(b-a) \tag{3.19}$$

This membership function, which is universally applicable, can be applied to the problem under consideration by considering the distance or similarity value as the input to the membership function. Assuming that $d_i = d(\mathbf{x}_i - \mathbf{x}_r)$

and $s_i = s_i(\mathbf{x}_i - \mathbf{x}_r)$ are appropriate distance or similarity measures between the vector under consideration and the ideal prototype, a membership function can be written as:

$$\mu_i = \frac{(1-\nu)^{\lambda-1}(d_{max}-d_i)^{\lambda}}{(1-\nu)^{\lambda-1}(d_{max}-d_i)^{\lambda}+\mu^{\lambda-1}(d_i-d_{min})^{\lambda}} \tag{3.20}$$

with $d_i \in [d_{max}, d_{min}]$. Alternatively a monotonically increasing function can be defined based on a similarity measure $s_i = s_i(\mathbf{x}_i - \mathbf{x}_r)$ as follows:

$$\mu_i = \frac{(1-\nu)^{\lambda-1}(s_i-s_{min})^{\lambda}}{(1-\nu)^{\lambda-1}(s_i-s_{min})^{\lambda}+\mu^{\lambda-1}(s_{max}-s-i)^{\lambda}} \tag{3.21}$$

with $s_i \in [s_{max}, s_{min}]$.

For the case of $\lambda = 1$ the linear form of the function is obtained. Namely;

$$\mu_i = \frac{s_i - s_{min}}{s_{max} - s_{min}} \tag{3.22}$$

for a monotonically increasing function, and

$$\mu_i = \frac{d_{max} - d_i}{d_{max} - d_{min}} \tag{3.23}$$

for a monotonically decreasing function, which corresponds to the nearest-neighbor rule used in [34–36] to define membership functions.

To summarize, a possible interpretation of the membership function has been outlined and how membership functions can be built based on similarity concepts has been discussed. A generalized model for building membership functions was utilized. The different similarity or distance measures discussed here can be used as input values to this membership function model.

3.2.4 Members of the Adaptive Fuzzy Filter Family

In this section, a number of color image filters derived from the generalized framework introduced is introduced. In the proposed filters, the distance d_i associated with the vector $\mathbf{y}_i$ inside the processing window is defined as the distance (or similarity) of this vector from a reference vector $\hat{\mathbf{y}}_r$. Therefore, assuming that the angle between the two vectors (see Sect. 2.10) is utilized to measure orientation difference, the scalar quantity:

$$d_i = cos^{-1}\left(\frac{\hat{\mathbf{y}}_r \mathbf{y}_i^t}{|\hat{\mathbf{y}}_r||\mathbf{y}_i|}\right) \tag{3.24}$$

is the directional (angular) distance associated with the noisy vector $\mathbf{y}_i$ inside the processing window of length n, with reference point $\hat{\mathbf{y}}_r$. Similar results can be obtained for all the different distance (or similarity) measures discussed in Chap. 2.

For example, measures such as the Canberra distance:

$$d_c = \sum_{k=1}^{p} \frac{|y^k - y_r^k|}{|y^k + y_r^k|} \tag{3.25}$$

or the Czekanowski coefficient:

$$d_z = 1 - \frac{2\sum_{k=1}^{p} \min(y^k, y_r^k)}{\sum_{k=1}^{p} (y^k, y_r^k)} \tag{3.26}$$

can also be used, with p as the dimension of the vectors $\mathbf{y}$ and y^k as the kth element of $\mathbf{y}$. In addition to distances, similarity measures such as:

$$s_7(\mathbf{y}, \mathbf{y}_r) = 1 - \frac{(|\mathbf{y}|^2 + |\mathbf{y}_r|^2 - 2|\mathbf{y}||\mathbf{y}_r|cos(\theta))^{\frac{1}{2}}}{(|\mathbf{y}|^2 + |\mathbf{y}_r|^2 + 2|\mathbf{y}||\mathbf{y}_r|cos(\theta))^{\frac{1}{2}}} \tag{3.27}$$

discussed in Chap. 2 can be used to quantify the similarity between the vector under consideration and the reference vector.

It is obvious that the value of the membership function and the performance of adaptive fuzzy filter depends not only on the distance or similarity measure selected to quantify the dissimilarity between the vector inputs, but on the choice of the appropriate reference vector (ideal prototype) as well. The ideal reference vector is the actual value of the multidimensional signal in the specific location under consideration. However, this signal is not available. Moreover, the noisy vector at the same location is not appropriate since any vector inside the window can be an outlier. Thus, a robust estimate of the location, usually evaluated in a smaller subset of the input vector set, is utilized as the reference vector. The selection of this robust reference vector depends on the signal characteristics. Usually the median is the preferable choice since it smoothes out impulsive noise and preserves edges and details. Moreover, unlike scalars, the center-most vector in a set of vectors can be defined in more than one way. Thus, the vector median filter (VMF), the basic vector directional filter (BVDF) or the marginal median filter (MAMF) operating in a processing window centered around the current vector input can be used to provide the requested reliable reference point [37].

The proposed adaptive fuzzy filter can be viewed as a double-window two stage estimator in which two operations can be distinguished between. First, the original signal is filtered by a multichannel median filter in a small processing window in order to reject possible outliers and then an adaptive fuzzy filter with data dependent weights to provide the final estimates. Thus, the overall filter can be viewed as a combined multichannel operator, which incorporates simple nonlinear statistical estimators, such as the VMF, into adaptive designs based on fuzzy membership functions.

Clearly, the filter depends on the reference point selected. To make the procedure robust and to make sure that the filter will provide accurate results, the need for a reference point is eliminated by evaluating the membership function used to weight each input vector $\mathbf{y}_i$ in (3.1) on the aggregate distance between the vector $\mathbf{y}_i$ under consideration and all the other vectors inside the processing window. Thus, the vector with the smallest overall distance (or maximum similarity) is now assigned the maximum membership value.

It is obvious that such a design does not depend on a reference point and thus is more robust to possible outliers. However, the computational complexity of the algorithm has increased as a result of the need to evaluate a number of distances (similarities) in the processing window. Any distance or similarity measure discussed in Chap. 2 can be used in the adaptive design. Needless to say the membership function selected is now evaluated on the aggregated distances and not on the distance between the vector and the ideal prototype.

Therefore, assuming, for example, that the Euclidean metric (L_2 norm):

$$d_2(i,j) = (\sum_{k=1}^{m} (y_i^k - y_j^k)^2)^{\frac{1}{2}} \tag{3.28}$$

has been selected by the designer as dissimilarity measure, the scalar quantity:

$$d_i = \sum_{j=1}^{n} L_2(\mathbf{y}_i, \mathbf{y}_j) \tag{3.29}$$

is the distance associated with the noisy vector $\mathbf{y}_i$, $\forall i = 1, 2, ..n$ inside the processing window of length n. This distance value is used as an input to the membership functions that determine the fuzzy weights in the multichannel filters. For such a distance an appropriate membership function is the exponential (Gaussian-like) form [22]:

$$\mu_i = exp[-\frac{d_p(i)^r}{\beta}] \tag{3.30}$$

where r is a positive constant, and β is a distance threshold. The actual values of the parameters vary with the application. The above parameters correspond to the denominational and exponential fuzzy generators controlling the amount of *fuzziness* in the fuzzy weight. It is obvious that since the distance measure is always a positive number, the output of this fuzzy membership function lies in the interval $[0, 1]$. The fuzzy transformation is such that the higher the distance value is, the lower the fuzzy weight becomes. It can easily be seen that the membership function is one (maximum value), when the distance value is zero and becomes zero (minimum value), when the distance value is infinite.

Similarly, the vector angle criterion (angular distance) [42] defines the scalar measure:

$$d_{ai} = \sum_{j=1}^{n} A(\mathbf{y}_i, \mathbf{y}_j) \tag{3.31}$$

as the distance associated with the noisy image vector $\mathbf{y}_i$ inside the processing window of length n, when the angle between two vectors:

$$A(\mathbf{y}_i, \mathbf{y}_j) = cos^{-1}(\frac{\mathbf{y}_i \mathbf{y}_j^t}{|\mathbf{y}_i||\mathbf{y}_j|}) \tag{3.32}$$

is used to measure dissimilarity.

For the case of the Canberra distance (3.25) the scalar measures:

$$d_{ac} = \sum_{j=1}^{n} d_c(i,j) \tag{3.33}$$

can be defined as the aggregated distance associated with the noisy image vector $\mathbf{y}_i$ inside the processing window of length n. On the other hand, if the measure of (3.27) is used to define similarity between two vectors, the quantity:

$$s_{a7} = \sum_{j=1}^{n} s_7(i,j) \tag{3.34}$$

is the corresponding aggregated similarity. Utilizing the general membership function of (3.12) and taking into account the fact that the relationship between a distance measured in units and perception is generally exponential, a sigmoidal (piece-wise) linear membership function was recommended [22]. In such a case the fuzzy weight μ_i associated with the vector $\mathbf{y}_j$ can take the following form:

$$\mu_i = \frac{\beta}{(1 + exp(\alpha_j))^r}, \tag{3.35}$$

where β, $r = \frac{1}{\sigma}$ are parameters to be determined and α_i is an angular distance measure.

The parameter β is a soft parameter used only to adjust the limit of the S-shaped membership function (weight scale threshold). Assigning β a value of 2, membership functions will deliver an output in the interval $[0,1]$. The parameter r is the smoothing parameter. Since, by definition, the angle distance measure in (3.35) delivers a positive number in the interval $[0,\pi]$ [77], the output of the fuzzy transformation introduced above produces a membership value in $[\frac{\beta}{(1+exp((n\pi)))^r}, \frac{\beta}{2}]$. It can easily be seen that for a moderated sized window, such as a (3x3) window, the above membership function can be considered as having values in the interval $(0,1]$, e.g. $[1.4x10^{-12}, 1]$ for the angular distance of (3.32) with parameters $r = 1$ and $\beta = 2$.

As one should expect, the function used here does not change too much when its values are around the minimum distance (region of confidence) and does not increase quickly when the membership function's input values are around ∞, the region of rejection. The membership functions defined through (3.30) and (3.35) can be used to derive the fuzzy weights introduced in the filter structure of (3.2). Their design parameters can be determined through an optimization procedure [24]. The general idea is to tune the shape and the parameters of the membership function using a training signal. Assuming that the fuzzy membership function is usually fixed ahead of time, a set of available training pairs (input, membership values) is used to tune its parameters. The most commonly used procedure exploits the mean square error (MSE) criterion. In addition, since most of the used shapes are nonlinear, iterative schemes, e.g. backpropagation, they are used in the calculations [15], [16], [23]. However, in an application such as image processing, in order for the membership function to be tuned adaptively, the original image or an image with properties similar to those of the original must be available. Unfortunately, this is seldom the case in real time image processing applications, where the uncorrupted original image or knowledge about the noise characteristics is not available. Therefore, alternative ways to obtain the best parameterization of the fuzzy transformation must be explored.

To this end, an approach is introduced here where instead of 'training' one membership function, a bank of candidate membership functions are determined in parallel using different distance measures [23], [39]. Then, a generalized nonlinear operator is used to determine the final optimized membership function, which is employed to calculate the fuzzy weights. This method of generating the overall function is closely related to the essence of computations with fuzzy logic. By choosing the appropriate operator, the generalized membership function can meet any specific objective requested by the design. As an example, if a minimum operator is selected, the designer pays more attention to the objectives that are satisfied poorly by the elemental functions and selects the overall value based on the worst of the properties. On the contrary, when using a maximum operator the positive properties of the alternative membership functions are emphasized. Finally, a mean-like operator provides a trade-off among different, possibly incompatible, objectives [40].

Using the previous setting, the problem of determining the overall function is transformed into a decision-making problem, where the designer has to choose among a set of alternatives after considering several criteria. Here only discrete solution spaces are discussed since distinct membership function alternatives are available. As in any decision problem, where satisfaction of an objective is required, two steps can be defined:

1. The determination of the efficient solutions
2. The determination of an optimal compromise solution

The optimal compromise solution is defined as the one which is preferred by the designer to all other solutions, taking into consideration the objective and all the constraints imposed by the design. The designer can specify the nonlinear operator used to combine elemental functions in advance and use this operator to single out the final value from the set of available different solutions. This is the approach followed in this section. An aggregator (fuzzy connective) whose shape is defined a-priori, will be used to combine the different elemental functions in order to produce the final weights at each position.

In fuzzy decision making, connectives or aggregators are defined as mappings from $[0,1]^m \rightarrow [0,1]$ and are often requested to be monotonic with respect to each argument. The subclass of aggregation operators which are continuous, neutral and monotonic is called the class of CNM operators [41]. An averaging operator is a member of the class of compensative CNM operators but different from *min* or *max* operators. Averaging operators can be characterized under several natural properties, such as monotonicity and neutrality [40]. It is widely accepted that an averaging operator verifies the following properties:

$\forall M\colon [0,1]^m \rightarrow [0,1]$

1. Idempotency: $\forall \alpha M(\alpha, \alpha, ..., \alpha) = \alpha$
2. Neutrality: The order of arguments is unimportant
3. M is non-decreasing in each place

The above implies that the averaging operator lies between *min* and *max*. However, aggregation operators are in general non-associative or decomposable since associativity may conflict with idempotence [41]. An example of averaging operators are the arithmetic mean, the geometric mean, the harmonic mean or the root-power mean. The problem of choosing operators for logical combination of criteria is a difficult one. Experiments in decision making indicate that aggregation among criteria are neither conjunctive or disjunctive. Thus, compensatory connectives which mix both conjunctive and disjunctive behavior were introduced in [30].

A compensative operator, first introduced in [28], is utilized to generate the final membership function. Following the results in [28], the operator is defined as the weighted mean of a (*logical AND*) and a (*logical OR*) operator:

$$A \bigodot_{\zeta} B = (A \bigcap B)^{1-\zeta} \cdot (A \bigcup B)^{\zeta} \tag{3.36}$$

where A, B are sets defined on the same space and represented by their membership functions. Different *t-norms* and *t-conorms* can be used to express a conjunctive or a disjunctive attitude. If product of membership functions is utilized to determine intersection (*logical AND*) and possibilistic sum for union (*logical OR*), the form of the operator for several sets is as follows [30]:

$$\mu_{ci} = \prod_{j=1}^{m} \mu_{ji}^{(1-\zeta)} \left(1 - \prod_{j=1}^{m} (1 - \mu_{ji})\right)^{\zeta} \tag{3.37}$$

where μ_{ci} is the overall membership function for the sample at pixel i, μ_{cj} is the jth elemental membership value and $\zeta \in [0,1]$. The weighting parameter ζ is interpreted as the *grade of compensation* taking values in the range of $[0,1]$. In this discussion a constant value of 0.5 is used for ζ.

The product and the possibilistic sum are not the only operators that can be used in (3.37). A simple and useful *t-norm* function is the *min* operator. In this section, this *t-norm* is also used to represent intersection. Subsequently, the *max* operator is the corresponding *t-conorm* [24]. In such a case, the compensative operator of (3.37) has the following form:

$$\mu_{ci} = (\min_{j=1}^{m} \mu_{ji})^{(1-\zeta)} (\max_{j=1}^{m} \mu_{ji})^{\zeta} \tag{3.38}$$

The form of the compensative operator is not unique. A number of other mathematical models can be used to represent the *AND* aggregation. An alternative operator, which combines the averaging properties of the arithmetic mean (member of the averaging operator class) with a *logical AND* operator (conjunctive operator) was also proposed in [30].

$$\mu_{ci} = \zeta \min_{j=1}^{m} \mu_{ji} + (1-\zeta)(m^{-1} \sum_{j=1}^{m} \mu_{ji}) \tag{3.39}$$

where μ_{ci} is the overall membership function for the sample at pixel i and the parameter $\zeta \in [0,1]$ is interpreted as the grade of compensation. In this equation the *min t-norm* stands for the *logical AND*. Alternatively, the product of membership functions can be used instead of the *min* operator in the above equation. The arithmetic mean is used to prevent higher elemental weights with extreme values to dominate the final outcome. The operator is computationally simple and possesses a number of desirable characteristics.

Compensatory operators are intuitively appealing but are based on ad-hoc definitions and properties, such as monotonicity, neutrality or idempotency, and cannot be proven always. However, despite these drawbacks, they are still an appealing and simple method for expressing compensatory effects or interactions between design objectives. For this reason, they are utilized in the next subsection to construct the overall fuzzy weights in our adaptive filter designs.

3.2.5 A Combined Fuzzy Directional and Fuzzy Median Filter

In the adaptive filter, its intended to assign higher weights to those samples that are more centrally located (inside the filter window). However, as observed in Sect. 3.2.4 in the case of multichannel data, the concept of vector ordering has more than one interpretation and the vector median inside the processing window can be defined in more than one way. Therefore, the determination of the most centrally positioned vector heavily depends on the distance measure used. Each distance measure described in Sect. 3.2.4 selects a different, most centrally located vector. Since multichannel ordering has no natural basis better filtering results combining ranking criteria which utilize different distances are expected.

Assuming that the adaptive multichannel filter of (3.1) has to be used and the weights μ_i, $\forall i$ inside the operational window have to be assigned consider the design objective: *'The* $\mathbf{y}_i$ *is centrally located as measured with the angle criterion and* $\mathbf{y}_i$ *is centrally located using the Minkowski distance'.* A fuzzy membership function will be established for this statement. The first step is to realize that this statement is a composition between two design objectives, which can be realized using elemental membership functions such as the ones discussed in the previous section. Then, utilizing the compensative operator, the overall function can be obtained. At this point, the effect of the compensatory operator in the filter has to be clarified. In the above design objective the same degree of attractiveness can be reached by having a less centrally located vector according to the Euclidean distance, but more central using the angle criterion and vice versa. That is, the higher value of *'the angle criterion'* compensates for the lower value of membership in *'using the Minkowski distance'.*

For the specific case of two elemental membership functions and equal exponents, the compensative operator defined in (3.37) has the form of a weighted membership product. Thus, depending on the *t-norm* or *t-conorm* used, the overall fuzzy function can be defined as:

$$\mu_{ci}^{a} = (\mu_{1i}\mu_{2i})^{0.5} \tag{3.40}$$

where μ_{ci}^{a} is the overall membership function for the sample at pixel i or:

$$\mu_{ci}^{a} = (\mu_{1i}\mu_{2i})^{0.5}(1 - ((1 - \mu_{1j})(1 - \mu_{2j})))^{0.5} \tag{3.41}$$

It can easily be seen from (3.40) that using the *min* and *max* operators and for equal powers, the operator in (3.37) has actually the form of the geometric mean, a member of the averaging operators family.

The alternative operator introduced in (3.39) has, for this specific case, the following form:

$$\mu_{ci}^{b} = 0.5\min_{j=1}^{2} \mu_{ji} + 0.25\sum_{j=1}^{2} \mu_{ji} \tag{3.42}$$

or

$$\mu_{ci}^{b} = 0.5(\mu_{1i} * \mu_{2i}) + 0.25\sum_{j=1}^{2} \mu_{ji} \tag{3.43}$$

In general, additional weighting factors which will absorb possible scale differences in the definition of the elemental membership functions must be used. However, since the two elemental functions used here run on the interval $[0,1]$, no such weighting factor is required.

The averaging operator defined in (3.38), and the two compensative operators defined in (3.42) and (3.43), respectively, can be used to define the fuzzy weights in (3.2) provided that the elemental fuzzy transforms discussed above have been used to construct the elemental weights. However, in order for the results to be meaningful, the nonlinear operator applied must satisfy some properties that will guarantee that its application will not alter, in any manner, the elemental decisions about the weights. In the literature, there are a number of properties that all the aggregation or compensative operators must satisfy. This subsection will try to examine if the operators that are used to calculate the adaptive weights satisfy these properties [30].

These properties are listed below:

1. *Convexity:* The mean operator in (3.38) is convex since it is known from statistics that:

$$\mu_{ci}{}^{a} = (\min_{k=1,2} \mu_{ki})(\max_{k=1,2} \mu_{ki})^{0.5} \tag{3.44}$$

$$\min_{k} \mu_{ki} \leq \mu_{ci}{}^{a} \leq \max_{k} \mu_{ki} \tag{3.45}$$

For the second operator, introduced in (3.39) the following holds:

$$\mu_{cj}i^{b} = 0.75\min_{k} \mu_{ki} + 0.25\max_{k} \mu_{ki} \tag{3.46}$$

Then, it can be concluded:

$$\min_{k} \mu_{ki} \leq \mu_{ci}^{b} \leq \max_{k} \mu_{ki} \tag{3.47}$$

The convexity of the operators allows for a compromise among the different elemental functions.

2. *Neutrality (Symmetry):* The operators introduced here are symmetric. The property guarantees that the order of presentation for the elemental functions does not affect the overall membership value.

3. *Monotonicity:* The property of monotonicity guarantees that the stronger piece of evidence (larger elemental membership value) generates a stronger support in the final membership function. By the definition in (3.40):

$$\mu_{ci}^{a*} \geq \mu_{ci}^{a} \tag{3.48}$$

where $\mu_{ci}^{a*} = (\mu_{1i}\mu_{ki})^{0.5}$, $\mu_{ci}^{a} = (\mu_{1i}\mu_{ji})^{0.5}$ and $\forall \mu_{ki} \geq \mu_{ji}$. Similarly, for μ_{1i} and $\forall \mu_{ki} \geq \mu_{ji}$ the $\min(\mu_{1i}, \mu_{ki}) \geq \min(\mu_{1i}, \mu_{ji})$, so using (3.42):

$$\mu_{ci}^{b*} \geq \mu_{ci}^{b} \tag{3.49}$$

4. *Idempotence:* The operators presented above are both idempotent. This property guarantees that the outcome of the overall function generates the same value with each elemental value if all of them report the same result. Thus:

$$\mu_{ci}^{a} = (\mu\mu)^{0.5} = \mu \tag{3.50}$$

$$\mu_{ci}^{b} = 0.5\mu + 0.25(\mu + \mu) = \mu \tag{3.51}$$

Examining (3.42) it can be seen that this operator is not idempotent. However, the operator is symmetric and satisfies the monotonicity requirement. Namely,

$$\mu_{ci}^{a*} \geq \mu_{ci}^{a} \tag{3.52}$$

where

$$\mu_{ci}^{a*} = (\mu_{1i}\mu_{ki})^{0.5}(1 - ((1 - \mu_{1i})(1 - \mu_{ki})))^{0.5} \tag{3.53}$$

and

$$\mu_{ci} = (\mu_{1i}\mu_{ji})^{0.5}(1 - ((1 - \mu_{1i})(1 - \mu_{ji})))^{0.5} \tag{3.54}$$

If $\forall \mu_{ki} \geq \mu_{ji}$

$$(1 - \mu_{ki}) \leq (1 - \mu_{ji}) \tag{3.55}$$

$$(1 - ((1 - \mu_{1i})(1 - \mu_{ki}))) \geq (1 - ((1 - \mu_{1i})(1 - \mu_{ji}))) \tag{3.56}$$

Combining (3.52)-(3.56) it can be concluded that the operator defined in (3.42) satisfies the monotonicity requirement. In summary, it was proven that the compensatory operators, that will be used for the fuzzy weights calculations in (3.1), correspond to an aggregation class which satisfies a number of natural properties, such as neutrality and monotonicity.

3.2.6 Comments

The decision to utilize a fuzzy aggregator to construct the overall weight is not arbitrary. On the contrary, it is anticipated that the operator will help accomplish the design objective. The introduction of a combination of different distances in the weight determination procedure is expected to enhance the filter performance. Each one of the above defined operators can generate a final membership function, which is sensitive to relative changes in the elemental membership values helping to accomplish our objective. A fuzzy filter, which utilizes this form of membership function for its fuzzy weights constitutes a fuzzy generalization of a combined vector median and vector directional filter. The following comments can be made regarding the combined fuzzy filter:

1. The computational efficiency of the proposed filter depends not only on the form of the membership function selected or the operator used for aggregation, but on both of them.
2. It must be emphasized that through this design the problem of determining the appropriate membership function is transformed into the problem of combining a collection of possible functions. This constitutes a problem of considerably reduced complexity, since admissible membership functions may be known from physical considerations or design specifications.

3. The shape of the membership function (e.g. sigmoidal or exponential) is not the only parameter that differentiates possible elemental fuzzy transformations. The designer may decide to use the same form for the elemental functions and assign different parameter values to them, e.g. different r or β. Then, an overall membership function can be devised using an appropriate combination of the individual functions.
4. The generalized membership function introduced here can include any weighting function already in use. In fact, the proposed methodology can be used to calculate in an efficient way possible parameters used in the function.
5. This parallel, adaptive on-line determination of the membership function allows for fast design of the membership function without time-consuming iterative processes. The filter's output is calculated in one-pass without any recursion. Thus, the filter does not depend on a 'good' initial estimate. On the contrary, it is well known that iterative 'adaptive' processes starting from an 'inappropriate' initial value are likely to be trapped in 'local optima' with profound consequences to the filter's performance.
6. The proposed adaptive design is a *scalable* one. The designer controls the complexity of the final membership function by determining the number and the form of the individual membership functions. Depending on the problem specification and the computational constraints, the designer can select the appropriate number of elemental functions to be used in the final weighting function.
7. No training signal is required in this adaptive design. Furthermore, the final fuzzy membership function is determined without any suboptimal local noise or signal statistic evaluation since such approaches usually lead to biased solutions. Thus, our adaptive color filters can be used in real-time image applications in contrast to other 'trainable' color filters, which are based on unrealistic assumptions about the availability of training sequences.

Although a number of different fuzzy designs were discussed here, all of them have some common design characteristics and exhibit similar behavior. A number of them are summarized in a series of comments.

- All of the adaptive vector processing filters discussed here perform smoothing of all vectors which are from the same region as the vector at the window center. It is reasonable to make their fuzzy weights proportional to the difference (similarity), in terms of a distance measure, between a given vector and its neighbors inside the operational window. At edges, or in areas with high details, the filter only smoothes inputs on the same side of the edge as the center-most vector, since vectors with relatively large distance values will be assigned smaller weights and will contribute less to the final filter output. Thus, through the utilization of the fuzzy adaptive designs a user is able to not only preserve the signal characteristics but also to reduce the computational effort by avoiding pre-filtering operations, such as edge or line detection operations. The proposed adaptive framework combines elements from almost all known classes of nonlinear filters. Namely, it combines Minkowski-type distances (used in order statistics based estimators) or non-metric 'Content-based' similarity measures (used in ranked type estimators), averaging outputs (used in linear filtering), with data dependent coefficients used in adaptive designs and membership functions used in fuzzy systems.

- In the framework described above, there is no requirement for fuzzy rules or local statistics estimates. Features extracted from local data, here in the form of distances or similarities, are used as inputs to the membership function. The fuzzy filters discussed in this section, do not utilize the distance measures to order the noisy input signals. Instead, they are used to provide selected features in a reduced space; features used as inputs for the adaptive weights.
- The adaptive vector filters can also be seen as directional filters since angular distances can be used in the derivation of the fuzzy weights. However, such an adaptive filter is not a pure chromaticity filter since it uses both the directional filtering information through the angular distances for its weights as well as the magnitude component of each one of the color vectors. This is a feature that differentiates the proposed design from the chromaticity filters with gray-level processing components introduced in [37]. The generalized chromaticity filters introduced there select a subset of the color vectors and then apply gray scale techniques only to the selected group of vectors. However, if important color information was eliminated due to errors in the chromaticity-based decision part, the filters in [42] are unable to compensate using their gray scale processing step. That is not the case in the new design. An adaptive filter based on the general form of (3.2) does not discard any magnitude information based on chromaticity analysis. All the vectors inside the operational window contribute to the final output. Simply stated, the filter assigns weights to the magnitude component of each color vector modifying in this way their contribution to the output. This natural blending of chromaticity-based weights with magnitude-based input contributions makes the filter appropriate for color image processing [38].
- The adaptive designs discussed here differ in their computational complexity. It should be noted at this point that the computational complexity of a given filter is a realistic measure of its practicality and usefulness, since it determines the required computing power and the associated processing time required for its implementation. The computational complexity analysis of the adaptive designs requires knowledge of the membership function used to calculate the adaptive weights and the exact form of the selected distance (similarity) measure used. The computationally intensive part of the adaptive scheme is the distance calculation part. This part, however, is

common to all vector processing designs. Thus, from a practical standpoint, the remarkably flexible structure of (3.2) yields realizations of different filters that can meet a number of design constraints including hardware and computational complexity.

3.2.7 Application to *1-D* Signals

It must be emphasized that the proposed filter framework can be applied in any multivariate signal with a spatial domain. However, since the main theme in the book is color image processing, this section outlines some basic properties that make the proposed filters appropriate in color image processing.

The use of nonlinear filters in color image processing is motivated primarily by the good performance of the filters near edges and other sharp signal transitions. Edges are basic image features which carry valuable information, useful in image analysis and object classification. Therefore, any nonlinear processing operator is required to preserve edges and smooth noise without altering sharp signal transitions.

Simple examples are introduced in this section to illustrate the effectiveness of the proposed algorithms in filtering operation near noisy edges. The goal is to determine the performance of the filter in terms of noise reduction and detail preservation. The fuzzy multivariate filter which utilizes the geometric mean operator to construct the final membership function is compared in terms of performance with the vector median filter (VMF) and the arithmetic (linear) mean filter (AMF).

To quantitatively evaluate the behavior of the algorithms one dimensional signals corrupted by noise are used. First, a step edge of height 2, for a two-variate, invariant signal corrupted by additive Gaussian mixture noise is used. The details are summarized below:

$$\mathbf{y}(t) = \mathbf{x} + \mathbf{w}(t) \tag{3.57}$$

with

$$\begin{cases} \mathbf{x} = \begin{pmatrix} 1.5 \\ 1.75 \end{pmatrix} & t \leq 45 \\ \mathbf{x} = \begin{pmatrix} 3.5 \\ 3.75 \end{pmatrix} & t > 45 \end{cases} \tag{3.58}$$

and

$$\mathbf{w}(t) = \mathbf{u}(t)\mathbf{v}_1(t) + (I - \mathbf{u}(t))\mathbf{v}_2(t) \tag{3.59}$$

where $\mathbf{u}(t) = u(t)I_{2x1}$. Here $u(t)$ is a random number uniformly distributed over the interval $[0,1]$, $\mathbf{v}_1(t)$ is from a Gaussian distribution with zero mean and covariance $0.05I_{2x2}$ and $\mathbf{v}_2(t)$ is from a Gaussian distribution with zero mean and covariance $0.25I_{2x2}$.

An operational window of size $N = 5$ was used in all the experiments reported here. The filtering results are shown below in Fig. (3.1)-(3.2). These figures depict the filter outputs for the first and second component.

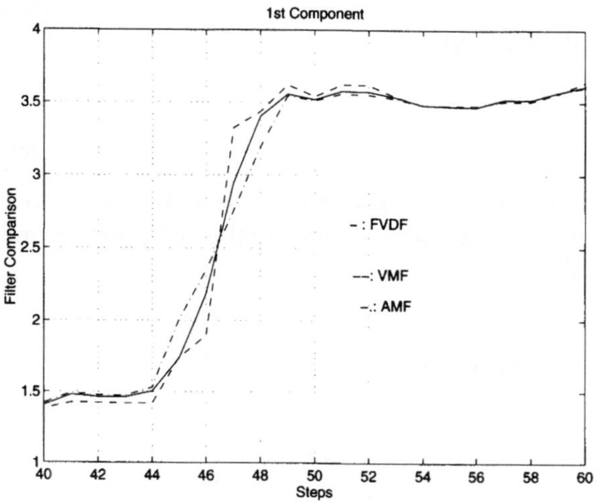

Fig. 3.1. Simulation I: Filter outputs (1st component)

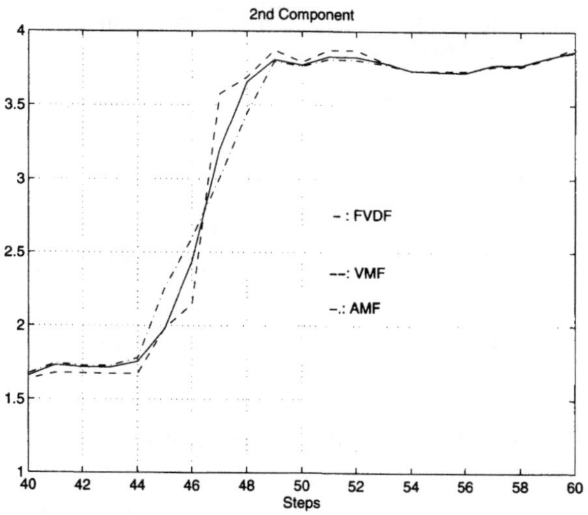

Fig. 3.2. Simulation I: Filter outputs (2nd component)

In order to evaluate the performance of the algorithms in the presence of mixed Gaussian and impulsive noise another simulation experiment was conducted. In this second experiment the actual signal is corrupted with mixed Gaussian and impulsive noise. The observed signal has the following form:

$$\mathbf{y}(t) = \mathbf{x} + \mathbf{w}(t) \tag{3.60}$$

with

$$\begin{cases} \mathbf{x} = \begin{pmatrix} 1.5 \\ 1.75 \end{pmatrix} & t{\leq}45,\ 35{\leq}t{\leq}55,\ t > 75 \\ \mathbf{x} = \begin{pmatrix} 3.5 \\ 3.75 \end{pmatrix} & 15leqt{\leq}35,\ 55{\leq}tleq75 \end{cases} \tag{3.61}$$

and

$$\mathbf{w}(t) = \mathbf{v}_1(t) + \mathbf{v}_2(t) \tag{3.62}$$

where $\mathbf{v_1}(t)$ is from a Gaussian distribution with zero mean and covariance $0.25 I_{2x2}$ and $\mathbf{v}_2(t)$ is impulsive noise with equal number of positive and negative spikes of height 0.25.

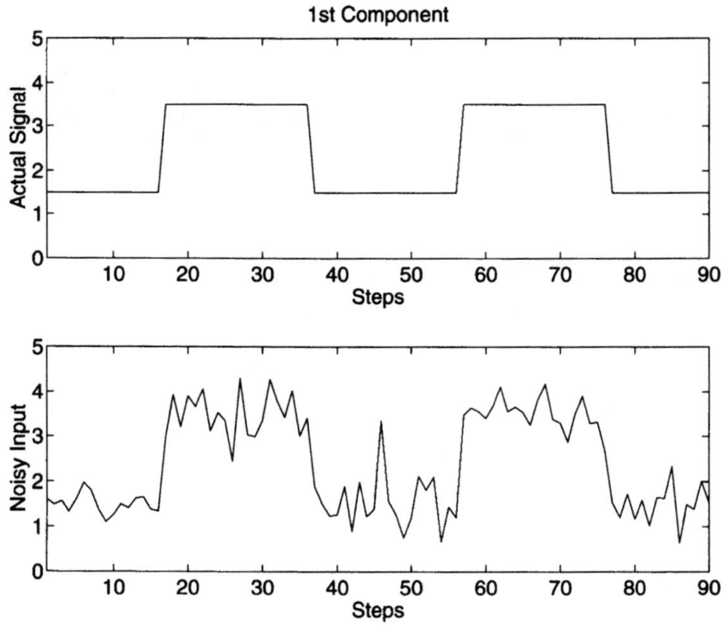

Fig. 3.3. Simulation II: Actual signal and noisy input (1st component)

Fig. (3.3) (i) denote the actual signal and (ii) the noisy input for the first component. Curves in Fig. (3.5) depicts (i) the output of the fuzzy adaptive filter, (ii) the output of the median filter and (iii) the output of the mean filter for the first vector component. Fig. (3.4) and Fig. (3.6) depict the corresponding signals for the second vector component with the same order.

From the above simulation experiments the following conclusions can be drawn:

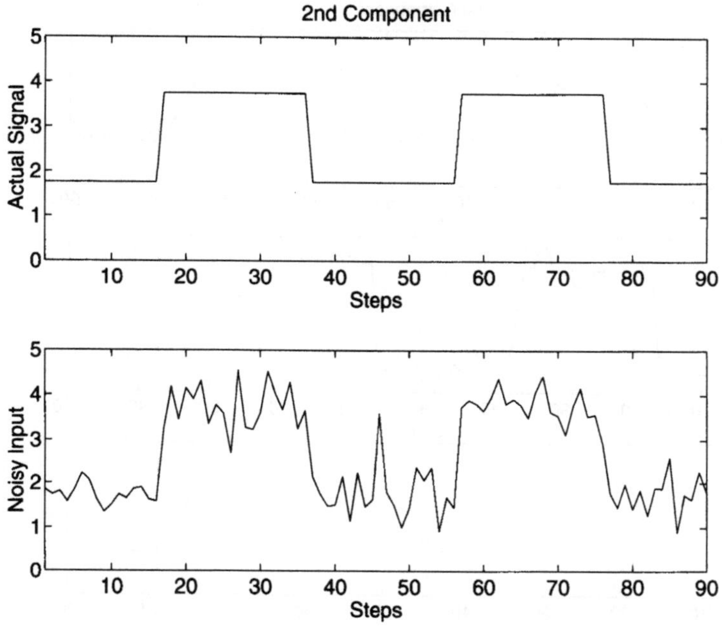

Fig. 3.4. Simulation II: Actual signal and noisy input (2nd component)

1. The vector median filter (VMF) works better near sharp edges.
2. The arithmetic mean (linear) filter works better for homogeneous signals with additive Gaussian-like noise.
3. The proposed adaptive filter can suppress the noise in homogeneous regions much better than the median filter and can preserve edges better than the simple averaging (arithmetic mean) filter.

3.3 The Bayesian Parametric Approach

In addition to fuzzy designs, statistical concepts can be used to devise adaptive color filters. In this section, adaptive filters based on generalized noise models and the principle of minimum variance estimation are discussed.

In all the adaptive schemes defined in this section, a 'loss function' which depends on the noiseless color vector and its filtered estimate is used to penalize errors during the filtering procedure [45]. It is natural to assume that if one penalizes estimation errors through a loss function then the optimum filter is that function of the measurements which minimizes the expected or average loss. In an additive noise scenario, the optimal estimator, which minimizes the average or expected quadratic loss, is defined as [46]:

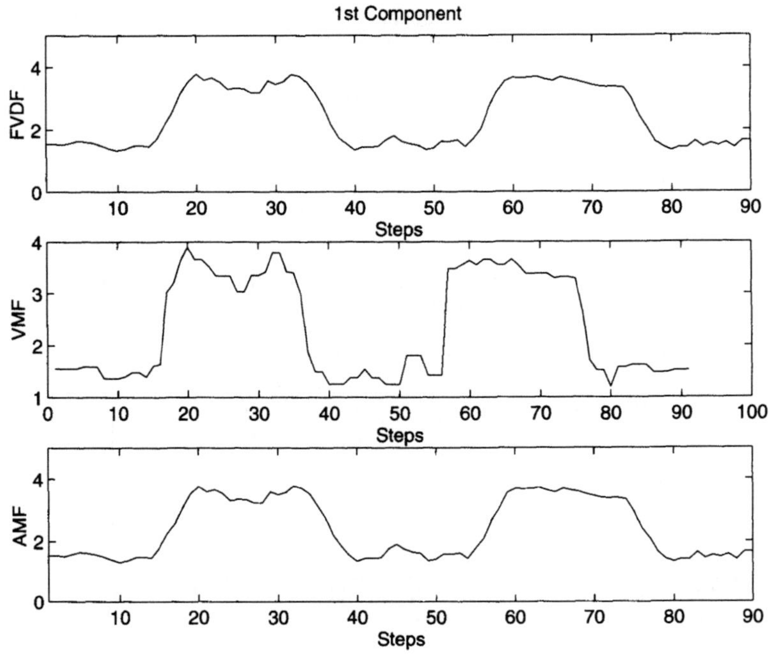

Fig. 3.5. Simulation II: Filter outputs (1st component)

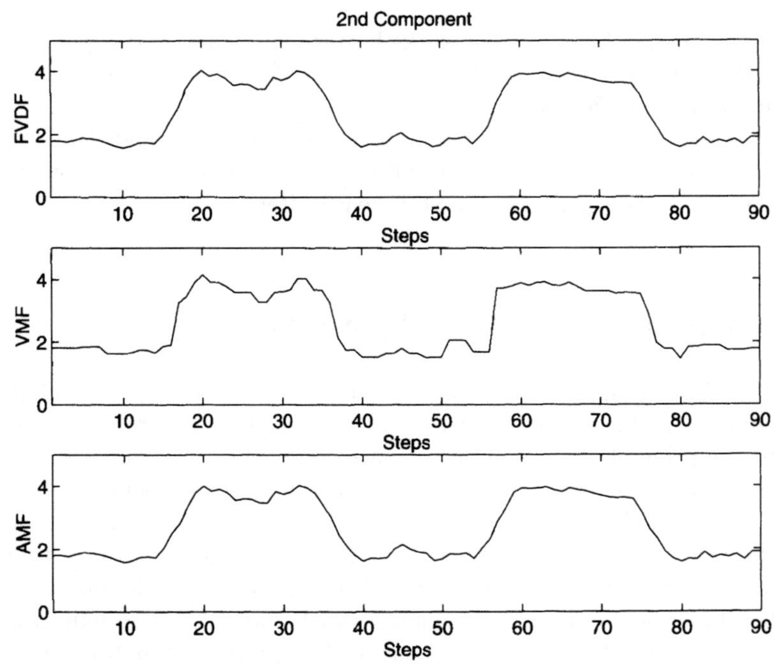

Fig. 3.6. Simulation II: Filter outputs (2nd component)

$$E(\mathbf{x}|\mathbf{y}) = \hat{\mathbf{y}}_{mv} = \int_{-\infty}^{\infty} \mathbf{x} f(\mathbf{x}|\mathbf{y})\, d\mathbf{x} \tag{3.63}$$

or

$$\hat{\mathbf{y}}_{mv} = \int_{-\infty}^{\infty} \frac{\mathbf{x} f(\mathbf{y},\mathbf{x})}{f(\mathbf{y})}\, d\mathbf{x} = \frac{\int_{-\infty}^{\infty} \mathbf{x} f(\mathbf{y},\mathbf{x})\, d\mathbf{x}}{f(\mathbf{y})} \tag{3.64}$$

with

$$f(\mathbf{y}) = \int_{-\infty}^{\infty} f(\mathbf{y},\mathbf{x})\, d\mathbf{x} \tag{3.65}$$

As in the case of order statistics based filters, a sliding window of size $W(n)$ is assumed. By assuming that the actual image vectors remain constant within the filter window, determination of the $\hat{\mathbf{x}}_{mv}$ at the window center corresponds to the problem of estimating the constant signal from n noisy observations present in the filter window [44]:

$$\hat{\mathbf{y}}_{mv} = E(\mathbf{x}|\mathbf{Y}) = \int_{-\infty}^{\infty} \mathbf{x} f(\mathbf{x}|\mathbf{Y})\, d\mathbf{x} \tag{3.66}$$

Central to the solution discussed above is the determination of the probability density function of the image vectors conditioned on the available noisy image data. If this a-posteriori density function is known, then the optimal estimate, for the performance criterion selected, can be determined. Unfortunately, in a realistic application scenario such a-priori knowledge about the process is usually not available. In our adaptive formulation, the requested probability density function is assumed to be of a known functional form but with a set of unknown parameters. This 'parent' distribution provides a partial description where the full knowledge of the underlying phenomenon is achieved through the specific values of the parameters. Given the additive nature of the noise, knowledge of the actual noise distribution is sufficient for the parametric description of the image vectors conditioned on the observations.

In image processing a certain family of noise models are often encountered. Thus, a symmetric 'parent' distribution can be introduced, which includes the most commonly encountered noise distributions as special cases [47]. This distribution function can be characterized by a *location* parameter, a *scale* parameter and a third parameter ζ which measures the degree of non-normality of the distribution [49]. The multivariate generalized Gaussian function, which can be viewed as an extension of the scalar distribution introduced in [48], is defined as:

$$f(\mathbf{m}|\theta,\sigma,\zeta) = k^M exp(-0.5\beta(\frac{|\mathbf{m}-\theta|}{\sigma})^{\frac{2}{1+\zeta}}) \tag{3.67}$$

where M is the dimension of the measurement space, σ, the variance, is an $M \times M$ matrix which can be considered as diagonal with elements σ_c with $c = 1, 2, ..., M$, while the rest of the parameters are defined as $\beta =$

$(\frac{\Gamma(1.5(1+\zeta))}{\Gamma(0.5(1+\zeta))})^{\frac{1}{1+\zeta}}$, $k = (\frac{(\Gamma(1.5(1+\zeta)))^{0.5}}{(1+\zeta)(\Gamma(0.5(1+\zeta)))^{0.5}})\sigma^{-1}$ with $\Gamma(x) = \int_0^{\infty} t^{x-1}e^{-t}\,dt$ and $x > 0$. This is a two-sided symmetric density, which offers great flexibility. By altering the 'shape' parameter ζ different members of the family can be derived. For example, a value of $\zeta = 0$ results in the Gaussian distribution. If $\zeta = 1$ the double exponential is obtained, and as $\zeta \rightarrow -1$ the distribution tends to the rectangular. For $-1 \leq \zeta \leq 1$ intermediate symmetrical distributions can be obtained [47]. Based in this generalized 'parent' distribution, an adaptive estimator can be devised utilizing Bayesian inference techniques. Assume, for example, that the image degradation process follows the additive noise model introduced in Chap. 2 and that the noise density function belongs to the generalized family of (3.67). Assuming that the shape parameter ζ and the location and scale parameters of this function are independent, $f(\mathbf{x}, \sigma, \zeta) \propto f(\mathbf{x}, \sigma) f(\zeta)$, the adaptively filtered result for a 'quadratic loss function' is given as:

$$E(\mathbf{x}|\mathbf{Y}) = \frac{\int\int\int \mathbf{x} f(\mathbf{Y}|\mathbf{x}, \sigma, \zeta) f(\mathbf{x}, \sigma) f(\zeta)\, d\mathbf{x}\, d\sigma\, d\zeta}{\int\int\int f(\mathbf{Y}|\mathbf{x}, \sigma, \zeta) f(\mathbf{x}, \sigma) f(\zeta)\, d\mathbf{x}\, d\sigma\, d\zeta} \tag{3.68}$$

$$E(\mathbf{x}|\mathbf{Y}) = \int (\frac{\int\int \mathbf{x} f(\mathbf{Y}|\mathbf{x}, \sigma, \zeta) f(\mathbf{x}, \sigma)\, d\mathbf{x}\, d\sigma}{\int\int f(\mathbf{Y}|\mathbf{x}, \sigma, \zeta) f(\mathbf{x}, \sigma)\, d\mathbf{x}\, d\sigma})(\frac{f(\zeta) f(\mathbf{Y}|\zeta)}{f(\mathbf{Y})})\, d\zeta \tag{3.69}$$

$$E(\mathbf{x}|\mathbf{Y}) = \int (E(\mathbf{x}|\mathbf{Y}, \zeta) f(\zeta|\mathbf{Y}))\, d\zeta \tag{3.70}$$

with

$$E(\mathbf{x}|\mathbf{Y}, \zeta) = \int \mathbf{x} f(\mathbf{x}|\mathbf{Y}, \zeta)\, d\mathbf{x} \tag{3.71}$$

The computational complexity of the adaptive filter depends on the information available about the shape parameter ζ. In applications such as image processing, where ζ is naturally discrete, the exact realization of the adaptive estimator can be obtained in a computationally efficient way. If the number of shape values is finite ($\zeta_1, ..., \zeta_{\Phi}$), then it is possible to obtain the overall adaptive filtered output by combining the conditional filtering results with the Bayesian learning of the unknown shape parameters. The form of the adaptive filter therefore becomes that of a weighted sum:

$$E(\mathbf{x}|\mathbf{Y}) = \sum_{\phi=1}^{\Phi} E(\mathbf{x}|\mathbf{Y}, \zeta_{\phi}) f(\zeta_{\phi}|\mathbf{Y}) \tag{3.72}$$

In cases where a continuous parameter space for the shape parameter is assumed, the a-priori density function can be quantized using the form $f(\zeta) = \sum_{\phi=1}^{\Phi} f(\zeta_{\phi}) \delta(\zeta - \zeta_{\phi})$ to obtain discrete values. Using the quantized values of the shape parameter, the approximate adaptive algorithm takes the form of (3.72).

Assume that for a given image location, a window W consisting of n noisy image vectors is available. Assume further that based on these $\mathbf{Y} = W(n)$ measurements, intermediate estimates, conditioned on various ζ, are available. For example, conditioned on $\zeta = 0$ the mean value of the $\mathbf{Y}$ measurements can be considered as the best estimate of the location. Alternatively, if $\zeta = 1$ the median value of the $\mathbf{Y}$ set is essentially accepted as the best estimator. In such a scenario, the main objective of the adaptive procedure is the calculation of the posterior densities which arise for different shape parameters. Assuming a uniform reference prior in the range of $-1 < \zeta \leq 1$ for $f(\zeta)$ the conditional densities are calculated through the following rule:

$$f(\zeta_\phi|\mathbf{Y}) = \frac{f(\mathbf{y}_n|\zeta_\phi, \mathbf{Y}_{n-1})f(\zeta_\phi|\mathbf{Y}_{n-1})}{\sum_{\mu=1}^{\Phi} f(\mathbf{y}_n|\zeta_\mu, \mathbf{Y}_{n-1})f(\zeta_\mu|\mathbf{Y}_{n-1})} \tag{3.73}$$

with

$$f(\mathbf{y}_n|\zeta_\phi, \mathbf{Y}_{n-1}) = k^M exp(-0.5(\frac{\mathbf{y} - \hat{\mathbf{x}}_\phi}{\sigma_\phi})^{\frac{2}{1+\zeta}}) \tag{3.74}$$

where $\mathbf{Y} = (\mathbf{y}_1, \mathbf{y}_2, ..., \mathbf{y}_{n-1}, \mathbf{y}_n)$, $\mathbf{Y}_{n-1} = (\mathbf{y}_1, \mathbf{y}_2, ..., \mathbf{y}_{n-1})$ are the observations obtained from the window and $\hat{\mathbf{x}}_\phi$ is the conditional filtered result for the image vector at the window center using a specific value of the shape parameter $\zeta = \zeta_\phi$.

The above result was obtained using Baye's rule:

$$f(\zeta_\phi|\mathbf{Y}) = \frac{f(\zeta_\phi, \mathbf{Y})}{f(\mathbf{Y})} = \frac{f(\zeta_\phi, \mathbf{y}_n, \mathbf{Y}_{n-1})}{f(\mathbf{Y}|\mathbf{Y}_{n-1})f(\mathbf{Y}_{n-1})}$$

$$= \frac{f(\zeta_\phi, \mathbf{y}_n|\mathbf{Y}_{n-1})}{\sum_{\mu=1}^{\Phi} f(\zeta_\mu, \mathbf{y}_n|\mathbf{Y}_{n-1})} \tag{3.75}$$

Further application of Baye's rule results with:

$$f(\zeta_\phi, \mathbf{y}_n|\mathbf{Y}_{n-1}) = \frac{f(\mathbf{y}_n, \zeta_\phi, \mathbf{Y}_{n-1})}{f(\mathbf{Y}_{n-1})} = \frac{f(\mathbf{y}_n|\zeta_\phi, \mathbf{Y}_{n-1})f(\zeta_\phi, \mathbf{Y}_{n-1})}{f(\mathbf{Y}_{n-1})} \tag{3.76}$$

or

$$f(\zeta_\phi, \mathbf{y}_n|\mathbf{Y}_{n-1}) = f(\mathbf{y}_n|\zeta_\phi, \mathbf{Y}_{n-1})(\frac{f(\zeta_\phi, \mathbf{Y}_{n-1})}{f(\mathbf{Y}_{n-1})})$$

$$f(\zeta_\phi, \mathbf{y}_n|\mathbf{Y}_{n-1}) = f(\mathbf{y}_n|\zeta_\phi, \mathbf{Y}_{n-1})f(\zeta_\phi|\mathbf{Y}_{n-1}) \tag{3.77}$$

To complete the adaptive determination of the a-posteriori density $f(\zeta_\phi|\mathbf{Y})$ in (3.76) the predictive density $f(\mathbf{y}_n|\zeta_\phi, \mathbf{Y}_{n-1})$ must be computed. Due to the additive nature of the noise:

$$f(\mathbf{y}_n|\zeta_\phi, \mathbf{Y}_{n-1}) = f(\mathbf{y}_n|\zeta_\phi, \mathbf{x}) = f_{\mathbf{n}|\mathbf{x}}(\mathbf{y}_n - \mathbf{x}|\zeta_\phi) = f_{\mathbf{n}}(\mathbf{y}_n - \mathbf{x}|\zeta_\phi) \tag{3.78}$$

where $f_{\mathbf{n}|\mathbf{x}}(.)$ denotes the conditional pdf of $\mathbf{n}$ given $\mathbf{x}$ and $f_{\mathbf{n}|\mathbf{x}}(.) = f_{\mathbf{n}}(.)$ when $\mathbf{n}$ and $\mathbf{x}$ are independent. Thus, the density $f(\mathbf{y}_n|\zeta_\phi, \mathbf{Y}_{n-1})$ can be considered to be generalized Gaussian with shape parameter ζ_ϕ and location estimate the conditional filter output.

The Bayesian inference procedure described above allows for the selection of the appropriate density from the family of densities considered. If the densities corresponding to the different shape values assumed are representative of the class of densities encountered in image processing applications, then the Bayesian procedure should provide good results regardless of the underlying density, resulting in a robust adaptive estimation procedure.

The adaptive filter described in this section can be viewed as a linear combination of specified, elemental filtered values. The weights in the adaptive design are nonlinear functions of the difference between the measurement vector and the elemental filtered values determined by conditioning on various ζ. In this context, the Bayesian adaptive filter can be viewed as a generalization of radial basis neural networks [50] or fuzzy basis functions networks [51].

If it is desired, the minimum mean square error of the unknown scalar shape parameter can be determined as:

$$E(\zeta|\mathbf{Y}) = \hat{\zeta}_{mmse}(\mathbf{Y}) = \sum_{\phi=1}^{\Phi} \zeta f(\zeta_\phi|\mathbf{Y}) \tag{3.79}$$

with the error in the shape parameter estimation calculated through:

$$E((\zeta - \hat{\zeta}_{mmse}(\mathbf{Y}))^2|\mathbf{Y}) = \sum_{\phi=1}^{\Phi} (\zeta - \hat{\zeta}_{mmse}(\mathbf{Y}))^2 f(\zeta_\phi|\mathbf{Y}) \tag{3.80}$$

In a similar manner, the maximum a-posteriori likelihood estimate of the shape parameter $\hat{\zeta}_{map}(\mathbf{Y}) = \hat{\zeta}_\mu(\mathbf{Y})$ can be obtained through the adaptive filter. The following comments can be made regarding the adaptive filter:

1. The adaptive filter of (3.72) is optimum in the Baye's sense every time it is used inside the window and its optimality is independent of the convergence. The weights that regulate the contribution of the elemental filters are not derived heuristically. Rather, the weights are determined through Baye's theorem using the assumptions on the noise density functions. The adaptive filter weights are dependent on the local image information and thus, as the filter window moves from one pixel to the next, a different adaptive filter output is obtained.
2. Through the adaptive design, the problem of determining the appropriate distribution for the noise is transformed into the problem of combining a collection of admissible distributions. This constitutes a problem of considerably reduced complexity since specific noise models, such as additive Gaussian noise, impulsive noise or a combination of both, are often encountered in image processing applications.

3. This adaptive design is also a scalable one. The designer controls the complexity of the procedure by determining the number and form of the individual filters. Depending on the problem specification and the computational constraints imposed by the design, an appropriate number of elemental filters can be selected. The filter requires no-prior training signals or test statistics and its parallel structure makes it suitable for real-time image applications.

The adaptive procedure is simple, computationally efficient, easy-to-use and reasonably robust. In the approach presented, the posterior probabilities are more important than the manner in which the designer can obtain the elemental estimates which are used in the procedure. Different methodologies can be utilized to obtain these estimates. Filters derived using the maximum likelihood principle, (e.g. VMF, BVDF), robust estimators, (e.g. α-mean filter) and estimators based on adaptive designs, such as the different fuzzy filters discussed in this chapter, can all be used to provide these needed elemental estimates.

From the large number of filters which can be designed using the adaptive procedure, a filter of great practical importance is the one which combines a vector median filter (VMF) with an arithmetic (linear) mean filter (AMF). Extensive experimentation in the past has proven that in the homogeneous regions of the image a mean filter is probably the most suitable estimator, whereas in areas where edges or fine details are present, a median filter is preferable. Through the adaptive design in (3.76), these two filters can be combined. By using local image information in the current processing window, such as an adaptive filter, which is called BFMA, can switch between the two elemental filters in a data dependent manner, offering enhanced filtering performance.

3.4 The Non-parametric Approach

The adaptive formulation presented in the previous section was based on the assumption that a certain class of densities can be used to describe the noise corrupting color images. Thus, a Bayesian adaptive procedure has been utilized to determine on-line the unknown parameters which are used to describe the noise density function. However, in a more general formulation, the functional form of the noise density may also be unknown. In such a case, the densities involved in the derivation of the optimal estimator of (3.64) cannot be determined through a parametric technique such as the one described in the previous section. Rather, they have to be estimated from available sample observations using a non-parametric technique.

Among the plethora of the different non-parametric schemes, the kernel estimator will be adopted here [52]. The notation of non-parametric estimation remains relatively unknown, therefore a brief overview is needed.

If the objective is the non-parametric determination of an unknown multivariate density $f(\mathbf{z})$ from a set of independent samples $\mathbf{Z} = \mathbf{z}_1, \mathbf{z}_2, ..., \mathbf{z}_n$ drawn from the unknown underlying density, the form of a data adaptive non-parametric kernel estimator is:

$$\hat{f}(\mathbf{z}) = (n^{-1})\sum_{l=1}^{n}(h_l)^{-p}K(\frac{\mathbf{z}-\mathbf{z}_l}{h_l}) \tag{3.81}$$

where $\mathbf{z}_l \in \mathbf{R}^p$, p is the dimensionality of the measurement space ($p = 3$ for color images), $K : \mathbf{R}^p \mapsto \mathbf{R}^1$ is a function centered at 0 that integrates to 1 and h_l is the smoothing term [53]-[56].

The form for the data-dependent smoothing parameter is of great importance for the non-parametric estimator. To this end, a new smoothing factor suitable for multichannel estimation is presented here. For the sample point defined in (3.81) a smoothing factor which is a function of the aggregate distance between the local observation under consideration and all the other vectors inside the $\mathbf{Z}$ set is defined, excluding the point at which the density is evaluated. The smoothing parameter is therefore given by:

$$h_l = n^{\frac{-k}{p}} A_l = n^{\frac{-k}{p}}(\sum_{j=1}^{n}|\mathbf{z}_j - \mathbf{z}_l|) \tag{3.82}$$

where $\mathbf{z}_j \neq \mathbf{z}_l$ for $\forall \mathbf{z}_j$, $j = 1, 2, ..., n$, $|\mathbf{z}_j - \mathbf{z}_l|$ is the absolute distance (L_1 metric) between the two vectors and k is a parameter to be determined. The resulting variable kernel estimator exhibits local smoothing which depends both on the point at which the density estimate is taken and information local to each sample observation in the $\mathbf{Z}$ set.

In addition to the smoothing parameter discussed above, the form of the kernel selected also affects the result. Usually, positive kernels are selected for the density approximation. The most common choices are kernels from symmetric distribution functions, such as the Gaussian or the double exponential. For the simulation studies reported in this section, the multivariate exponential kernel $K(\mathbf{z}) = exp(-|\mathbf{z}|)$ and the multivariate Gaussian kernel $K(\mathbf{z}) = exp(-0.5\mathbf{z}^{\tau}\mathbf{z})$ were selected [55].

As for any estimator, the behavior of the non-parametric estimator of (3.81) is determined through the study of its statistical properties. Certain restrictions should apply to the design parameters, such as the smoothing factor, in order to obtain an asymptotically unbiased and consistent estimator. According to the analysis introduced in [55], if the conditions $(\lim_{n\to\infty}(nh_l^{2p}(n)) = \infty$ (asymptotic consistency), $(\lim_{n\to\infty}(nh_l^p(n)) = \infty$ (uniform consistency), and $(\lim_{n\to\infty}(h_l^p(n)) = 0$ (asymptotic unbiasedness) are satisfied then $\hat{f}(\mathbf{z})$ becomes an asymptotically unbiased and consistent estimate of $f(\mathbf{z})$. The multiplier $n^{\frac{-k}{p}}$ in (3.82) with $(0.5 > k > 0)$ guarantees the satisfaction of the conditions for an asymptotically unbiased and consistent estimator [55]. The selection of the A_l for the same design parameter does not affect the asymptotic properties of the estimator in (3.81). However, for a finite number of samples, as in our case, the function A_l is the dominant parameter which determines the performance of the non-parametric estimator.

After this brief introduction to the problem of non-parametric evaluation of the densities involved in the derivation of the optimal estimator in (3.72) will be considered. This time, no assumption regarding the functional form of the noise present in the image is made.

It is only assumed that n pairs of image vectors $(\mathbf{x}_l, \mathbf{y}_l)$, $l = 1, 2, ..., n$ are available through a sliding window of length n centered around the noisy observation $\mathbf{y}$. Based on this sample, the densities $f(\mathbf{y})$ and $f(\mathbf{y}, \mathbf{x})$ will be approximated using sample point adaptive non-parametric kernel estimators.

The first task is to approximate the joint density $f(\mathbf{y}, \mathbf{x})$. As a non-parametric density approximation the following may be chosen:

$$\hat{f}(\mathbf{x}, \mathbf{y}) = n^{-1} \sum_{l=1}^{n} (h_{lx})^{p} (h_{ly})^{-p} K((\frac{(\mathbf{x} - \mathbf{x}_l)}{h_{lx}}), (\frac{(\mathbf{y} - \mathbf{y}_l)}{h_{ly}})) \tag{3.83}$$

Assuming a product kernel estimator [57], the non-parametric approximation of the joint density $f(\mathbf{y}, \mathbf{x})$ has is follows:

$$\hat{f}(\mathbf{x}, \mathbf{y}) = n^{-1} \sum_{l=1}^{n} (h_{lx})^{-p} (h_{ly})^{-p} K(\frac{(\mathbf{x} - \mathbf{x}_l)}{h_{lx}}) K(\frac{(\mathbf{y} - \mathbf{y}_l)}{h_{ly}}) \tag{3.84}$$

The marginal density $f(\mathbf{y})$ in the denominator of (3.65) can then be approximated using the results in (3.83) as follows:

$$\int \hat{f}(\mathbf{y}, \mathbf{x})\, d\mathbf{x}$$

$$= n^{-1} \sum_{l=1}^{n} (h_{ly})^{-p} K(\frac{(\mathbf{y} - \mathbf{y}_l)}{h_{ly}}) (\int (h_{lx})^{-p} K(\frac{(\mathbf{x} - \mathbf{x}_l)}{h_{lx}})\, d\mathbf{x}) \tag{3.85}$$

$$\int \hat{f}(\mathbf{y}, \mathbf{x})\, d\mathbf{x} = n^{-1} \sum_{l=1}^{n} (h_{ly})^{-p} K(\frac{(\mathbf{y} - \mathbf{y}_l)}{h_{ly}}) \tag{3.86}$$

since $\int K(\mathbf{z})\, d\mathbf{z} = 1$ assuming that the kernel results from a real density.

The determination of the numerator is now feasible. The assumption that $\int z_1^1 ... z_p^1 K(\mathbf{z})\, d\mathbf{z} = 0$ implies that [57]:

$$\int \mathbf{x} K(\mathbf{x} - \mathbf{x}_l)\, d\mathbf{x} = \mathbf{x}_l \tag{3.87}$$

Thus, the numerator of (3.67) becomes:

$$\int \mathbf{x}\hat{f}(\mathbf{y},\mathbf{x})\,d\mathbf{x} = n^{-1}\sum_{l=1}^{n}\mathbf{x}_l(h_{ly})^{-p}K(\frac{(\mathbf{y}-\mathbf{y}_l)}{h_{ly}}) \tag{3.88}$$

Utilizing (3.82)-(3.88) the sample point adaptive non-parametric estimator of $\hat{\mathbf{y}}$ can be defined as:

$$\begin{aligned}\hat{\mathbf{y}}_{np} &= \int_{-\infty}^{\infty}\frac{\mathbf{x}\hat{f}(\mathbf{x},\mathbf{y})}{\hat{f}(\mathbf{y})}\,d\mathbf{x} = \frac{\int_{-\infty}^{\infty}\mathbf{x}\hat{f}(\mathbf{x},\mathbf{y})\,d\mathbf{x}}{\int_{-\infty}^{\infty}\hat{f}(\mathbf{x},\mathbf{y})\,d\mathbf{x}} \\ &= \frac{\sum_{l=1}^{n} x_l((n^{-1})h_l^{-p}K(\frac{\mathbf{y}-\mathbf{y}_l}{h_l}))}{\sum_{l=1}^{n}((n^{-1})h_l^{-p}K(\frac{\mathbf{y}-\mathbf{y}_l}{h_l}))} \\ &= \sum_{l=1}^{n} x_l(\frac{h_l^{-p}K(\frac{\mathbf{y}-\mathbf{y}_l}{h_l})}{\sum_{l=1}^{n} h_l^{-p}K(\frac{\mathbf{y}-\mathbf{y}_l}{h_l})}) = \sum_{l=1}^{n}\mathbf{x}_l w_l(\mathbf{y})\end{aligned} \tag{3.89}$$

where $\mathbf{y}_l \in W$ and $w_l(\mathbf{y})$ is a weighting function defined in the interval [0,1]. From (3.89), it can be seen that the non-parametric estimator, often called the Nadaraya-Watson estimator [58], [59], is given as a weighted average of the samples in the window selected. The inputs in the mixture are the noise-free vectors $\mathbf{x}_l$. This estimator is linear with respect to the $\mathbf{x}_l$ and can therefore be considered as a linear smoother. The basis functions on the other hand, determined by the kernel function $K(.)$ and the smoothing parameter $h(.)$, can be either linear or nonlinear on the noisy measurements $\mathbf{y}_l$. It is easy to recognize the similarity between the Bayesian adaptive parametric filter discussed in this chapter with the Nadaraya-Watson estimator. The Bayesian adaptive filter is also linearly smoother with respect to a function of the $\mathbf{x}_l$ (the elemental filtered results) and utilizes nonlinear basis functions which are determined by the unknown 'shape' (ζ) parameter from the generalized 'parent' distribution assumed.

Although the existence of a consistent estimate in mean square has been proven, there are no a-priori guidelines on the selection of design parameters, such as the smoothing vector or the kernel, on the basis of a finite set of data. Smoothing factors, other than the aggregated distance introduced here, such as the minimum distance or the maximum distance between the $\mathbf{y}_l$ and the kth nearest neighbors, constitute valid solutions and can be used instead [60].

The adjustable parameters of the proposed filter are $\mathbf{x}$, $\mathbf{y}$, $K(.)$ and $h(.)$. The degree of the smoothness is controlled by the smoothing factor $h(.)$. It can easily be seen that by appropriately modifying the smoothing factor the non-parametric estimator can be forced to match a given sample arbitrarily close. To accomplish this the kernel is modified by adjusting, through an exponent, the effect of the smoothing parameter $h(.)$. In this case, the form of the estimator is as follows:

$$\hat{\mathbf{y}} = \sum_{l=1}^{n}\mathbf{x}_l(\frac{h_l^{-p}K(\frac{\mathbf{y}-\mathbf{y}_l}{h_l^{r}})}{\sum_{l=1}^{n} h_l^{-p}K(\frac{\mathbf{y}-\mathbf{y}_l}{h_l^{r}})}) \tag{3.90}$$

where, the parameter r regulates the smoothness of the kernel. Since the non-parametric filter is a regression estimator which provides a smooth interpolation among the observed vectors inside the processing window, the r parameter can provide the required balance between smoothing and detail preservation. Because r is a one-dimensional parameter, it is usually not difficult to determine an appropriate value for a practical application. By increasing the value of the r the non-parametric estimator can be forced to approximate arbitrarily close any one of the vectors inside the filtering window. To this end, suppose that a non-parametric estimator with given value $r = r^*$ exists, given the available input set $\mathbf{Y}$. Then the following relation holds:

$$\hat{\mathbf{y}} = \frac{\mathbf{x}_j + \sum_{l=1_{l \neq j}}^{n} \mathbf{x}_l (h_l^{-p} K(\frac{\mathbf{y}-\mathbf{y}_l}{h_l{}^r}))}{1 + \sum_{l=1_{l \neq j}}^{n} (h_l^{-p} K(\frac{\mathbf{y}-\mathbf{y}_l}{h_l{}^r}))} \tag{3.91}$$

with $l = 1, 2, ..., n$. Assuming that $\mathbf{x}_j \neq \mathbf{x}_l$ for $j \neq l$. Then for arbitrary $\epsilon > 0$ and any l, j with $l = 1, 2, ..., n$ and $j \neq l$ you can force $K(\frac{\mathbf{y}-\mathbf{y}_l}{h_l{}^r}) < \epsilon$ since by properly choosing a value $r^* = \frac{1}{r_v}$ the kernel $K(\frac{\mathbf{y}-\mathbf{y}_l}{h_l{}^{r^*}}) \mapsto 0$ as $r_v \mapsto 0$ if $\mathbf{y} \neq \mathbf{y}_l$. Thus, it can be concluded that there exists some value of r such that the non-parametric regressor approaches arbitrarily close to an existing vector.

To obtain the final estimate it is assumed that, in the absence of noise, the actual image vectors $\mathbf{x}_l$ are available. As is the case for the adaptive/trainable filters, a training record can be obtained in some cases during some calibration procedure in a controlled environment. In a real time image processing application however, that is not always possible. Therefore, alternative suboptimal solutions are introduced. In a first approach, each vectors $\mathbf{x}_l$ in (3.89) is replaced with its noisy measurement $\mathbf{y}_l$. The resulting suboptimal estimator, called adaptive multichannel non-parametric filter (hereafter AMNF), is solely based on the available noisy vectors and the form of the data-adaptive kernel selected for the density approximations. Thus, the AMNF form is as follows:

$$\hat{\mathbf{y}}_{AMNF} = \sum_{l=1}^{n} \mathbf{y}_l (\frac{h_l^{-p} K(\frac{\mathbf{y}-\mathbf{y}_l}{h_l})}{\sum_{l=1}^{n} h_l^{-p} K(\frac{\mathbf{y}-\mathbf{y}_l}{h_l})}) \tag{3.92}$$

A different form of the adaptive non-parametric estimator can be obtained if a reference vector $\mathbf{x}_{lr}$ is used instead of the actual color vector $\mathbf{x}_l$ in (3.92). A robust estimate of the location, usually evaluated in a smaller subset of the input set, is utilized instead of the $\mathbf{x}_l$. Usually the median is the preferable choice since it smoothes out impulsive noise and preserves edges and details. However, unlike scalars, the most central vector in a set of vectors can be defined in more than one way. Thus, the vector median filter (VMF) or the marginal median filter (MAMF) operating in a (3×3) window centered around the current pixel can be used to provide the requested reliable reference. In this paper, the VMF evaluated in a (3×3) window was selected to provide the reference vector. The new adaptive multichannel non-parametric filter, (hereafter AMNF2), has the following form:

$$\hat{\mathbf{y}}_{AMNF} = \sum_{l=1}^{n} \mathbf{x}_l^{VM} \left(\frac{h_l^{-p} K(\frac{\mathbf{y}-\mathbf{y}_l}{h_l})}{\sum_{l=1}^{n} h_l^{-p} K(\frac{\mathbf{y}-\mathbf{y}_l}{h_l})} \right) \tag{3.93}$$

The AMNF2 can be viewed as a double-window two stage estimator. First the original image is filtered by a multichannel median filter in a small processing window in order to reject possible outliers and then the adaptive filter of (3.93) is utilized to provide the final filtered output. The AMNF2 filter can be viewed as an extension to the multichannel case of the double-window (DW) filtering structures extensively used for gray scale image processing. As in gray scale processing, with this adaptive filter, the user can distinguish between two operators: (i) the computation of the median in a smaller window; and (ii) the adaptive averaging in a second processing window.

A kernel estimator designed specifically for directional data, such as color vectors, can be devised based on the properties of the color samples on the sphere [63]. When dealing with directional data, a kernel other than the exponential (Gaussian-like) often used in non-parametric density approximation should be utilized. In [62] the following kernel is recommended for the evaluation of the density at point $\mathbf{Y}$ given a set of n available data points $\mathbf{y}_1, \mathbf{y}_2, ..., \mathbf{y}_n$:

$$K_n(\mathbf{Y}) = \frac{C_n}{4\pi sinh(C_n)} exp(C_n(\mathbf{Y}^{\tau}\mathbf{y}_i)) \tag{3.94}$$

where $A(C_n) = \frac{C_n}{4\pi sinh(C_n)}$ normalizes the kernel to a probability density, C_n is the reciprocal of h_l used in the definition of the data-adaptive non-parametric estimator used in [11], and $\mathbf{Y}$ and $\mathbf{y}_i$ are the Cartesian representations of the color vectors with the term $(\mathbf{Y}^{\tau}\mathbf{y}_i)$ denoting the inner product between the two vectors. Alternatively, the term $1 - (\mathbf{Y}^{\tau}\mathbf{y}_i)$ which is the cosine of the distance between the two vectors $(\mathbf{Y}, \mathbf{y}_i)$ along the surface of the sphere can be used. Comparing the adaptive kernel estimator of (3.81) with the proposed [62] kernel estimator of (3.94), it can be seen that the first is based on the Euclidean distance among the available data points, with the latter based on the inner product (the cosine of the distance) between the data points available.

In [64] a simple and computationally attractive alternative to the non-parametric directional estimator was proposed. The proposed non-parametric directional density estimator is of the following form:

$$K_n(\mathbf{Y}) = \frac{1}{n}\left(\frac{1}{A_m} cos^{2m}\left(\frac{a_l}{2}\right)\right) \tag{3.95}$$

where A_m is a normalization constant with m a user defined smoothing factor approaching infinity to be determined as a function of the data record n and $\frac{a_l}{2}$ denotes the angle between the point $\mathbf{Y}$ and the vector with spherical coordinates $(0,0)$. The normalization factor A_m is given to be $A_m = \int (cos^{2m}(\frac{a_l}{2}))\, d\mathbf{x}$ so that $\frac{1}{A_m} cos^{2m}(\frac{a_l}{2})$ integrates to 1 over the sphere S^2. Using direct evaluation of the integral and the Stirling's formula the approximate normalized factor of $A_m = \frac{4\pi}{m+1}$ was proposed in [64]. It can be seen by inspection that $A(C_n)$ is independent of the coordinate system since it is only a function of the angle between two vectors $(\mathbf{Y}, \mathbf{y}_i)$.

This new vector directional kernel estimator is then utilized to assist in the development of an adaptive multivariate non-parametric filter (hereafter AMNFD) which can provide reliable filtered estimates without any prior knowledge regarding signal or noise characteristics.

Given the form of the optimal minimum variance estimator and the adaptive non-parametric kernel, the resulting non-parametric filter can be defined as:

$$\hat{\mathbf{y}}_{np} = \sum_{l=1}^{n} \frac{\mathbf{x}_l (n^{-1})(\frac{1}{A_m}) cos^{2m}(\frac{a_l}{2})}{\sum_{l=1}^{n} (n^{-1})(\frac{1}{A_m}) cos^{2m}(\frac{a_l}{2})} \tag{3.96}$$

where $\frac{a_l}{2}$ denotes the angle between the point $\mathbf{y}_l$ and the vector with spherical coordinates $(0,0)$.

If it is not possible to access the noise free color vectors $\mathbf{x}$, the noisy measurements $\mathbf{y}$ can be used instead. The resulting filter is solely based on the available noisy vectors and the form of the minimum variance estimator.

$$\hat{\mathbf{y}}_{np} = \sum_{l=1}^{n} \frac{\mathbf{y}_l (n^{-1})(\frac{1}{A_m}) cos^{2m}(\frac{a_l}{2})}{\sum_{l=1}^{n} (n^{-1})(\frac{1}{A_m}) cos^{2m}(\frac{a_l}{2})} \tag{3.97}$$

In the derivation of the adaptive non-parametric estimators presented in (3.91), (3.92), (3.93) and (3.97) a number of design parameters have been introduced. Namely:

- the window size, and therefore, the number of noisy measurement vectors available for the evaluation of the approximate density,
- the form of the smoothing factor h_l where decisions about the multiplier and the distance measure utilized can greatly affect the performance of the density estimator,
- the type of kernel used in (3.89),
- the vectors used instead of the actual, unavailable color vectors $\mathbf{x}_l$ in the derivation of (3.89).

All of these elements affect the filtering process since they determine the output of the estimator. In an adaptive formulation, Θ can be defined as the parameter vector, which is the abstract representation of all elements listed above. It is not necessary that all of these elements be treated as parameters in a specific design. Problem specifications and design objectives can be used to determine the elements included in the parameter vector Θ.

By varying the different parameters on the design of the non-parametric kernel, different results $\hat{\mathbf{x}}_{\Theta}(\mathbf{Y}) = \hat{m}(\mathbf{x})$ can be obtained. Suppose that

$\hat{\mathbf{m}}(\mathbf{x}_i)$, $i = 1, 2, ..., P$ are different non-parametric estimators, all based on the same sample record $\mathbf{Y} = (\mathbf{y}_1, \mathbf{y}_2, ..., \mathbf{y}_n)$ but possibly with different kernels $K_1, K_2, ..., K_p$ and different smoothing factors $h_1, h_2, ..., h_p$. An overall estimator based on these values can be obtained as the expected value $\hat{\mathbf{x}}_{np} = E(\hat{\mathbf{m}}(\mathbf{x})|\mathbf{y})$ calculated over the given non-parametric values determined by the different techniques.

Assuming that the different estimated values $\hat{m}(\mathbf{x})$ are available and that they are related to the observed sample through the model

$$\mathbf{y} = \hat{\mathbf{m}}(\mathbf{x}) + \xi \tag{3.98}$$

with ξ additive corruption noise, it can be claimed that the minimization of the expected error leads to a solution for $\hat{\mathbf{y}}_{np}$ as:

$$\hat{\mathbf{y}}_{np} = \frac{\sum_{i=1}^{P_\xi} \hat{\mathbf{m}}_i(\mathbf{y}) f_\xi(\mathbf{y} - \hat{\mathbf{m}}_i(\mathbf{y}))}{\sum_{i=1}^{P_\xi} f_\xi(\mathbf{y} - \hat{\mathbf{m}}(\mathbf{y}))} = \sum_{i=1}^{P_\xi} \hat{\mathbf{m}}_i(\mathbf{y}) w_{npi} \tag{3.99}$$

with

$$w_{npi} = \frac{f_\xi(\mathbf{y} - \hat{\mathbf{m}}_i(\mathbf{y}))}{\sum_{j=1}^{P_\xi} f_\xi(\mathbf{y} - \hat{\mathbf{m}}_j(\mathbf{y}))} \tag{3.100}$$

To calculate the exact value of the multiple non-parametric estimator, the function $f_\xi(.)$ must be evaluated. Since it is generally unknown, it is approximated in a non-parametric fashion based on the set of the elemental values $\hat{\mathbf{m}}(\mathbf{y})$ available. If P_ξ elemental estimates $\hat{\mathbf{m}}_i(\mathbf{y})$ are available, with $i = 1, 2, ..., P_\xi$ the nominal parameter $\xi_i = \mathbf{y} - \hat{\mathbf{m}}_i(\mathbf{y})$ is introduced. Therefore, our objective is the non-parametric evaluation of the density $f_\xi(.)$ using the set of the available data points $\Xi = \xi_1, \xi_2, ..., \xi_{P_\xi}$. The approximation task can be carried out by using any standard non-parametric approach, such as the different kernel estimators discussed in (3.90). For the simulation studies discussed in Sect. 3.6, the sample point adaptive kernel estimator of (3.82)is used. Thus, the following estimate of the density $f_\xi(\mathbf{y} - \hat{\mathbf{m}}_i(\mathbf{y}))$ is used:

$$\hat{f}_\xi(\mathbf{y} - \hat{\mathbf{m}}_i(\mathbf{y})) = \hat{f}_\xi(\xi) = (P_\xi^{-1}) \sum_{l=1}^{P_\xi} (h_l)^{-p} K(\frac{\xi - \xi_l}{h_l}) \tag{3.101}$$

with the smoothing parameter calculated as:

$$h_l = P_\xi^{\frac{-k}{p}} A_l = P^{\frac{-k}{p}} (\sum_{j=1}^{P_\xi} |\xi_j - \xi_l|) \tag{3.102}$$

where $\xi_j \neq \xi_l$ for $\forall \xi_j$ $j = 1, 2, ..., P_\xi$, and $|\xi_j - \xi_l|$ is the absolute distance (L_1 metric) between the two vectors.

From (3.101) it can be claimed that $\hat{f}_\xi(\xi)$ integrates to 1, given the form of the approximation and the fact that the kernel $K(.)$ results from a real density. Thus, the set of weights w_{npi} has the following properties:

1. Each weight is a positive number, $w_{npi} \geq 0$,
2. The summation of all the weights is equal to one, $\sum_{j=1}^{P_\xi} w_{npi} = 1$.

These properties can be interpreted as posterior probabilities used to incorporate prior information concerning local smoothness. Thus, each weight in (3.99) regulates the contribution of the associated filter by its posterior component density.

The following comments can be made regarding the multiple non- parametric filter:

- The general form of the filter is given as a linear combination of nonlinear basis functions. The weights in the above mixture are the elemental filtered values obtained by the different non-parametric estimators applied to the problem. The non-linear basis function is determined by the form of the approximate density $\hat{f}_\xi$ and can take many different forms, such as Gaussian, exponential or triangular. It is not hard to see that in the case of a Gaussian kernel the multiple estimator of (3.99) can be viewed as a radial basis function (RBF) neural network.
- The adaptive procedure can be used to combine a variety of non-parametric estimators each one of them developed for a different value set of the parameter vector Θ. For example, such a structure can be used to combine elemental non-parametric filters derived for different window sizes W. The number of color vector samples utilized in the development of the non-parametric estimator depends on the window W centered around the pixel under consideration. Usually a square (3×3) or (5×5) window is selected. However, such a decision affects the filter's performance. In smooth areas or when Gaussian noise is anticipated, a larger window (e.g. (5×5)) is preferable. On the other hand, near edges or when impulsive noise is assumed a smaller window (usually (3×3) window) is more appropriate. An adaptive filter which utilizes elemental filters with different window sizes, hereafter MAMNF35, is probably a better choice in an unknown or mixed noise environment. Using the same approach other practical adaptive filters, such as the MAMNFEG which utilizes two elemental non-parametric filters with an exponential and a Gaussian kernel respectively, can be devised. Due to the specific form of the kernel, it is anticipated that a non-parametric filter with a Gaussian kernel is probably a better choice for Gaussian noise smoothing. Similarly, a filter with an exponential kernel will provide better filtering results when impulsive or long tailed noise is present. An adaptive design which allows for both filters to be utilized simultaneously is of paramount importance in an unknown noise environment. Such examples emphasize the versatility of the proposed adaptive approach which can provide a wide range of different practical filters.
- Although the filter in (3.99) has been derived as a generalization of a non-parametric estimator it can be used to combine different heterogeneous estimators applied to the same task. Specifically, the designer can utilize a number of different elemental filters, such as order statistics based filters,

the Bayesian adaptive filter, nearest neighbor filters and non-parametric estimators and then combine all the different results using (3.99)-(3.102). The effectiveness of the adaptive scheme is determined by the validity of the elemental filtered results and the approximation required in (3.100). However, due to the different justification of the elemental filters, extensive simulation results are the only way to examine the performance of the filter in practical applications. Experimentation with color images will be used to demonstrate the effectiveness of the multiple filter and to access the improvement in terms of the performance achieved using a multiple non-parametric filter vis. a vis. a simple non-parametric filter. The multiple filter can be a powerful design tool since it allows the combination of filters designed using different methodologies and different design parameters. This is of paramount importance in practical applications since it allows for the development of efficient adaptive filters when no indication for the selection of a suitable filtering approach is available.

3.5 Adaptive Morphological Filters

3.5.1 Introduction

In recent years, a great deal of work has been reported on the development of geometrical based image processing techniques, especially on transformations based on the morphological operations of erosion, dilation, opening and closing. Mathematical morphology can be described geometrically, in terms of the actions of the operators on binary, monochrome or color images. The geometric description depends on small synthetic images called structuring elements. This form of mathematical morphology, often called structural morphology, is highly useful in the analysis and processing of images [65]-[70]. Since objects in nature are generally random in their shape, size and location the notion of a random set provides the means of studying the geometrical parameters of naturally occurring objects.

Mathematical morphology was first introduced for the case of binary images. The objects within a binary image are easily viewed as sets. The interaction between an image set and a second set, the structural element, produces transformations in the image. Measurements taken of the image set, the transformation set, and the difference between the two provide information describing the interaction of the set with the structuring element. The interactions between the image set and the structuring element are set-based transformations. The intersection or union of translated, transposed or complimented versions of the image set and structuring element filter out information. Through the utilization of the umbra, an n-dimensional function described in terms of an $(n+1)$-dimensional set, morphological transformations can be applied to monochrome images [67]. Thresholding of a monochrome image into a group of two-dimensional sets representing the three-dimensional umbra also provides a method of transforming gray scale images using the original definitions of mathematical morphology.

Throughout this book, color images are considered as vector signals and it is well known that the correlation of the color components is of paramount importance in the development of efficient color image processing techniques. In this section another aspect of color image processing is studied. That is, the effect of geometrical adaptivity in processing color images. Morphological techniques developed for use with monochrome images can be extended to color images by applying the algorithm to each of the color component separately. Since morphology is readily defined in terms of scalar signals, the individual color channels can be processed separately as three monochrome images [74]. The idea is to introduce new types of opening and closing operators that allow for the development of fast processing algorithms. The new operators utilize structuring elements that adapt their shape according to the local geometrical structures of the processed images. Although the proposed algorithm process color images in quite different ways from those of the vector based filters it can improve the quality of the processed color images by smoothing out noise while preserving fine details.

Mathematical morphology is based in set theory. The reason for using set theory is to consider objects as being part of a space S. The description of an object is therefore reduced to describing an element or subset X is S. This section summarizes several definitions related to mathematical morphology and relates them to the application of morphological transformations used in image processing, starting with the simplest case of binary images and expanding to the case of monochrome images. Consider initially a two-dimensional space. This space may be pictured as a binary image where each object in the image is a subset of the digital space $\mathbf{Z^2}$. The mathematical definition of an object X in a binary image in terms of a set is:

$$X = \{(i,j) : f(I,j) = 1\} \tag{3.103}$$

where $f(.)$ is called the characteristic function of X. The remaining space in $\mathbf{Z^2}$ is the background or complement of X and is denoted by the set X^c and defined as:

$$X^c = \{(i,j) : f(I,j) = 0\} \tag{3.104}$$

The above definitions may also be defined in terms of vectors. If a vector $\mathbf{x}$ is the vector from the origin to the point (x, y) then (3.103) and (3.104) may become:

$$\begin{aligned} X &= \{\mathbf{x} : f(\mathbf{x}) = 1\} \\ X^c &= \{\mathbf{x} : f(\mathbf{x}) = 0\} \end{aligned} \tag{3.105}$$

The set X also has associated with it its translate and transposition. The translate of X by a vector $\mathbf{b}$ is denoted as X_b. The transposition of X or the symmetric set of X is denoted by $\check{X}$.

$$X = \{\mathbf{x} : \mathbf{x} - \mathbf{b} \in X\}$$
$$\check{X} = \{\mathbf{x} : (-\mathbf{x}) \in X\} \tag{3.106}$$

Consider two sets, X and B. The set B is a set to be included in X if every element of B is also an element of X. If B hits X then the intersection of X and B produces a non-empty set. The opposite of B hitting X is B missing X. The intersection of these two sets is an empty set in this case. If the set of all possible subsets of S, denoted by $F(S)$, are considered and supposing that X and B are elements of $F(S)$ then the following definitions may be made:

- B is included in X, $B \subset X \Rightarrow b \in X, \forall b \in B$
- B hits X, $B \uparrow X \Rightarrow B \cap X \neq \emptyset$
- B misses $X \Rightarrow B \subset X^c$

Mathematical morphology is concerned with the interaction between the set X, the second set B, their complements, translates and transposes. The set X is usually associated with an image while the set B is the structural element. To relate mathematical morphology to other image processing techniques, the structuring element B corresponds to a mask used in linear FIR filtering. The interaction between image X and the structuring element B transforms the image to a new 'filtered' image. Depending on the size and shape of both X and B and the type of interaction looked at different transformations will result. It is these transformations which enable information to be extracted from an image for use in various applications.

There exist two basic transformations in mathematical morphology. These are erosion and dilation . These two transformations form the basis over which all other morphological transformations exist. The erosion of X by B is defined as being the set of all translation vectors so that when B is translated by any of them, its translate is included in X. Assume that Y is the eroded set of X, then in mathematical terms:

$$Y = \{\mathbf{x} : B_{\mathbf{x}} \subset X\} \tag{3.107}$$

The operation resembles the definition of the Minkowski subtraction:

$$X \ominus B = \bigcap_{\mathbf{b} \in B} X_{\mathbf{b}} \tag{3.108}$$

in the sense that morphological erosion of a set X by the structuring element B is the Minkowski subtraction of X and $\check{B}$, the symmetric set of B.

$$Y = \{\mathbf{x} : B_{\mathbf{x}} \subset X\} = \bigcap_{\mathbf{b} \in B} X_{-\mathbf{b}} = \bigcap_{-\mathbf{b} \in B} X_{-\mathbf{b}} = X \ominus \check{B} \tag{3.109}$$

To derive the definition of the second basic morphological transformation, dilation, consider an operation which is dual to erosion. A transformation which is the dual of another is defined as being the resultant transformation when a known operator is applied to the complement of a set and the complement of the result is taken. Assuming that dilation is the dual translation of erosion the following equation can be obtained:

$$X \oplus \check{B} = [X^c \ominus \check{B}^c]^c \tag{3.110}$$

The erosion determines all of $\check{B}_\mathbf{b}$ which are included in X^c. This is equivalent to determining all the $\check{B}_\mathbf{b}$ which do not hit X. The complement of the set which this statement produces must therefore be the set of all $\check{B}_\mathbf{b}$ which hit X. This is the definition of the morphological transformation of dilation.

$$X \oplus \check{B} = \{\mathbf{x} : B_\mathbf{x} \subset X \neq \emptyset\} = \{\mathbf{x} : B_\mathbf{x} \uparrow X\} \tag{3.111}$$

Erosion and dilation respectively shrink and expand an image object. However, they are not inverses of each other. That is, these transformations are not information preserving. A dilation after an erosion will not necessarily return the image to its original state nor will an erosion of a dilated object necessarily restore the original object. This loss of information from these two operators and the results obtained when cascading the two transformations one after the other provide the basis for the definition of another pair of morphological transformations called morphological opening and closing.

The morphological transformations of opening and closing are not exactly defined as being the cascading of an erosion followed by a dilation or a dilation followed by an erosion. The symmetric set of B is not used in both steps. A morphological erosion or dilation as defined above is first performed on an image X using a structuring element B. The second transformation is, respectively, a dilation or an erosion not by the structuring element B but by the symmetric set $\check{B}$. Formally, a mathematical opening of a set X by a structuring element B, denoted by $X \circ B$, is defined in terms of the Minkowski operators as the following:

$$X \circ B = (X \ominus \check{B}) \oplus B \tag{3.112}$$

This transformation is an erosion of X using a structuring element B followed by a dilation using a structuring element equal to $\check{B}$. In other words the set is first shrunk and then re-expanded, not necessarily to its original state.

The definition of the morphological closing of a set X by a structuring element b may be derived in a similar fashion. Morphological closing, denoted by $X \bullet B$, is the dual operation of morphological opening and may also be defined in terms of the Minkowski operators as follows:

$$X \bullet B = (X \oplus \check{B}) \ominus B \tag{3.113}$$

The discussion of mathematical morphology has until now been restricted to binary images and the space Z^2. Extending the definitions enables the use of mathematical morphology on monochrome (gray scale) images. In the binary case the two dimensions were the (i, j) coordinates of the binary image. If the gray scale values of an image $X(i, j)$ are taken as the third dimension then an image becomes a surface in Z^3. The term umbra $U[X]$ was defined in [67] as a set which extends unbroken indefinitely downward in the negative Z direction below the two dimensional function's surface. A point $p = (i, j, k)$ is an element of an image's umbra if and only if $k \leq X(i, j)$. An image's umbra is a set in Z^3. Once this definition of a set in a three-dimensional space representing a monochrome image is made, the extension of morphological transformations to monochrome images is quite simple. Structuring elements also become two dimensional functions defined over a domain. The set associated with the two dimensional structuring element function is defined as all points (i, j, k) such that k is non-negative and (i, j) lies in the domain over which B is defined. If the structuring element is restricted so that $B(i, j)$ is uniformly equal to zero over the entire domain of B, then B is considered to be a flat structuring element [69]. Once the assumption of B being flat is made, the set associated with the structuring element becomes a set in two dimensions. This set is simply the set of all points (i, j) over which B is defined. Therefore, the definitions of monochrome erosion and dilation simplify to:

$$\begin{aligned}(X \ominus \check{B})(i,j) &= \min_{(t_1,t_2)\in B_{i,j}} X(t_1,t_2)\\ (X \oplus \check{B})(i,j) &= \max_{(t_1,t_2)\in B_{i,j}} X(t_1,t_2)\end{aligned} \tag{3.114}$$

where B now represents the set of all points in the structuring element, $\check{B}$ is the symmetric set of B, and $B_{i,j}$ is the translation of the set B by the two dimensional vector (i, j). $(X \ominus \check{B})(i, j)$ and $(X \oplus \check{B})(i, j)$ represent the two dimensional functions corresponding to the resultant images obtained after eroding or dilating an image X by a flat structuring element B.

Using (3.107) and (3.108) and the above definitions of gray scale erosion and dilation, the monochrome morphological transformations of opening and closing may be defined as:

$$\begin{aligned}(X \circ B)(i,j) &= \max_{(s_1,s_2):(i,j)\in B_{s_1,s_2}} \left[\min_{(t_1,t_2)\in B_{s_1,s_2}} (X(t_1,t_2))\right]\\ (X \bullet B)(i,j) &= \min_{(s_1,s_2):(i,j)\in B_{s_1,s_2}} \left[\max_{(t_1,t_2)\in B_{s_1,s_2}} (X(t_1,t_2))\right]\end{aligned} \tag{3.115}$$

The definitions of monochrome morphological transformations describe transformations which enable noisy images to be filtered. Combinations of several transformations may be used to improve the noise removal of morphological monochrome filters which combine the results from several openings and closings of an image X by a set of several structuring elements.

The conventional morphological operators have one structuring element with fixed size and fixed shape. It has been known that the fixed structuring element may cause the lose of image details, the loss of the detailed parts of large objects and may cause distortion in smooth areas in noise filtering.

In fact many other operators with one fixed operational window may share the same problems. Many approaches have been suggested to deal with those problems. Among them, a type of new opening operators (NOP) and closing operators (NCP) were introduced in [72]. The structuring element of NOP and NCP adapts its shape according to the local geometric structures of the proposed images, and can be any shape formed by connecting a given number of N pixels. The NOP can be developed on the basis of (3.112)-(3.115). The opening definition in (3.115) states that for a flat structuring element acting like a moving window fitting over the features around the pixel (i,j) from the inside of the surface, the output value for the pixel (i,j) is the minimum value in the fitted window B. The group opening defined in (3.115) computes the maximum over all the minima obtained from the opening by each individual G^k. To achieve a larger degree of freedom in manipulating the geometric structures in the images than that of (3.115) a large set of group openings is required before selecting the maximum as output. Denoting the set of all possible structuring elements formed by connecting N points as Ξ^N, the NOP is defined as:

$$(X \circ \Xi^N)(i,j) = \max_{B^k \in \Xi^N} [(X \circ B^k)](i,j) \tag{3.116}$$

In practice it is impossible to compute (3.116) by the conventional opening operations, since there are too many elements in Ξ^N. To alleviate the problem, a way to directly search for the resulting structuring element obtained from the maximum of group openings was devised in [72]. In essence, the algorithm finds a connected structuring element that best fits the local structure then flattens the feature by assigning the minimum value within the domain of the selected element to the pixel (i,j). A structuring element that fits best to the local feature is the trace of the maxima containing the point (i,j). In other words, an adaptive structuring element which maximizes the minimum is searched. In this way, information extracted from an image is biased only by the size, not the shape of the structuring element. The size N is usually chosen according to the requirement of a specific image processing task. According to the above, the objective of the NOP operator is to find N connected points that trace the maxima of the local feature along and include the point (i,j), then assign the minimum value in the window of these chosen N points as the output for the pixel (i,j). Based on the above interpretation of NOP it is now possible to develop a fast algorithm for computing it.

The NCP, as the dual of the NOP, has its structuring element follow the trace of the minima of the local feature containing the point (i,j). The output (i,j) is assigned the maximum value of those pixels within the domain of the adaptive structuring element. The NCP at point (i,j) is essentially a search for an adaptive structuring element which minimizes the maximum. Thus, it is easy to complement the NOP definition to that of the NCP:

$$(X \bullet \Xi^N)(i,j) = \max_{B^k \in \Xi^N} [(X \bullet B^k)](i,j) \tag{3.117}$$

Based on (3.117) NCP at point (i, j) has to find N connected points that trace the minima of the local feature along and include the point (i, j) then assign the maximum value in the window of these chosen N points as the output for the pixel (i, j). NCP fills any valley smaller than N points to form a larger basin of at least N points, whose shape contains the adaptive structuring element. If the area of a uniform basin is larger than or equal to N pixels, its surface structure will not be altered. Other points of the surfaces, such as the slopes and the peaks, will remain intact under the NCP operation. It should be noted that the NOP (NCP) cannot be decomposed into an erosion (dilation) followed by a dilation (erosion).

3.5.2 Computation of the NOP and the NCP

Since NOP and NCP are derived from the conventional opening and closing operators, they share many of their properties, such as translation invariance, increasing, ordering and idempotency. The new operators also attain some distinct properties that exploit the geometric structures. The intuitive geometric operations are the most distinguishing characteristics of the NOP and the NCP. They differ from most of the existing linear and nonlinear processing techniques discussed in this book.

The definition and the properties of the NOP and NCP show a great potential to develop fast algorithms. To fully develop the potential is a complicated problem that requires considerable effort. The basic algorithm structure proposed in [73] is only a straight forward realization of the definition of the NOP and NCP. Study has shown that from the basic structure, there are many ways for further development. In this section, a fast and computationally efficient algorithm for the computation of NOP and NCP is reviewed. The core of the NOP and NCP is the search for the adaptive structuring element which follows the shape of the local features. An essential requirement in the search is connectivity. That is, the N-point structuring element must be connected via the current pixel (i, j). The search procedure is iterative until all N points in accordance to the NOP or the NCP definition are found.

The NOP algorithm can be divided into five steps, of which the middle three are repeated in finding the N points which trace the local feature with the largest values. The five steps are:

1. **Initialization:** All buffers, counters, registers and flags are initialized.
2. **Search:** Immediate neighborhoods of those newly included points in the structuring element during the previous iteration are searched, flagged and identified as possible candidates. The candidates $\mathbf{b} = (b_i, b_j)$ are arranged in descending order of their pixel values $f(\mathbf{b})$ as follows:

   $$f(\mathbf{b}_1) \geq \cdots f(\mathbf{b}_k) \geq \cdots$$

   Since the best structuring element for the opening follows the maxima of the local feature, the largest K numbers of the ordered candidates are singled out for the decision in step 3. In other words, K is the lesser of the number of points to be found and the number of possible candidates. The rest of the candidates are purged while their flags remain set to indicate exclusion from any further iteration.

3. **Decision:** The K candidates are examined for inclusion in the set of the structuring element, by comparing to the minimum value f_{MIN} of the points chosen. Initially, $f_{MIN} = f(i,j)$. There are three possible cases:
   a) All K candidates have pixel values larger than or equal to f_{MIN}:
      $f\mathbf{b}_K \geq f_{MIN}$
      In this case, the coordinates of all K candidates in the set are assigned in the set of the structuring element.
   b) Some of the K candidates have pixel values smaller than f_{MIN}.
      $f\mathbf{b}_K \geq f_{MIN}$
      $f\mathbf{b}_k < f_{MIN} \quad 1 \leq k \leq K$
      Only those coordinates with pixel values not smaller than f_{MIN} are assigned in the set of the structuring element. Others are left as candidates for the next search cycle.
   c) All K candidates have pixel values smaller than f_{MIN}.
      $f\mathbf{b}_K < f_{MIN}$
      In this case, the coordinates, $\mathbf{b}_1$ of the largest pixel value is included in the set of the structuring element as a connecting point to the larger outer points. f_{MIN} is also replaced by $f\mathbf{b}_1$.
4. **Update:** Buffers, counters and registers are updated according to the decision made in step (3). If less than N points have been located, steps (2) to (4) are repeated. Otherwise, step (5) is followed to output.
5. **Output:** Assign the minimum pixel value f_{MIN} in the window of the N point structuring element as output for the pixel (i,j). The search is complete.

To ensure the search progress is smooth, there are two buffers, three counters, and two registers to keep track of the records in each iteration:

1. Buffer $\{\mathbf{a}\}$ stores the pixel coordinates chosen to be in the set of the adaptive structuring element. Initially, $\{\mathbf{a}\}$ contains (i,j).
2. Buffer $\{\mathbf{b}\}$ stores the coordinates of all possible candidates for the current iteration. These include the rejected $\mathbf{b}_k$, and the immediate neighbors of those added to $\{\mathbf{a}\}$ during the previous iteration. Initially, $\{\mathbf{b}\}$ contains the eight immediate neighbors of (i,j).
3. Counter M keeps count of the number of pixels that have been located for the structuring element. Initially, $M = 1$ since (i,j) is always included in the set.
4. Counter BN stores the number of all possible candidates.
5. Counter K stores the number of pixels to be decided upon for the current iteration. If $N \leq 9$, K is usually set to $N - M$. In the case of $N > 9$, K is set to the lesser of $N - M$ and BN.

6. Register L holds the position of the last entry in $\{\mathbf{a}\}$. This ensures that only the neighbors of those points added to the structuring element during the current iteration will be searched in the next cycle.
7. A register to store the smallest pixel value f_{MIN} in the domain of the structuring element.

In addition, every pixel is associated with a flag which will be set once the pixel is chosen to be included in the search. The flag guarantees that a pixel will not be searched twice. The area of possible domain for the structuring element is the rectangle bounded exclusively by $((i-N, j-N), (i-N, j+N), (i+N, j-N), (i+N, j+N))$. To ensure that the search will not go beyond the image frame, the original image is augmented with a one-pixel wide frame whose values equal the smallest pixel value. A flowchart of the NOP algorithm is shown in Fig. 3.7.

The NCP algorithm can be derived directly from its NOP dual with the following changes:

1. Reverse the ordering so that the coordinates are put in ascending order of the pixel value:

   $$f(\mathbf{b}_1) \leq \cdots f(\mathbf{b}_k) \leq \cdots \leq f(\mathbf{b})_{BN}$$

   The K smallest numbers of the ordered candidates are included for decision.
2. f_{MIN} is changed to f_{MAX}, such that the maximum pixel value in the set of the structuring element is stored and output.
3. The candidates $\mathbf{b}_k$ is chosen if $f(\mathbf{b})_k$ for $1 \leq k \leq K$ is smaller than, or equal to f_{MAX}. That is, all comparison inequalities between $f(\mathbf{b}_k)$ and f_{MAX} are reversed.

Moreover, the original image is augmented with one-pixel wide frame whose values equal the largest pixel value.

3.5.3 Computational Complexity and Fast Algorithms

In the algorithms discussed above, the major computational efforts lie in the ordering and the comparison of the pixel values. With a good and efficient ordering algorithm, it is fair to say that the number of comparisons used for an image represents the computational complexity to a certain degree. Therefore, reducing the number of comparisons required per pixel will result in less computational effort. Fast algorithms can be obtained by exploiting the nature of basic search. Keeping the connectivity in mind and the fact that the search of maximum path in NOP is performed outward from the current computation point (i, j), close examination reveals that those pixels included in the final structuring element before locating the minimum pixel for the first time, will share the same output f_{MIN} as (i, j).

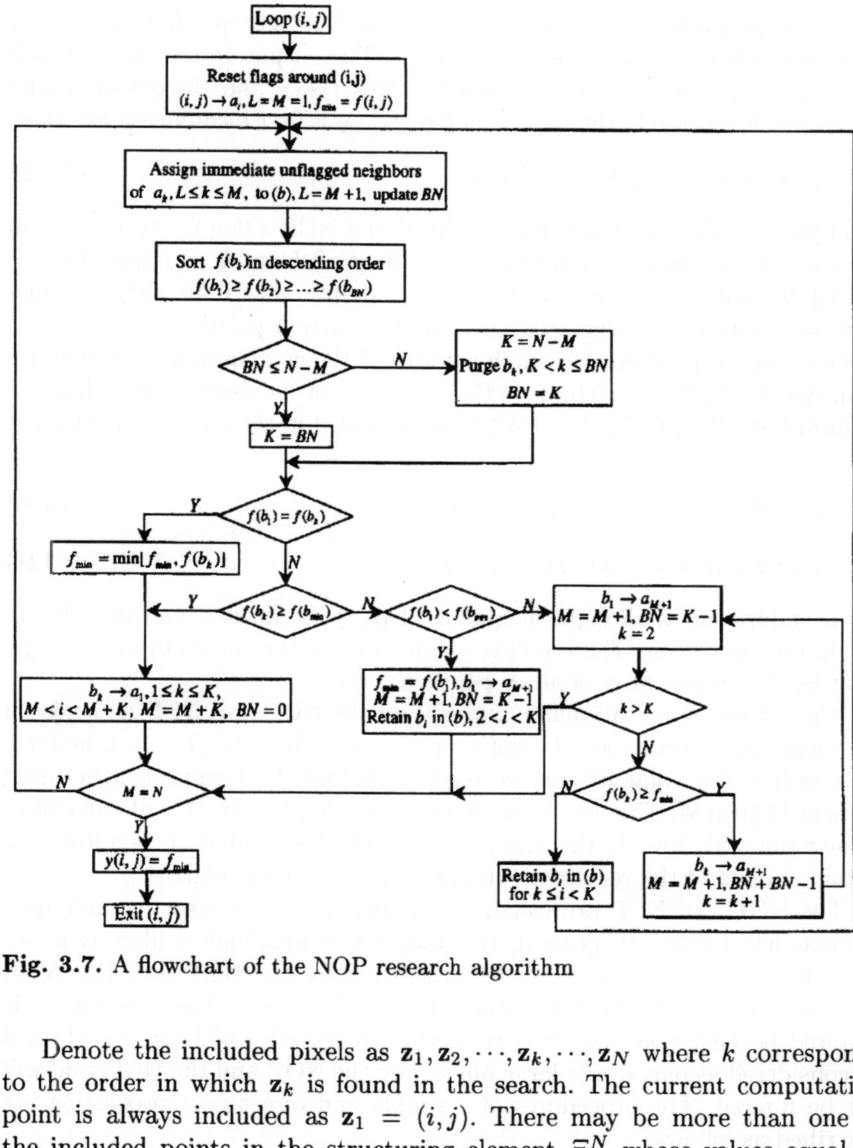

Fig. 3.7. A flowchart of the NOP research algorithm

Denote the included pixels as $\mathbf{z}_1, \mathbf{z}_2, \cdots, \mathbf{z}_k, \cdots, \mathbf{z}_N$ where k corresponds to the order in which $\mathbf{z}_k$ is found in the search. The current computation point is always included as $\mathbf{z}_1 = (i, j)$. There may be more than one of the included points in the structuring element Ξ^N whose values equal to the minimum value f_{MIN} for the output of NOP at (i, j). Assume that $\mathbf{z}_{k_1}, \cdots, \mathbf{z}_{k_t}$, where $1 \le k_1 \le k_2 \le \cdots \le k_t \le \cdots \le N$ and $1 \le t \le N$, as the pixels whose values equal f_{MIN}, then for $1 \le k \le k_1$ the following is true:

$$(X \circ \Xi_N)(\mathbf{z}_k) = f(\mathbf{z}_{k_1}) = f_{MIN_{ij}} \tag{3.118}$$

That is, the pixels $\mathbf{z}_k$ for $1 \le k \le k_1$ and the pixels $\mathbf{z}_{k_i}$ for $i = 2, \cdots, t$ do not require a search for the structuring element of their own, since they share the same output as (i, j).

The same property can be applied to the NCP except that $\mathbf{z}_{k_1}, \cdots, \mathbf{z}_{k_t}$ are now pixels whose values equal to f_{MAX}. The output values for the pixels $\mathbf{z}_{k2}, \cdots, \mathbf{z}_{kt}$ are the same as their input values f_{MAX} and the output values for the pixels located in the set Ξ^N before f_{MAX} is first located, are assigned:

$$(X \bullet \Xi_N)(\mathbf{z}_k) = f(\mathbf{z}_{k_1}) = f_{MAX_{ij}} \tag{3.119}$$

To implement the fast algorithms for NOP and NCP a flag for every pixel in the image only needs to be included. The flags of those $\mathbf{z}_k$ satisfying (3.118) or (3.119), and $\mathbf{z}_{k2}, \cdots, \mathbf{z}_{kt}$, will be set to signify that their output values have been determined when they become the current position.

One way to speed up the search is to test if the neighborhood is a uniform area, that is, if $f(\mathbf{b}_1) = f(\mathbf{b}_K)$, at the beginning of the decision step. If it is a uniform area, then all $\mathbf{b}_k$ for $1 \leq k \leq K$ are included in the structuring element and

$$(X \circ \Xi_N)(i,j) = \min(f_{MIN}, f(\mathbf{b}_K)) \tag{3.120}$$

$$(X \bullet \Xi_N)(i,j) = \max(f_{MAX}, f(\mathbf{b}_K)) \tag{3.121}$$

For a uniform area, all eight neighbors of (i,j) are located and included in Ξ^N in one operation. These points will also share the same output as (i,j). That is, the output flags of all N points are set.

The actual computational complexity of the NOP and NCP depends on the image to be processed. In the simplest case where (i,j) is in a uniform area, only a few comparisons are required before the resultant structuring element is located. The worst case happens is the pixel (i,j) is at the end of a one point wide line. In this case, only one pixel is located in each iteration of the search and the resultant computational burden is high.

The NOP and NCP are usually used together to construct an adaptive morphological filter. In general, the adaptive morphological filter is a two stage filter. The first stage is the processing by the NOP and the NCP. The second stage is the post processing of the image. Post processing is required because noise patterns connected to the edge of large objects will be considered as part of the large objects by the NOP and the NCP, and will not be filtered. The procedures of a simple and direct post processing are described as follows:

1. Denote the image filtered by an adaptive morphological filter as $y(i,j)$. Decompose $y(i,j)$ into a coarse image $z_1(i,j)$ and a detailed image $z_2(i,j)$ by a conventional opening or closing morphological filter with a small structuring element, such as a (2×2) element. This decomposition separates the noise patterns from the large objects:

$$z_1(i,j) = (y \circ B) \bullet B \tag{3.122}$$

$$z_2(i,j) = y(i,j) - z_1(i,j) \tag{3.123}$$

As a result, $z_1(i,j)$ contains only objects not smaller than B, leaving $z_2(i,j)$ with the isolated noise pixels and the fine details with size smaller than B.

2. Remove the noise patterns in $z_2(i,j)$ by the same adaptive morphological filter but with a smaller size structuring element, which will remove the noise but leave the fine detail intact. The filtered image $z_3(i,j)$ will contain only the fine details. Thus, for $N_1 < N$:

$$z_3(i,j) = (z_2 \circ \Xi^N) \bullet \Xi_{N_1} \tag{3.124}$$

3. Output the final image $y^*(i,j)$ by adding the noise free details back to the coarse image. The post process image $y^*(i,j)$ has sharper edges than $y(i,j)$:

$$y^*(i,j) = z_1(i,j) + z_3(i,j) \tag{3.125}$$

The main drawback of this simple post processing method is that it cannot remove noise pixels connected to one-pixel wide details. Although more sophisticated post processing methods can be used to deliver better results, these remaining noise pixels are usually negligible since the human eye is more tolerable to small amounts of noise in the neighborhood of an edge.

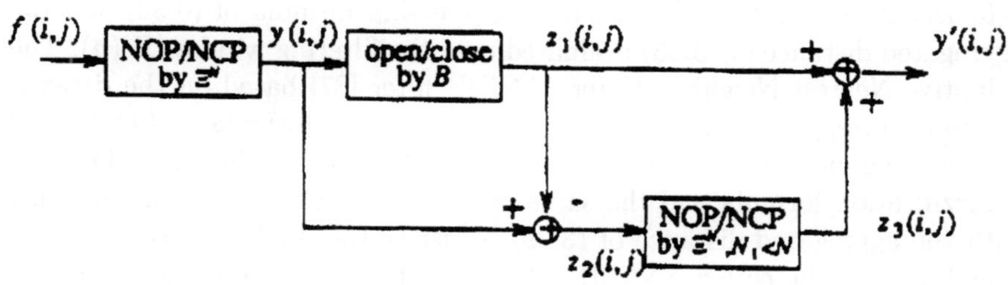

Fig. 3.8. The adaptive morphological filter

3.6 Simulation Studies

A set of experiments has been conducted in order to evaluate the adaptive designs presented. In this first part, the performance of adaptive designs based on fuzzy, statistical and non-parametric techniques with that of commonly used filters, such as the vector median filter (VMF), the generalized vector directional filter (GVDF), the distance-direction filter (DDF) and the hybrid filters of [79] are compared. The noise attenuation properties of the different filters are examined by utilizing the color images 'Lenna' and 'Peppers'. The test images have been contaminated using various noise source models in order to assess the performance of the filters under different scenarios (see Table 3.3). The original images as well as their noisy versions are represented in the RGB color space. The filters operate on the images in the RGB color space.

Table 3.1. Noise Distributions

Number	Noise Model
1	Gaussian ($\sigma = 30$)
2	impulsive (4%)
3	Gaussian ($\sigma = 15$), impulsive (2%)
4	Gaussian ($\sigma = 30$), impulsive (4%)

Since it is impossible to discuss all the fuzzy adaptive filters resulting from the theory introduced in this chapter, five different filters based on the designs are constructed. These filters are compared, in terms of performance, with other widely used multichannel filters (see Table 3.2). In particular, a simple rank-order filter is introduced, based on the distance measure of [36], hereafter content-based rank filter (CBRF), which can be seen as an adaptive fuzzy system with the defuzzification rule of (3.27). The simulation studies also include the fuzzy vector directional filter (FVDF) which is based on the defuzzification strategy of (3.2), the membership formula of (3.32) and the aggregated distance of (3.35) evaluated over the filtering window $W(n)$. The Adaptive Nearest Neighbor Filter (ANNF) filter [37] based on the defuzzification strategy of (3.2), the membership function formula of (3.23), and the distance measure of (3.33) is also included in the set. Further, the same defuzzification formula and the same membership function is utilized along with the aggregated distance of (3.29) to derive the double window nearest neighbor filter, hereafter ANNMF. By using the Canberra distance and the distance measure of (3.27) instead of the angular distance, four other filters have been devised, named CANNF, CANNMF, CBANNF and CBANNMF respectively (see Table 3.2).

A number of different objective measures can be utilized to assess the performance of the different filters. All of them provide some measure of closeness between two digital images by exploiting the differences in the statistical distributions of the pixel values [1], [11]. The most widely used measure is the normalized mean square error (NMSE) defined as:

$$NMSE = \frac{\sum_{i=0}^{N1} \sum_{j=0}^{N2} ||(\mathbf{y}(i,j) - \hat{\mathbf{y}}(i,j)||^2}{\sum_{i=0}^{N1} \sum_{j=0}^{N2} ||(\mathbf{y}(i,j)||^2} \tag{3.126}$$

where $N1$, $N2$ are the image dimensions, and $\mathbf{y}(i,j)$ and $\hat{\mathbf{y}}(i,j)$ denote the original image vector and the estimation at pixel (i,j), respectively.

In many application areas, such as multimedia, telecommunications (e.g. HDTV), production of motion pictures, printing industry and graphic arts,

Table 3.2. Filters Compared

Notation	Filter	Reference
AMF	Arithmetic (Linear) Mean Filter	[1]
VMF	Vector Median Filter	[76]
BVDF	Basic Vector Directional Filter	[42]
CBRF	Content-based Rank Filter, eq.	[36]
GVDF	Generalized Vector Directional Filter with an α-trimmed magnitude module, ($\alpha = 1.5$)	[43]
DDF	Directional-Distance Filter	[78]
HF	Hybrid Directional Filter	[79]
AHF	Adaptive Hybrid Directional Filter	[79]
FVDF	Fuzzy Vector Directional Filter with structure/weights determined through (3.2), (3.32), (3.35) $r = 1$, $\beta = 2$	[22]
ANNF	Adaptive Nearest Neighbor Filter with (3.2), (3.23), (3.32)	[34]
ANNMF	Double window Adaptive Nearest Neighbor Filter with (3.2), (3.23), (3.24)	[37]
CANNF	Adaptive Nearest Neighbor Filter eqs. (3.2) (3.23), (3.33)	
CANNMF	Double window adaptive nearest neighbor filter (3.2), (3.23), (3.25)	
CBANNF	Adaptive Nearest Neighbor Filter, (3.2) (3.22), (3.34)	
CBANNMF	Double window adaptive nearest neighbor filter with (3.2), (3.22) and (3.27)	
AMNFE	Adaptive Non-parametric Filter with an exponential kernel, (3.93)	[11]
AMNFE	Adaptive Non-parametric Filter with an Gaussian kernel, (3.93)	[11]
AMNFD	Adaptive Non-parametric Filter with a directional kernel, (3.97)	
BFMA	Bayesian Adaptive filter with median and mean sub-filters	[11]

greater emphasis is given to perceptual image quality. Consequently, the perceptual closeness (alternatively the perceptual difference or error) of the filtered image to the uncorrupted original image is ultimately the best measure of the efficiency of any color image filtering method. There are basically two major approaches used for assessing the perceptual error between two color images. In order to make a complete and thorough assessment of the performance of the various filters, both approaches are used in this section.

The first approach is to make an objective measure of the perceptual error between two color images. This leads to the question of how to estimate the perceptual error between two color vectors. Precise quantification of the perceptual error between two color vectors is one of the most important and open research problem. RGB is the most popular color space used conventionally to store, process, display, and analyze color images. However, the human perception of color cannot be described using the RGB model. Therefore, measures such as the normalized mean square error (NMSE) defined in the RGB color space are not appropriate to quantify the perceptual error between images. Thus, it is important to use color spaces which are closely related to the human perceptual characteristics and suitable for defining appropriate measures of perceptual error between color vectors. A number of such color spaces are used in areas such as computer graphics, motion pictures, graphic arts, and printing industry. Among these, perceptually uniform color spaces are the most appropriate to define simple yet precise measures of perceptual error. As seen in Chap. 1 the Commission Internationale de l'Eclairage (CIE) standardized two color spaces, the L*u*v* and L*a*b*, as perceptually uniform. The L*u*v* color space is chosen for this analysis because it is simpler in computation than the L*a*b* color space, without any sacrifice in perceptual uniformity.

The conversion from the non-linear RGB color space (the non-linear RGB values are the ones stored in the computer and applied to the CRT of the monitor to generate the image) to the L*u*v* color space is explained in detail in Chap. 1 and elsewhere [80]. Non-linear RGB values of both, the uncorrupted original image and the filtered image, are converted to corresponding L*u*v* values for each of the filtering methods under consideration. In the L*u*v* space, the L^* component defines the lightness and the u^* and v^* components together define the chromaticity. In a uniform color space, such as the L*u*v*, the perceptual color error between two color vectors is defined as the Euclidean distance between them given by :

$$\Delta E_{Luv} = [(\Delta L^*)^2 + (\Delta u^*)^2 + (\Delta v^*)^2]^{\frac{1}{2}} \tag{3.127}$$

where ΔE_{Luv} is the color error and ΔL^*, Δu^*, and Δv^* are the difference in the L^*, u^*, and v^* components, respectively, between the two color vectors under consideration. Once the ΔE_{Luv} for each pixel of the image under consideration is computed, the normalized color distance (NCD) is estimated according to the following formula:

$$NCD = \frac{\sum_{i=0}^{N1} \sum_{j=0}^{N2} \|\Delta E_{Luv}\|}{\sum_{i=0}^{N1} \sum_{j=0}^{N2} \|E^*_{Luv}\|} \tag{3.128}$$

where $E^*_{Luv} = [(L^*)^2 + (u^*)^2 + (v^*)^2]^{\frac{1}{2}}$ is the *norm* or *magnitude* of the uncorrupted original image pixel vector in the $L^*u^*v^*$ space.

Although quantitative measures, such as ΔE_{Luv} and NCD are close approximations to the perceptual error they cannot exactly characterize the quite complex attributes of the human perception. Therefore, an alternative subjective approach is commonly used by researchers [81] for estimating the perceptual error.

The second approach, the easiest and simplest, is the subjective evaluation of the two images to be compared in which both images are viewed, simultaneously, under identical viewing conditions by a set of observers. A set of color image quality attributes can be defined for the subjective evaluation [81]. The evaluation must take into consideration important factors in image filtering.

For the results presented, the performance is ranked subjectively in five categories: excellent (5), very good (4), good (3), fair (2) and bad (1) using the following subjective criteria (see Table 3.4).

1. Detail preservation: which corresponds to edge and fine detail preservation.
   One of the most important criteria in the subjective examination of a filter performance is edge preservation. Color edges in an image may be defined as a discontinuity or abrupt change in the color attributes. Edges are important features since they provide an excellent indication of the shape of the objects in the image. Maintaining the sharpness of the edges is as important as removing the noise in the image. The same holds true for fine details in the image. An image void of details looks plain and unpleasant. Therefore, it is important to distinguish the fine elements from the noise, so that they can be preserved during the filtering process.
2. Color appearance: which refers to color sharpness, the distinctness of boundaries among colors, and color uniformity which refers to the consistency of the color in uniform areas.
   The human eye is very sensitive to small changes in color. Therefore, it is important to keep the chromaticity (namely hue and saturation) constant while removing noise. The natural appearance of the color features in the scene must be preserved while artificial contrast, color drift and other abberations that make the filtered image to look unpleasant should be avoided.
3. Defects: classify any imperfection such as blocking artifacts that was not present in the original (noise free) image.

Table 3.3. Subjective Image Evaluation Guidelines

Score	Overall Evaluation	Noise Removal Evaluation
1	Very disruptive distortion	poor
2	Disruptive distortion	fair
3	Destructive but not disruptive distortion	good
4	Perceivable but not destructive distortion	very good
5	Imperceivable distortion	excellent

Table 3.4. Figure of Merit

a	Overall Subjective Evaluation
b	Additive Gaussian noise
c	Impulsive noise
d	Moderate mixed (Gaussian/impulsive) noise
e	Mixed (Gaussian/impulsive) noise

In this study, the color images under consideration were viewed in parallel, on a SUN Sparc 20 with a 24-bit color monitor, and the observers were asked to mark scores on a printed evaluation sheet following the guidelines summarized in Table3.3 [82]. To subjectively evaluate the noise removal capabilities of the algorithms a similar procedure was followed. Observers were instructed to assign a lower number if noise was still present in the filtered output (Table 3.3).

The second approach, the easiest and simplest, is the subjective evaluation of the resulting images when they are viewed, simultaneously, under identical viewing conditions by a set of observers. To this end, the performance of the different filters in noise attenuation using the test RGB image 'Peppers' is compared. The image is corrupted by outliers (4% impulsive noise (Fig. 3.9)). The RGB color image 'Lenna' is also used. The test image is corrupted with Gaussian noise $\sigma = 15$ mixed with 2% impulsive noise (Fig. 3.10). All the filters considered in this section operate using a square 3×3 processing window.

Filtering results using different estimators are depicted in (Fig. 3.18) and (Fig. 3.26). A visual comparison of the images clearly favors the adaptive designs over existing techniques.

One of the obvious observations from the results in Tables 3.5-3.12 is the effect of window size on the performance of the filter. In the case of rank-type filters, such as the VMF, BVDF, CBVF, DDF as well as the HF and the AHF, the bigger window size (5×5) gives considerably better results for the removal of Gaussian noise (noise model 1), while decreasing the performance for the removal of impulsive noise (noise model 2). Although a similar pattern follows for the adaptive filters, fuzzy, Bayesian or non-parametric the effect of the window size on performance is less dramatic as compared to the rank-type of filters.

Analysis of the results summarized here reveals the effect that the distance (or similarity) measure can have on the filter output. Even filters which are based on the same concept, such as VDF, CVDF and CBVF, or ANNF and CANNF have different performance simply because a different distance measure is utilized to quantify dissimilarity among the color vectors. Similarly, double window adaptive filters have better smoothing abilities, outperforming the other filters, when a Gaussian noise or mixed noise model is assumed.

For the case of impulsive noise, the VMF gives the best performance among the rank-type filters according to the results, as well as the theory, and is thus used as a benchmark to evaluate the fuzzy adaptive designs. The proposed fuzzy filters perform close to the VMF and outperform existing adaptive designs, such as the HF or the AHF with respect to NMSE and NCD, and for both window sizes. For the case, of pure Gaussian noise, the VMF gives the worst results. The results summarized in Tables 3.5-3.12 indicate that the adaptive filters perform exceptionally well in this situation.

The arithmetic mean filter (AMF) is theoretically the best non-adaptive filter for the removal of pure Gaussian noise (noise model 1). In other words, the NMSE, NCD, and the subjective measure all indicate the best performance by AMF. So the performance of the AMF filter is used as a benchmark to compare the performance of the new adaptive filters in the same noise environment. The results indicate that the adaptive filters, both fuzzy and non-parametric, perform better or close enough to the AMF and outperform existing adaptive filters, such as the AHF in NMSE, NCD and subjective sense. Clearly, the new AMNFG adaptive filter is the best for Gaussian noise and performs exceptionally well, outperforming the existing filters, both adaptive and non-adaptive, with respect to all three error measures and for both window sizes.

For the mixture of Gaussian and impulsive noise (noise models 3 and 4), the adaptive fuzzy filters consistently outperform any of the existing listed filters, both rank type or adaptive with respect to NMSE and NCD. This is demonstrated by the simple fact that, for noise models 3 and 4 (see Table 3.1), the highest error among the new adaptive filters is comparable to the lowest error among the existing rank type, non-adaptive filters. Herein lies the real advantage of the adaptive designs, such as the fuzzy, Bayesian or non-parametric filters introduced here. In real applications, the noise model is unknown a-priori. Nevertheless, the most common noise types encountered in real situations are Gaussian, impulsive or a mixture of both. Therefore, the use of the proposed fuzzy adaptive filters guarantees near optimal performance for the removal of any kind of noise encountered in practical applications. On the contrary, application of a 'noise-mismatched' filter, such as the VMF for Gaussian noise can have profound consequences leading to unacceptable results.

In conclusion, from the results listed in the tables, it can easily be seen that the adaptive designs provide consistently good results in all types of noise, outperforming the other multichannel filters under consideration. The adaptive designs discussed here attenuate both impulsive and Gaussian noise. The versatile design of (3.1) allows for a number of different filters, which can provide solutions to many types of different filtering problems. Simple adaptive fuzzy designs, such as the ANNF or the CANNF can preserve edges and smooth noise under different scenarios, outperforming other widely used multichannel filters. If knowledge about the noise characteristics is available, the designer can tune the parameters of the adaptive filter to obtain better results. Finally, considering the number of computations, the computationally intensive part of the adaptive fuzzy system is the distance calculation part. However, this step is common in all multichannel algorithms considered here. In summary, the adaptive design is simple, does not increase the numerical complexity of the multichannel algorithm and delivers excellent results for complicated multichannel signals, such as real color images.

Table 3.5. NMSE($x10^{-2}$) for the RGB 'Lenna' image, 3×3 window

Filter	Noise Model			
	1	2	3	4
None	4.2083	5.1694	3.6600	9.0724
AMF	0.6963	0.8186	0.6160	1.2980
BVDF	2.8962	0.3448	0.4630	1.1354
CBRF	1.3990	0.1863	0.5280	1.5168
GVDF	1.4600	0.3000	0.6334	1.9820
DDF	1.5240	0.3255	0.6483	1.6791
VMF	1.6000	0.1900	0.5404	1.6790
FVDF	0.7335	0.2481	0.4010	1.0390
ANNF	0.8510	0.2610	0.3837	1.0860
ANNMF	0.6591	0.1930	0.3264	0.7988
HF	1.3192	0.2182	0.5158	1.6912
AHF	1.0585	0.2017	0.4636	1.4355
CANNF	0.8360	0.2497	0.3471	1.0481
CANNMF	0.6001	0.1891	0.3087	0.7137
CBANNF	0.8398	0.2349	0.3935	1.0119
CBANNMF	0.6011	0.1894	0.3087	0.7149
AMNFE	0.5650	0.1710	0.3020	0.6990
AMNFG	0.8417	0.2006	0.3578	1.0070
AMNFD	0.8045	0.2350	0.3537	1.0101
BFMA	0.7286	0.3067	0.4284	1.0718

Table 3.6. NMSE(x10^{-2}) for the RGB 'Lenna' image, 5×5 window

Filter	Noise Model 1	2	3	4
None	4.2083	5.1694	3.6600	9.0724
AMF	0.5994	0.6656	0.5702	0.8896
BVDF	2.800	0.7318	0.6850	1.3557
CBRF	0.9258	0.3180	0.4890	1.0061
GVDF	1.0800	0.5400	0.4590	1.1044
DDF	1.0242	0.5126	0.6913	1.3048
VMF	1.1700	0.5800	0.5172	1.0377
FVDF	0.7549	0.3087	0.4076	0.9550
ANNF	0.6260	0.4210	0.4360	0.7528
ANNMF	0.5445	0.2505	0.3426	0.6211
HF	0.7700	0.3841	0,4890	1.1417
AHF	0.6762	0.3772	0.4367	0.7528
CANNF	0.5950	0.4028	0.4091	0.7380
CANNMF	0.5208	0.3017	0.3671	0.5802
CBANNF	0.5925	0.3943	0.4045	0.7111
CBANNMF	0.5201	0.3014	0.3662	0.5795
AMNFE	0.5180	0.3010	0.3710	0.5830
AMNFG	0.5140	0.3070	0.3620	0.5810
AMNFD	0.4587	0.3492	0.4258	0.8211
BFMA	0.5809	0.3146	0.3799	0.6637

Table 3.7. NMSE(x10^{-2}) for the RGB 'peppers' image, 3×3 window

Filter	Noise Model 1	2	3	4
None	5.0264	6.5257	3.2890	6.5076
AMF	1.0611	4.8990	3.4195	4.8970
BVDF	3.9267	1.5070	0.8600	1.4911
CBRF	1.9622	0.4650	0.4354	0.4711
GVDF	1.8640	0.4550	0.3613	0.4562
DDF	3.5090	0.5886	0.5336	0.5893
VMF	1.8440	0.3763	0.3260	0.3786
FVDF	1.4550	0.4246	0.3412	0.4046
ANNF	1.1230	0.5110	0.3150	0.5180
ANNMF	0.9080	0.3550	0.3005	0.3347
HF	1.5892	0.4690	0.3592	0.4781
AHF	1.4278	0.4246	0.3566	0.4692
CANNF	1.1382	0.4696	0.3492	0.4699
CANNMF	0.8994	0.4526	0.4284	0.4545
CBANNF	1.2246	0.4546	0.4566	0.4548
CBANNMF	0.8964	0.4546	0.4300	0.4548
AMNFE	1.1489	0.4976	0.4779	0.4996
AMNFG	1.1130	0.4984	0.4786	0.5084
AMNFD	1.1495	0.4584	0.3700	0.4583
BFMA	1.4118	0.4887	0.4494	0.4876

Table 3.8. NMSE(x10^{-2}) for the RGB 'peppers' image, 5×5 window

Filter	Noise Model 1	2	3	4
None	5.0264	6.5257	3.2890	6.5076
AMF	0.9167	1.7341	2.1916	1.1706
BVDF	4.2698	2.7920	1.6499	4.1350
CBRF	1.4639	0.7090	0.6816	0.7161
GVDF	1.2534	0.6977	0.6600	0.7030
DDF	2.1440	0.7636	0.7397	0.7612
VMF	1.3390	0.6740	0.6563	0.6812
FVDF	2.1120	0.7310	0.6971	0.7178
ANNF	1.0027	0.5230	0.5200	0.6210
ANNMF	0.8050	0.4471	0.4047	0.4458
HF	1.0040	0.9970	0.7684	0.9970
AHF	1.1167	0.9841	0.7632	0.9841
CANNF	1.0281	0.7393	0.6718	0.7426
CANNMF	0.8687	0.6355	0.6405	0.6420
CBANNF	1.0145	0.7281	0.6677	0.7310
CBANNMF	0.8634	0.6338	0.6313	0.6371
AMNFE	1.0001	0.6665	0.6527	0.6686
AMNFG	0.09945	0.6671	0.6533	0.6693
AMNFD	0.9889	0.6540	0.6155	0.6555
BFMA	1.1972	0.48577	0.4524	0.4817

Table 3.9. NCD for the RGB 'Lenna' image, 3×3 window

Filter	Noise Model 1	2	3	4
None	0.1149	0.0875	0.7338	0.1908
AMF	0.0334	0.0284	0.0295	0.0419
BVDF	0.0508	0.0082	0.0210	0.0708
CBRF	0.0467	0.0051	0.0169	0.0524
GVDF	0.0462	0.0079	0.0191	0.0489
DDF	0.0398	0.0073	0.0179	0.0426
VMF	0.0432	0.0053	0.0238	0.0419
FVDF	0.0377	0.0049	0.0144	0.0394
ANNF	0.0338	0.0061	0.0149	0.0412
ANNMF	0.0316	0.0047	0.01374	0.0402
HF	0.03824	0.0061	0.0147	0.0486
AHF	0.0347	0.0593	0.0139	0.0442
CANNF	0.0222	0.0057	0.0090	0.0255
CANNMF	0.0175	0.0046	0.0081	0.0193
CBANNF	0.0229	0.0055	0.0089	0.0250
CBANNMF	0.0175	0.0046	0.0081	0.01934
AMNFE	0.0311	0.0151	0.0213	0.0331
AMNFG	0.0301	0.0169	0.0213	0.0325
AMNFD	0.0218	0.0054	0.0091	0.0283
BFMA	0.0360	0.0201	0.0250	0.0404

Table 3.10. NCD for the RGB 'Lenna' image, 5×5 window

Filter	Noise Model 1	2	3	4
None	0.1149	0.0875	0.7338	0.1908
AMF	0.0275	0.0270	0.0252	0.0338
BVDF	0.0408	0.0084	0.0267	0.0631
CBRF	0.0284	0.0070	0.0130	0.0310
GVDF	0.0220	0.0089	0.0189	0.0474
DDF	0.0279	0.0079	0.0171	0.0368
VMF	0.0193	0.0062	0.0236	0.0344
FVDF	0.0218	0.0057	0.0129	0.0339
ANNF	0.0202	0.0071	0.0120	0.0329
ANNMF	0.0181	0.0059	0.0123	0.0318
HF	0.0199	0.0097	0.0123	0.01205
AHF	0.0188	0.0941	0.0120	0.0322
CANNF	0.0129	0.0078	0.0085	0.0153
CANNMF	0.0126	0.0063	0.0080	0.0134
CBANNF	0.0130	0.0077	0.0084	0.0150
CBANNMF	0.0126	0.0063	0.0080	0.0134
AMNFE	0.0261	0.0173	0.0212	0.0281
AMNFG	0.0279	0.0177	0.0216	0.0294
AMNFD	0.0140	0.0070	0.0086	0.0168
BFMA	0.0309	0.0192	0.0228	0.0339

Table 3.11. NCD for the RGB 'peppers' image, 3×3 window

Filter	Noise Model 1	2	3	4
None	0.2414	0.0854	0.0831	0.0859
AMF	0.1042	0.1296	0.1144	0.1298
BVDF	0.1916	0.0774	0.0668	0.0775
CBRF	0.1579	0.0560	0.0541	0.0561
GVDF	0.1463	0.0631	0.0596	0.0639
DDF	0.2113	0.0678	0.0657	0.0679
VMF	0.1624	0.0559	0.0533	0.0558
FVDF	0.1217	0.0585	0.0558	0.0591
ANNF	0.1135	0.0642	0.0578	0.0643
ANNMF	0.0997	0.0575	0.0565	0.0579
HF	0.1406	0.0609	0.0553	0.0605
AHF	0.1346	0.0605	0.0557	0.0601
CANNF	0.1137	0.0610	0.0561	0.0610
CANNMF	0.1009	0.0571	0.0560	0.0574
CBANNF	0.1132	0.0605	0.0558	0.0606
CBANNMF	0.1007	0.0569	0.0559	0.0570
AMNFE	0.1003	0.0597	0.0585	0.0598
AMNFG	0.1007	0.0597	0.0584	0.0597
AMNFD	0.109	0.0621	0.0584	0.0623
BFMA	0.1311	0.0583	0.0566	0.0582

Table 3.12. NCD for the RGB 'peppers' image, 5×5 window

Filter	Noise Model 1	2	3	4
None	0.2414	0.0854	0.0831	0.0859
AMF	0.0916	0.1029	0.0944	0.1028
BVDF	0.186235	0.1056	0.0867	0.1047
CBRF	0.1281	0.0657	0.0646	0.0659
GVDF	0.1384	0.0941	0.0870	0.0946
DDF	0.1613	0.0706	0.0695	0.0706
VMF	0.1301	0.0662	0.0648	0.0663
FVDF	0.1310	0.0658	0.0644	0.0659
ANNF	0.0917	0.0760	0.0698	0.0760
ANNMF	0.0895	0.0657	0.0652	0.0658
HF	0.1118	0.0798	0.0697	0.0798
AHF	0.1070	0.0795	0.0699	0.0792
CANNF	0.0896	0.0652	0.0651	0.0659
CANNMF	0.0896	0.0652	0.0651	0.0659
CBANNF	0.0988	0.07246	0.06837	0.0725
CBANNMF	0.0893	0.0649	0.06452	0.0651
AMNFE	0.0915	0.0671	0.0660	0.0672
AMNFG	0.0917	0.0670	0.0659	0.0671
AMNFD	0.0917	0.0687	0.0672	0.0688
BFMA	0.1191	0.0579	0.0563	0.0577

Table 3.13. Subjective Evaluation

Filter	Figure of Merit				
Filter	a	b	c	d	e
AMF	2	5	1	3	4
BVDF	3	3	3	2	1
CBRF	3	3	3	3	3
GVDF	3	3	4	3	3
DDF	3	3	3	3	2
VMF	4	2	5	3	3
FVDF	3	3	4	3	3
ANNF	4	3	3	3	3
ANNMF	4	3	3	3	3
HF	3	3	3	3	3
AHF	3	3	4	3	3
CANNF	4	3	3	3	3
CAMMMF	4	4	3	3	3
CBANNF	4	3	3	3	3
CBANNMF	4	4	3	3	3
AMNFE	4	4	5	5	4
AMNFG	4	4	5	5	5
MAMNFD	4	4	5	4	5
BFMA	4	5	4	4	4

Fig. 3.9. 'Peppers' corrupted by 4% impulsive noise

Fig. 3.10. 'Lenna' corrupted with Gaussian noise $\sigma = 15$ mixed with 2% impulsive noise

The effect of geometric adaptability in color image processing is also studied in this section through experimental analysis. The processing task on hand is color image filtering. The noise is distributed randomly in the RGB image planes. The adaptive morphological filter processes each of the three color components separately. Since there is no generally accepted quantitative difference measure for image quality, the normalized mean square error (NMSE) is used to quantify the performance of the filters. As in the experiments discussed previously, four subjective criteria are introduced to quantify the performance of the different filters under consideration.

1. Edge preservation
2. Detail preservation
3. Color appearance
4. Smoothness of uniform areas

Fig. 3.11. *VMF* of (3.9) using 3x3 window

Fig. 3.12. *BVDF* of (3.9) using 3x3 window

Fig. 3.13. *HF* of (3.9) using 3x3 window

Fig. 3.14. *AHF* of (3.9) using 3x3 window

Fig. 3.15. *FVDF* of (3.9) using 3x3 window

Fig. 3.16. *ANNMF* of (3.9) using 3x3 window

Fig. 3.17. *CANNMF* of (3.9) using 3x3 window

Fig. 3.18. *BFMA* of (3.9) using 3x3 window

Fig. 3.19. *VMF* of (3.10) using 3x3 window

Fig. 3.20. *BVDF* of (3.10) using 3x3 window

Fig. 3.21. *HF* of (3.10) using 3x3 window

Fig. 3.22. *AHF* of (3.10) using 3x3 window

Fig. 3.23. *FVDF* of (3.10) using 3x3 window

Fig. 3.24. *ANNMF* of (3.10) using 3x3 window

Fig. 3.25. *CANNMF* of (3.10) using 3x3 window

Fig. 3.26. *BFMA* of (3.10) using 3x3 window

The performance is ranked subjectively in five categories: excellent (5), very good (4), good (3), fair (2) and bad (1). The RGB color image 'Mandrill' is used for comparison purposes. The size and the shape of the window and the operation sequence for each filter is chosen such that the best result can be obtained from the filter. In particular, the three filters compared here are:

1. Non-adaptive morphological close-opening filter with four 5-point one-dimensional structuring elements oriented in 0°, 45°, 90°, and 135°.
2. NOP-NCP filter with a 9-point ($N = 9$) adaptive structuring element. The structuring element B used in the open-closing of the post processing in obtaining the coarse image z_1 is a (2×2) window. A 5-point ($N_1 = 5$) adaptive structuring element is used in the NOP-NCP filtering of the isolated noise patterns in z_2.
3. Vector median filter (VMF) with a (3×3) square window.

The performance of the filters on the noise corrupted Mandrill image is illustrated in Table 3.14.

Table 3.14. Performance measures for the image Mandrill

Filter	NMSE 10^{-2}	Edge preserv.	Detail preserv.	Color appear.	CPU time
Close-opening	1.7234	2	2	2	6.26
NOP-NCP	1.2823	5	5	4	14.39
VMF	2.31	5	4	4	8.70

The adaptive morphological filter is the most effective in providing a good quality image in terms of detail, edge and color preservation. The close-opening morphological filter is quite effective in removing the noise, but it tends to blur the image, cause unnatural color, and smooth the texture. Its performance also depends on the patterns in the image. It provides good filtering only if the object patterns are aligned in the same directions as the four structuring elements, and hence limiting its effectiveness in general cases. The experiment demonstrates the way that the morphological filter deals with color distortion. Because of its adaptive structuring element, the morphological filter can best preserve the detailed structures in each color component, and thus achieves a better final result. By comparison, VMF utilizes the color correlation, and thus shows a strong performance along the edges of large blocks. But the fixed window of the filter cannot fit into many detailed structures of the image. Thus, the vector median filter may alter those structures and create artificial patterns. The sudden change of color vectors at those patterns appears as artificial color changes. According to the experimental results reported here it can be concluded that to best preserve the color quality, not only does the correlation between the three color channels have to be utilized, but the images have to be processed according to the

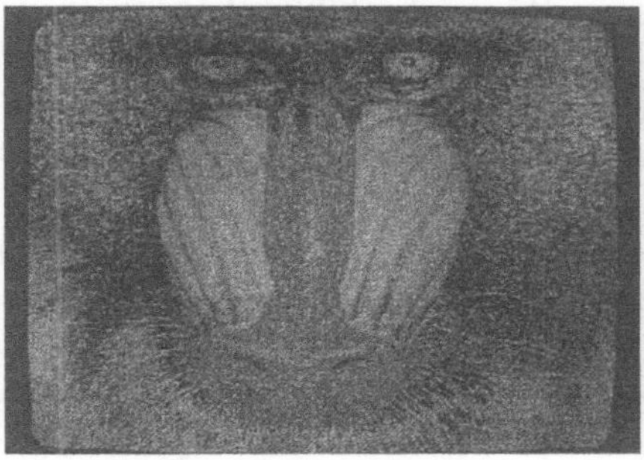

Fig. 3.27. 'Mandrill' - 10% impulsive noise

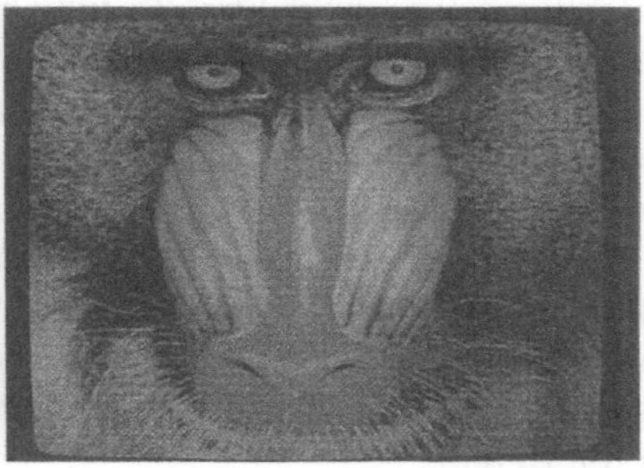

Fig. 3.28. NOP-NCP filtering results

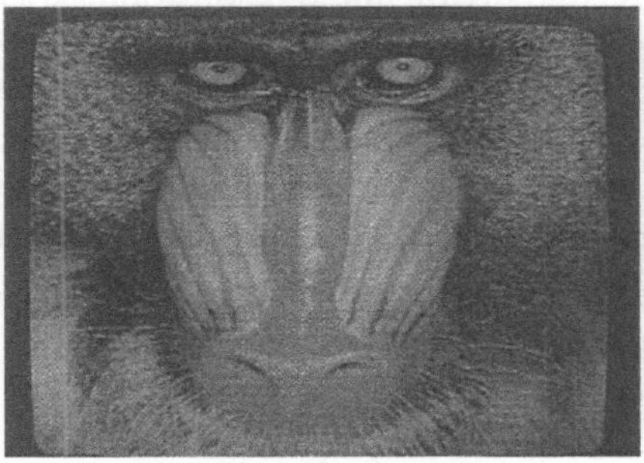

Fig. 3.29. VMF using 3x3 window

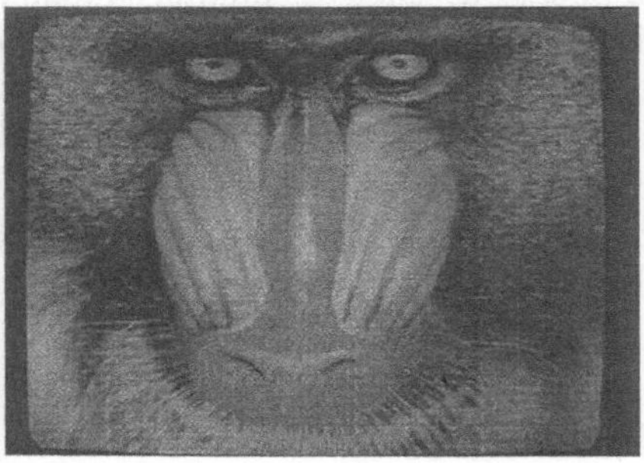

Fig. 3.30. Mutistage Close-opening filtering results

image structure. The computation time taken by the close-opening filter and the VMF is relatively short compared to that of the NOP-NCP filter. The adaptive morphological filter provides the best detail and highlight preservation at the expense of more CPU time. Although research is needed towards more efficient algorithms that can provide a faster search for the structuring elements, the geometric approach to color image processing can be proven valuable in applications where no prior knowledge of the image statistics is available.

3.7 Conclusions

In this chapter adaptive filters suitable for color image processing have been discussed. The behavior of these adaptive designs was analyzed and their performance was compared to that of the most commonly used nonlinear filters. Particular emphasis was given to the formulation of the problem and the filter design procedure. To fully assess the applicability of the adaptive techniques, further analysis is required on algorithms and architectures which may be used for the realization of the adaptive designs. Issues, such as speed, modularity, the effect of finite precision arithmetic, cost and software transportability should be addressed.

The adaptive designs not only have a rigid theoretical foundation but promising performance in a variety of noise characteristics. Indeed, the simulation results included and the subjective evaluation of the filtered color images indicate that adaptive filters compare favorably with other techniques in use to date.

The rich and expanding area of color signal processing underline the importance of the tools presented here. In addition to color image processing, application areas, such as multi-modal signal processing, telecommunication applications, such as channel equalization and digital audio restoration, satellite imagery, multichannel signal processing for seismic deconvolution and applications in biomedicine, such as multi electrode ECG/EEG and CT scans, to name a few, are potential application fields of the adaptive methodologies discussed in this chapter. Problems motivated by the new applications demand investigations into algorithms and methodologies which may result in even more effective adaptive filtering structures.

References

1. Pitas, I., Venetsanopoulos, A .N. (1990): Nonlinear Digital Filters: Principles and Applications. Kluwer Academic Publishers, Boston, MA.
2. J.S. Lee, J.S. (1980): Digital image enhancement and noise filtering by local statistics. IEEE Trans. on Pattern Recognition and Machine Intelligence, **2**, 165-168.
3. Sun, X.Z., Venetsanopoulos, A.N. (1988): Adaptive schemes for noise filtering and edge detection by use of local statistics. IEEE Trans. on Circuit and Systems, **35**(1), 59-69.
4. Cotropoulos, C., Pitas, I (1994): Adaptive nonlinear filter for digital signal/image processing. (Advances In 2D and 3D Digital Processing, Techniques and Applications, edited by C.T. Leondes), Academic Press, **67**, 263-317.
5. Kosko, B. (1991): Neural Networks for Signal Processing. Prentice Hall, Englewood Cliffs, N.J., USA.
6. Yin, L., Astola, J., Neuvo, Y., (1993): A new class of nonlinear filters: Neural filters. IEEE Trans. on Signal Processing. **41**, 1201-1222.
7. Russo, F. (1996): Nonlinear fuzzy filters: An overview. Proceedings European Signal Processing Conference, **VIII**, 1709-1712.
8. Y. Choi, Y., Krishnapuram, R., A robust approach to image enhancement based on fuzzy logic. IEEE Trans. on Image Processing, **6**(6), 808-825.
9. Yu, P.T., Chung Chen, R. (1996): Fuzzy stack filters: Their definitions, fundamental properties and application in image processing. IEEE Trans. on Image Processing, **5**(6), 838-854.
10. Russo, F., Ramponi, G. (1996): A fuzzy filter for images corrupted by impulsive noise. IEEE Signal Processing Letters, **3**(6), 168-170.

11. Plataniotis, K.N., Androutsos, D., Vinayagamoorthy, S., Venetsanopoulos, A.N. (1997): Color image processing using adaptive multichannel filters. IEEE Trans. on Image Processing, **6**(7), 933-950.
12. Russo, F., Ramponi, G. (1994): Nonlinear fuzzy operators for image processing. Signal Processing, **38**(4), 429-440.
13. Yang, X., Toh, P.S. (1995): Adaptive fuzzy multilevel median filter. IEEE Trans. on Image Processing, **4**(5), 680-682.
14. Taguchi, A., Kimura, T. (1996): Data-dependent filtering based on if-then rules and else rules. Proceedings of European Signal Processing Conference, **VIII**, 1713-1716.
15. Arakawa, K., Arakawa, Y. (1991): Digital signal processing using fuzzy clustering. IEICE Transactions, **E 74**(11), 3554-3558.
16. Arakawa, K., Arakawa, Y. (1993): Proposal of median-type fuzzy filter and its optimum design. Electronics and Communications in Japan: part 3, **76**(7), 27-35.
17. Taguchi, A., Izawa, N. (1996): Fuzzy center weighted median filters. Proceedings of European Signal Processing, **VIII**, 1721-1724.
18. Russo, F. (1997): Nonlinear filtering of noisy images using neuro-fuzzy operators. Proceedings of the IEEE Conference on Image Processing, 1997.
19. Tsai, H-H., Yu, Pao-Ta (1999): Adaptive fuzzy hybrid multichannel filters for color image restoration. Proceedings of the 1999 IEEE Workshop on Nonlinear Signal and Image Processing, **I**, 134-138.
20. Tsai, H-H., Yu, Pao-Ta (1999): Adaptive fuzzy hybrid multichannel filters for removal of impulsive noise from color images. Signal Processing, **74**(20, 127-152.
21. Kosko, B., (1992): Neural Networks and Fuzzy Systems: A Dynamic Systems Approach to Machine Intelligence. Prentice Hall, Englewood Cliffs, N.J., USA.
22. Plataniotis, K.N., Androutsos, D., Venetsanopoulos, A.N. (1996): Fuzzy adaptive filters for multichannel image processing. Signal Processing Journal, **55**(1), 93–106.
23. Plataniotis, K.N., Androutsos, D., Venetsanopoulos, A.N. (1996): Multichannel filters for image processing. Signal Processing: Image Communications, **9**(2), 143-158.
24. Mendel, J.M. (1995): Fuzzy logic systems for engineering: A tutorial. Proceedings of the IEEE, **26**(3), 345-377.
25. Bilgic, t., Turksen, I.B., (1996): Measurement of membership functions: Theoretical and empirical work. Technical report, Department of Mechanical and Industrial Engineering, University of Toronto, Toronto, Canada.
26. Zysno, P. (1981): Modelling membership functions. (Empirical Semantics, B. Rieger Editor), Brockmeyer, Bochum, Germany, 350-375.
27. Zimmerman, H.J., Zysno, P. (1996): Quantifying vagueness in decision models. European Journal of Operation Research, **22**, 148-154.
28. H.J. Zimmermann, P. Zysno, Latent connectives in human decision making, Fuzzy Sets and Systems, vol. 4, pp. 37-51, 1980.
29. F.S. Roberts, F.S. (1979): Measurement Theory with Applications to Decision-Making, Utility and the Social Sciences. Addison-Wesley, Reading, Massachusetts.
30. Zimmermann, H.J. (1987): Fuzzy Sets, Decision Making and Expert System. Kluwer Academic, Boston, Massachusetts.
31. Zadeh, L.A. (1965): Fuzzy sets. Information control, **8**, 338-353.
32. Shepard R.N. (1981): Towards a universal law of generalization for psychological science Science, **237**, 1317-1323.
33. Dombi, J. (1990): Membership function as an evaluation. Fuzzy Sets and Systems, **35**, 1-21.

34. Plataniotis, K.N., Androutsos, D., Sri, V., Venetsanopoulos, A.N. (1995): 'A Nearest Neighbour Multichannel Filter,' Electronic Letters, **31**, 1910-1911.
35. Plataniotis, K.N., Androutsos, D., Vinayagamoorthy, S., Venetsanopoulos, A.N. (1996): An adaptive nearest neighbor multichannel filter. IEEE Trans. on Circuits and Systems for Video Technology, **6**(6), 699-703.
36. Plataniotis, K.N., Androutsos, D., Venetsanopoulos, A.N. (1997): Content-based colour image filters. Electronic Letters, **33**(3), 202-203.
37. Plataniotis, K.N., Androutsos, D., Venetsanopoulos, A.N. (1997): Color image filters: The vector directional appoach. Optical Engineering, **36**(9), 2375-2383.
38. Plataniotis, K.N., Androutsos, D., Venetsanopoulos, A.N. (1996): Color image processing using adaptive vector directional filters. IEEE Trans. on Circuits and Systems II: Analog and Digital Signal Processing, **45**(10), 1414-1419.
39. Plataniotis, K.N., Androutsos, D., Venetsanopoulos, A.N. (1996): An adaptive multichannel filters for color image processing. Canadian Journal of Electrical & Computer Engineering, **21**(4), 149-152.
40. Grabisch, M., Nguyen, H.T., Walker, E.A. (1996): Fundamentals of Uncertainty Calculi with Applications to Fuzzy Inference. Kluwer Academic Publishers, Dordrecht.
41. Fodor, J., Marichal, J., Raibens, M. (1995): Characterization of the ordered weighted averaging operators. IEEE Trans. on Fuzzy Systems, **3**(2), 231-240.
42. Trahanias, P.E., Venetsanopoulos, A.N. (1993): Vector directional filters. A new class of multichannel image processing filters. IEEE Trans. on Image Processing, **2**, 528-534.
43. Trahanias, P.E., Karakos D., Venetsanopoulos, A.N. (1996): Directional processing of color images: theory and experimental results. IEEE Trans. on Image Processing, **5**(6), 868-880.
44. Plataniotis, K.N., Androutsos, D., Vinayagamourthy, S., Venetsanopoulos, A.N. (1997): Color image processing using adaptive multichannel filters. IEEE Trans. on Image Processing, **6**(7), 933-950.
45. Bickel, P.J. (1982): On adaptive estimation. Annals of Statistics, **10**, 647-671.
46. Sage, A.P., Melsa, J.L. (1979): Estimation Theory with Applications to Communication and Control, R.E. Krieger Publishing Co., Huntington N.Y.
47. Box, G.E., Tiao, G.C. (1964): A note on criterion robustness and inference robustness. Biometrika, **51**(2), 169-173.
48. Box, G.E., Tiao, G.C. (1973): Bayesian Inference in Statistical Analysis. Addison-Wesley publishing Co, Toronto, Canada.
49. Pan, W., Jeffs, B.D. (1995): Adaptive image restoration using a generalized Gaussian model for unknown noise. IEEE Trans. on Image Processing, **4**(10) 1451-1456.
50. Plataniotis, K.N. (1994): Distributed Parallel Processing State Estimation Algorithms, Ph.D Dissertation, Florida Institute of Technology, Melbourne, Florida, USA.
51. Kim, H.M., Mendel, J.M., Fuzzy basis functions: Comparisons with other basis functions. IEEE Trans. on Fuzzy Systems, **3**(2), 158-169.
52. Plataniotis, K.N., Androutsos, D., Venetsanopoulos, A.N. (1998): Adaptive multichannel filters for color image processing. Signal Processing: Image Communications, **11**(3), 1998.
53. Cacoullos, T. (1966): Estimation of a multivariate density. Annals of Statistical Mathematics, **18**(2), 179-189.
54. Epanechnikov, V.K. (1969): Non-parametric estimation of a multivariate probability density. Theory Prob. Appl., **14**, 153-158.
55. Fukunaga, K. (1990): Introduction to Statistical Pattern Recognition, Academic Press, Second Edition, London, UK.

56. Breiman, L., Meisel, W., Purcell, E. (1977): Variable kernel estimates of multivariate densities. Technometrics, **19**(2), 135-144.
57. Rao, Prasaka B.L.S. (1983): Non-parametric functional estimation Academic Press, N.Y.
58. Nadaraya, E.A. (1964): On estimating regression. Theory Probab. Applic., **15**, 134-137.
59. Watson, G.S. (1964): Smooth regression analysis. Sankhya Ser. A, **26**, 359-372.
60. T.J. Wagner, T.J. (1975): Nonparametric estimates of probability density. IEEE Trans. on Information Theory, **21**(4), 438-440.
61. Prat, W.K. (1991): Digital Image Processing. Second Edition, John Wiley, N.Y.
62. Fisher, N.I., Lewis, T., Embleton, B.J.J. (1993): Statistical Analysis of Spherical Data. Cambridge University Press, Paperback Edition, Cambridge.
63. Plataniotis, K.N., Androutsos, D., Venetsanopoulos, A.N. (1998): Processing color images using vector directional filters: extensions and new results. Proceedings, Nonlinear Image Processing IX, **3304**, 268-276.
64. Srinivasan, A. (1996): Computational issues in the solution of liquid crystalline polymer flow problems. Ph.D Dissertation, Department of Computer Science, University of California, Santa Barbara, CA.
65. Matheron, G. M. (1975): Random Sets of Integral Geometry. Wiley, New York, N.Y.
66. Serra, J. (1982): Image Analysis and Mathematical Morphology. Academic Press, London, U.K.
67. Sternberg, S.R. (1986): Greyscale morphology. Computer Vision, Graphics and Image Processing, **35**: 333-355.
68. Serra, J. (1986): Introduction to mathematical morphology. Computer Vision, Graphics and Image Processing, **35**: 283-305.
69. Smith, D. G. (1992): Fast Adaptive Video Processing: A Geometrical Approach. M.A.Sc. thesis, University of Toronto, Toronto, Canada.
70. Maragos, P.A. (1990): Morphological systems for multidimensional signal processing. Proceedings of the IEEE, **78**(4): 690-709.
71. Serra, J. (1988): Image Analysis and Mathematical Morphology: Theoretical Advances. Academic Press, London, U.K.
72. Cheng, F., Venetsanopoulos, A.N. (1992): An adaptive morphological filter for image processing. IEEE Trans. on Image Processing, **1**(4), 533-539.
73. Cheng, F., Venetsanopoulos, A.N. (1999): Adaptive morphological operators, fast algorithms and their applications. Pattern Recognition, forthcoming special issue on Mathematical Morphology and its applications.
74. Deng-Wong, P., Cheng, F., Venetsanopoulos, A.N. (1996): Adaptive morphological filters for color image enhancement. Journal of Intelligence and Robotic Systems, **15**: 181-207.
75. Maragos, P. (1996): Differential morphology and image processing. IEEE Trans. on Image Processing, **5**(6), 922-937.
76. Astola, J., Haavisto, P., Neuvo, Y. (1990): Vector median filters. Proceedings of the IEEE, **78**, 678-689.
77. Trahanias, P.E., Pitas, I., Venetsanopoulos, A.N. (1994): Color Image Processing. (Advances In 2D and 3D Digital Processing: Techniques and Applications, edited by C.T. Leondes), Academic Press, **67**, 45-90.
78. Karakos, D., Trahanias, P.E. (1997): Generalized multichannel image filtering structures. IEEE Trans. on Image Processing, **6**(7), 1038-1045.
79. Gabbouj, M., Cheickh, F.A. (1996): Vector median-vector directional hybrid filter for color image restorartion, Proceedings of the European Signal Processing Conference, **VIII**, 879-881.

80. Poynton, C.A. (1996): A Technical Introduction to Digital Video. (*http : //www.inforamp.net/~poynton/Poynton – T – I – Digital – Video.html*), Prentice Hall, Toronto.
81. Engeldrum, P.G. (1995): A framework for image quality models. Journal of Imaging Science and Technology, **39**(4), 312-318.
82. Narita, N. (1994): Consideration of subjective evaluation method for quality of image coding, Electronics and Communications in Japan: Part 3, **77**(7) 84-97.

4. Color Edge Detection

4.1 Introduction

One of the fundamental tasks in image processing is edge detection. High level image processing, such as object recognition, segmentation, image coding, and robot vision, depend on the accuracy of edge detection. Edges contain essential information about an image. Most edge detection techniques are based on finding maxima in the first derivative of the image function or zero-crossings in the second derivative of the image function. This concept is illustrated for a gray-level image in Fig. 4.1 [4]. The figure shows that the first derivative of the gray-level profile is positive at the leading edge of a transition, negative at the trailing edge, and zero in homogeneous areas. The second derivative is positive for that part of the transition associated with the dark side of the edge, negative for that part of the transition associated with the light side of the edge, and zero in homogeneous areas. In a monochrome image an edge usually corresponds to object boundaries or changes in physical properties such as illumination or reflectance. This definition is more elaborate in the case of color (multispectral) images since more detailed edge information is expected from color edge detection. According to psychological research on human visual system [1], [2], color plays a significant role in the perception of boundaries. Monochrome edge detection may not be sufficient for certain applications since no edges will be detected in gray-level images when neighboring objects have different hues but equal intensities [3]. Objects with such boundaries are treated as one big object in the scene. Since the capability of distinguishing between different objects is crucial for applications such as object recognition and image segmentation, the additional boundary information provided by color is of paramount importance. Color edge detection also outperforms monochrome edge detection in low contrast images [3]. There is thus a strong motivation to develop efficient color edge detectors that provide high quality edge maps.

Despite the relatively short period of time, numerous approaches of different complexities to color edge detection have been proposed. It is important to identify their strength and weaknesses in choosing the best edge detector for an application. In this chapter particular emphasis will be given to color edge detectors based on vector order statistics. If the color image is considered as three dimensional vector space, a color edge can be defined as a significant

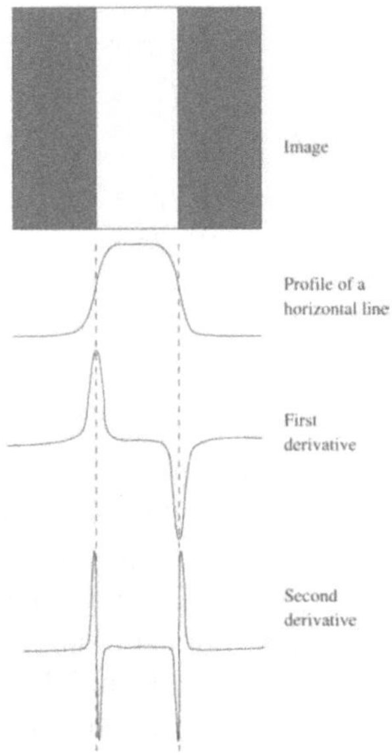

Fig. 4.1. Edge detection by derivative operators

discontinuity in the vector field representing the color image function. An abrupt change in the vector field characterizes a color step edge, whereas a gradual change characterizes a color ramp edge. It should be noted that the above definitions are not intended as formal definitions that can lead to edge detectors. Rather they are intuitive descriptions of the notion of color edges in order to facilitate discussion on order statistics based edge detectors. Edge detectors based on order statistics operate by detecting local minima and maxima in the color image function and combining them in an appropriate way in order to produce a positive response for an edge pixel. Since there is no unique way to define order for multivariable signals, such as color images, the reduced ordering (R-ordering) scheme discussed in Chap. 2 will be used to sort vector samples. A class of color edge detectors will then be defined using linear combinations of the sorted vector samples. The minimum over the magnitudes of these linear combinations defines this class of edge operators. Different coefficients in the linear combinations result in different edge detectors that vary in simplicity and in efficiency.

The major performance issues concerning edge detectors are their ability to extract edges accurately, their robustness to noise, and their computational efficiency. In order to provide a fair assessment, it is necessary to have a set of effective performance evaluation methods. Though numerous evaluation methods for edge detection have been proposed, there has not been any standardized method. In image processing, the evaluation methods can usually be categorized into objective and subjective evaluation. While objective evaluation can provide analytical data for comparison purpose, it is not sufficient to represent the complexity of human visual systems. In most image processing applications, human evaluation is the final step, as noted by Clinque [7]. The subjective evaluation, which takes into account the human perception, seems to be very attractive in this perspective. The visual assessment method proposed by Heath [8], [9] is entirely based on subjective evaluation. In this chapter, both types of evaluation methods are utilized for comparing various edge detectors.

The chapter is organized as follows: An overview of the methodology for color edge detection is presented first. Early approaches extended from monochrome edge detection, as well as the more recent approaches of vector space are addressed. The edge detectors illustrated in this section, among others, are the Sobel operator [3], [10], Laplacian operator [3], Mexican Hat operator [3], [11], Vector Gradient operator [12], Directional operator [13], Entropy operator [14], and Cumani operator [15]. In the sequence, two families of vector based edge detection operators, Vector Order Statistic operators [16] [17] and Difference Vector operators [25] are studied in detail. A variety of edge detectors obtained as special cases of the two families are introduced and their performances are evaluated. Evaluation results from both objective and subjective tests as well as conclusion from the tests performed are also listed here.

4.2 Overview Of Color Edge Detection Methodology

4.2.1 Techniques Extended From Monochrome Edge Detection

In a monochrome image, an edge is defined as an intensity discontinuity. In the case of color images, the additional variation in color also needs to be considered. Early approaches to color edge detection comprise extensions from monochrome edge detection. These techniques are applied to the three color channels independently and then the results are combined using a certain logical operation [3].

Sobel operator. The first derivative at any point in an image is obtained using the magnitude of the gradient at that point. This can be done using various operators including the Sobel, Prewitt, and Roberts operators [4],[5]. The Sobel operators have the advantage of providing both a differencing and a smoothing effect. Because derivatives enhance noise, the smoothing effect is a particularly attractive feature of the Sobel operators. A number of edge detectors including the Sobel operator are compared in [3]. The Sobel operator is implemented by convolving a pixel and its eight neighbors with the following two 3×3 convolution masks [4], [5]:

$$M_x = \begin{pmatrix} 1 & 0 & -1 \\ 2 & 0 & -2 \\ 1 & 0 & -1 \end{pmatrix}, \qquad M_y = \begin{pmatrix} 1 & 2 & 1 \\ 0 & 0 & 0 \\ -1 & -2 & -1 \end{pmatrix} \tag{4.1}$$

The two masks are applied to each color channel independently and the sum of the squared convolution results states an approximation of the magnitude of the gradient in each channel. A pixel is regarded as an edge point if the mean of the gradient magnitude values in the three color channels exceed a given threshold. According to [3] the Sobel operator produces very thick edges that have to be thinned.

Laplacian operator. The second derivative at any point in an image is obtained by using the Laplacian operator. The basic requirement in defining the Laplacian operator is that the coefficient associated with the center pixel be positive and the coefficients associated with the outer pixels be negative [4]. The sum of the coefficients has to be zero. An eight-neighbor Laplacian operator can be defined using the following convolution mask:

$$M = \begin{pmatrix} 10 & 22 & 10 \\ 22 & 128 & 22 \\ 10 & 22 & 10 \end{pmatrix} \tag{4.2}$$

The Laplacian mask is applied to the three color channels independently and the edge points are located by thresholding the maximum gradient magnitude. The methodology is simple, easy to implement and is very successful in located edges. However, there are problems when the Laplacian methodology is applied to color images. First, many of the Laplacian zero crossings are spurious edges which really correspond to local minima in gradient magnitude. It is well known that zero crossings in a second order derivative indicates an extremum in the first order derivative but not necessary a local maximum. To improve the performance of the Laplacian operator and differentiate between global and local minima, the sign of the third derivative may have to be examined. Performance, however, can be hampered due to the noise usually corrupting real images. Since differentiation amplifies noise acting like a high pass filter, the Laplacian zero crossings for an image may have numerous false edges caused by noise. It is therefore recommended that a smoothing operator is applied to the image prior to the detection module.

Mexican Hat operator. Another group of edge detectors commonly used in monochrome edge detection is based on second derivative operators and they can also be extended to color edge detection in the same way. A second derivative method can be implemented based on the above operator. The Mexican Hat operator uses convolution masks generated based on the negative Laplacian derivative of the Gaussian distribution:

$$-\nabla^2 G(x,y) = \frac{x^2+y^2-2\sigma^2}{2\pi\sigma^6} exp(-\frac{x^2+y^2}{2\sigma^2}) \tag{4.3}$$

Edge points are located if zero-crossing occurs in any color channel. The gradient operators proposed for gray scale images [26] can also be extended to color images by taking the vector sum of the gradients for individual components [12], [14]. Similar to Sobel and Laplacian operators, the gradient operator also employs first derivative-like mask patterns. Other approaches consider performing operations in alternative color spaces. The Hueckel edge operator [27] operates in the luminance, chrominance color space. The edges in the three color components are also assumed to be independent under the constraints that they must have the same orientation. In studying the application of the compass gradient edge detection method to color images, Robinson [28] also utilized different color coordinates.

One common problem with the above approaches is that they fail to take into account the correlation among the color channels, and as a result, they are not able to extract certain crucial information conveyed by color. For example, they tend to miss edges that have the same strength but are in opposite direction in two of their color components. Consequently, the approach to treat the color image as vector space has been proposed.

4.2.2 Vector Space Approaches

Various approaches proposed consider the problem of color edge detection in vector space. Color images can be viewed as a two-dimensional three-channel vector field [29] which can be characterized by a discrete integer function $\mathbf{f}(x, y)$. The value of this function at each point is defined by a three dimensional vector in a given color space. In the RGB color space, the function can be written as $\mathbf{f}(x,y) = (R(x,y), G(x,y), B(x,y))$, where (x,y) refers to the spatial dimensions in the 2-D plane. Most existing edge detection algorithms use either first or second differences between neighboring pixels for edge detection. A significant change gives rise to a peak in the first derivative and zero-crossing in the second derivative, both of which can be identified fairly easily. Some of these operators are considered here.

Vector gradient operators. The vector gradient operator employs the concept of gradient operator with modifications such that instead of a scalar space the operator performs on a two-dimensional three channel color vector space. There are several ways of implementing the vector gradient operator [12]. One simple approach is to employ a (3x3) window centered on each pixel and then obtain eight distance values, $(D_1, D_2, \ldots, D_8)$, by computing the Euclidean distance between the center vector and its eight neighboring vectors. The vector gradient (g) is then chosen as:

$$g = max(D_1, D_2, \ldots, D_8) \tag{4.4}$$

Another approach employs directional operators . Let the image be a vector function $\mathbf{f}(x,y) = (R(x,y), G(x,y), B(x,y))$, and let $\boldsymbol{r}, \boldsymbol{g}, \boldsymbol{b}$ be the unit vectors along the R, G, B axes, respectively. The horizontal and vertical directional operators can be defined as:

$$\boldsymbol{u} = \frac{\partial R}{\partial x}\boldsymbol{r} + \frac{\partial G}{\partial x}\boldsymbol{g} + \frac{\partial B}{\partial x}\boldsymbol{b} \tag{4.5}$$

$$v = \frac{\partial R}{\partial y} r + \frac{\partial G}{\partial y} g + \frac{\partial B}{\partial y} b \tag{4.6}$$

$$g_{xx} = u \cdot u = |\frac{\partial R}{\partial x}|^2 + |\frac{\partial G}{\partial x}|^2 + |\frac{\partial B}{\partial x}|^2 \tag{4.7}$$

$$g_{yy} = v \cdot v = |\frac{\partial R}{\partial y}|^2 + |\frac{\partial G}{\partial y}|^2 + |\frac{\partial B}{\partial y}|^2 \tag{4.8}$$

$$g_{xy} = \frac{\partial R}{\partial x}\frac{\partial R}{\partial y} + \frac{\partial G}{\partial x}\frac{\partial G}{\partial y} + \frac{\partial B}{\partial x}\frac{\partial B}{\partial y} \tag{4.9}$$

Then the maximum rate of change of **f** and the direction of the maximum contrast can be calculated as:

$$\theta = \frac{1}{2} \arctan \frac{2g_{xy}}{g_{xx} - g_{yy}} \tag{4.10}$$

$$F(\theta) = \frac{1}{2}\{(g_{xx} + g_{yy}) + cos2\theta(g_{xx} - g_{yy}) + 2g_{xy}sin\theta\} \tag{4.11}$$

The edges can be obtained by thresholding $\sqrt{F(\theta)}$.

The image derivatives along the x and y directions can be computed by convolving the the vector function **f** with two spatial masks as follows:

$$\frac{\partial f_i}{\partial x} \simeq \frac{1}{6} \begin{bmatrix} -1 & 0 & 1 \\ -1 & 0 & 1 \\ -1 & 0 & 1 \end{bmatrix} * f_i, \quad \frac{\partial f_i}{\partial y} \simeq \frac{1}{6} \begin{bmatrix} 1 & 1 & 1 \\ 0 & 0 & 0 \\ -1 & -1 & -1 \end{bmatrix} * f_i \tag{4.12}$$

Unlike the gradient operator extended from monochrome edge detection, the vector gradient operator can extract more color information from the image because it considers the vector nature of the color image. On the other hand, the vector gradient operator is very sensitive to small texture variations [17]. This may be undesirable in some cases since it can cause confusion in identifying the real objects. The operator is also sensitive to Gaussian and impulse noise.

Directional operators. The direction of an edge in color images can be utilized in a variety of image analysis tasks [18]. A class of directional vector operators was proposed to detect the location and orientation of edges in color images [13]. In this approach, a color $\mathbf{c}(r, g, b)$ is represented by a vector $\mathbf{c}$ in color space. Similar to the well known Prewitt operator [20] shown below,

$$\Delta H = \frac{1}{3} \begin{pmatrix} -1 & 0 & 1 \\ -1 & 0 & 1 \\ -1 & 0 & 1 \end{pmatrix}, \quad \Delta V = \frac{1}{3} \begin{pmatrix} -1 & -1 & -1 \\ 0 & 0 & 0 \\ 1 & 1 & 1 \end{pmatrix} \tag{4.13}$$

the row and column directional operators (i.e. in the horizontal and vertical directions), each have one positive and one negative component. For operators of size $(2w + 1) \times (2w + 1)$ the configuration is the following:

$$\boldsymbol{\Delta H} = \left[\boldsymbol{H}_{-} \; \mathbf{0} \; \boldsymbol{H}_{+} \right] \,, \quad \boldsymbol{\Delta V} = \begin{bmatrix} \boldsymbol{V}_{-} \\ \mathbf{0} \\ \boldsymbol{V}_{+} \end{bmatrix} \tag{4.14}$$

where the parameter w is a positive integer. These positive and negative components are *convolution kernels*, denoted by $\boldsymbol{V}_{-}$, $\boldsymbol{V}_{+}$, $\boldsymbol{H}_{-}$ and $\boldsymbol{H}_{+}$, whose outputs are vectors corresponding to the local *average colors*. In order to estimate the color gradient at the pixel (x_o, y_o), the outputs of these components are calculated as follows:

$$\boldsymbol{H}_{+}(x_o, y_o) = \frac{1}{w(2w+1)} \sum_{y=y_o-w}^{y=y_o+w} \sum_{x=x_o+1}^{x=x_o+w} \mathbf{c}(x,y)$$

$$\boldsymbol{H}_{-}(x_o, y_o) = \frac{1}{w(2w+1)} \sum_{y=y_o-w}^{y=y_o+w} \sum_{x=x_o-1}^{x=x_o-w} \mathbf{c}(x,y)$$

$$\boldsymbol{V}_{+}(x_o, y_o) = \frac{1}{w(2w+1)} \sum_{y=y_o+1}^{y=y_o+w} \sum_{x=x_o-w}^{x=x_o+w} \mathbf{c}(x,y)$$

$$\boldsymbol{V}_{-}(x_o, y_o) = \frac{1}{w(2w+1)} \sum_{y=y_o-1}^{y=y_o-w} \sum_{x=x_o-w}^{x=x_o+w} \mathbf{c}(x,y) \tag{4.15}$$

where $\mathbf{c}(x,y)$ denotes the RGB color vector (r,g,b) at the image location (x,y). Local colors and local statistics affect the output of the operator components ($\boldsymbol{V}_{+}(x,y)$, $\boldsymbol{V}_{-}(x,y)$, $\boldsymbol{H}_{+}(x,y)$ and $\boldsymbol{H}_{-}(x,y)$). In order to estimate the local variation in the vertical and horizontal directions, the following vector differences are calculated:

$$\boldsymbol{\Delta H}(x_o, y_o) = \boldsymbol{H}_{+}(x_o, y_o) - \boldsymbol{H}_{-}(x_o, y_o)$$

$$\boldsymbol{\Delta V}(x_o, y_o) = \boldsymbol{V}_{+}(x_o, y_o) - \boldsymbol{V}_{-}(x_o, y_o) \tag{4.16}$$

The scalars $||\boldsymbol{\Delta H}(x_o, y_o)||$ and $||\boldsymbol{\Delta V}(x_o, y_o)||$ give the variation rate at (x_o, y_o) in orthogonal directions (i.e. are the amounts of color contrast in the horizontal and vertical directions). The local changes in the color channels (i.e. R, G and B) can not be combined properly by simply adding the R, G and B components of $\boldsymbol{\Delta H}$ and $\boldsymbol{\Delta V}$. This approach leads to a mutual cancel out effect in several situations (e.g. when contrast is in phase opposition in different channels). Instead, the local changes in R, G and B are assumed to be independent (i.e. orthogonal), and the intensity of the local color contrast is obtained as the magnitude of the resultant vector in the RGB space (using the Euclidean norm), as shown in (4.17) and (4.18). Therefore, the magnitude B of the maximum variation rate at (x_o, y_o) is estimated as the magnitude of the resultant vector:

$$B(x_o, y_o) = \sqrt{||\boldsymbol{\Delta V}(x_o, y_o)||^2 + ||\boldsymbol{\Delta H}(x_o, y_o)||^2} \tag{4.17}$$

and the direction θ of the maximum variation rate at (x_o, y_o) is estimated as:

$$\theta = arctan(\frac{\Delta V'(x_o, y_o)}{\Delta H'(x_o, y_o)}) + k\pi \tag{4.18}$$

where k is an integer and:

$$\Delta V'(x_o, y_o) = \begin{cases} ||\boldsymbol{\Delta V}(x_o, y_o)|| & \text{if } ||\boldsymbol{V}_+(x_o, y_o)|| \geq ||\boldsymbol{V}_-(x_o, y_o)|| \\ -||\boldsymbol{\Delta V}(x_o, y_o)|| & \text{otherwise} \end{cases}$$

$$\Delta H'(x_o, y_o) = \begin{cases} ||\boldsymbol{\Delta H}(x_o, y_o)|| & \text{if } ||\boldsymbol{H}_+(x_o, y_o)|| \geq ||\boldsymbol{H}_-(x_o, y_o)|| \\ -||\boldsymbol{\Delta H}(x_o, y_o) & \text{otherwise}|| \end{cases}$$

where $||.||$ denotes the Euclidean norm. In this formulation, the color contrast has no sign. In order to obtain the direction of maximal contrast, a convention is adopted to attribute signs to the quantities $\Delta V'(x_o, y_o)$ and $\Delta H'(x_o, y_o)$ in (4.18). These quantities are considered positive if the luminance increases in the positive directions of the image coordinate system. The luminance quantities are estimated here by the norms: $||\boldsymbol{H}_+||$, $||\boldsymbol{H}_-||$, $||\boldsymbol{V}_+||$ and $||\boldsymbol{V}_-||$. Typically the luminance has been estimated by the luminance quantity, using the norm $||\mathbf{c}||_1 = r + g + b$. However, the norm $||\mathbf{c}||_2 = \sqrt{r^2 + g^2 + b^2}$ also has been used to estimate luminance [21]. Another possibility would be to consider the local color contrast with respect to a reference (e.g. the central portion of the operator $\mathbf{c}_o$), instead of the luminance quantity. However, this last possibility could present some ambiguities. For example, in vertical ramp edges $||\boldsymbol{H}_- - \mathbf{c}_o|| = ||\boldsymbol{H}_+ - \mathbf{c}_o||$, then $\Delta H'(x_o, y_o)$ would have positive sign, irrespective to the actual sign of the ramp slope [13].

Note the similarity between the color gradient formulated above, and a Prewitt-type $(2w+1) \times (2w+1)$ monochromatic gradient [20]. The larger the parameter w, the smaller the operator sensitivity to noise, and also to sharp edges. This happens because there is a smoothing (low pass) effect associated with the convolution mask. Therefore, the larger the size of the convolution mask, the stronger the low pass effect, and the less sensitive to high spatial frequencies becomes the operator. Also note that $\boldsymbol{H}_-, \boldsymbol{H}_+, \boldsymbol{V}_-$ and $\boldsymbol{V}_+$ are in fact convolution masks and could easily implement the latest vector order statistics filtering approaches.

Compound edge detectors. The simple color gradient operator can also be used to implement compound gradient operators [13]. A well known example of a compound operator is the derivative of a Gaussian (ΔG) operator [20]. In this case, each channel of the color image is initially convolved with a Gaussian smoothing function $G(x, y, \sigma)$, where σ is the standard deviation, and, then, this gradient operator is applied to the smoothed color image to detect edges. Torre and Poggio [22] stated that differential operations on sampled images require the image to be first smoothed by filtering. The Gaussian filtering has the advantage that guarantees the bound-limitedness of the signal, so the derivative exists everywhere. This is equivalent to regularizing the signal using a low pass filter prior to the differentiation step.

The low pass filtering (regularization) step is done by the convolution of a Gaussian $G(x, y, \sigma)$ with the image signal. In a multi-spectral image, each pixel is associated with a vector $\mathbf{c}(x, y)$ whose components are denoted by $c_i(x, y)$ where $i = 1, 2, 3$. This convolution is expressed as follows:

$$G(x, y, \sigma) \otimes \boldsymbol{I} = G(x, y, \sigma) \otimes I_i \ , \forall \, \mathrm{i} \tag{4.19}$$

where $\boldsymbol{I}$ and I_i denote the image itself and the image component i. The image edges are then detected using the operator described before and at each pixel the edge orientation $\theta(x, y)$ and magnitude $B(x, y)$ are obtained. The filtering operation introduces an arbitrary parameter, the scale of the filter, e.g., the standard deviation for the Gaussian filter. A number of authors have discussed the relationship existing between multiresolution analysis, Gaussian filtering and zero-crossings of filtered signals [22] [23].

The actual edge locations are detected by computing the zero-crossings of the second-order differences image, obtained by applying first-order difference operators twice. Once the zero-crossings are found, they still must be tested for maximality of contrast. Let the zero-crossing image elements denote $ZC(x, y)$. In practice, the image components are only known at the nodes of a rectangular grid of sampling points (x, y), and the zero-crossing condition $ZC(x, y) = 0$ often does not apply. The simple use of local minima conditions leaves a margin for uncertainty. The zero-crossing image locations can be located by identifying how the sign of $ZC(x, y)$ varies in the direction of maximal contrast, near the zero-crossing location [15]. Therefore, the condition $ZC(x, y) = 0$ must then be substituted by the more practical condition:

$$ZC(x_i, y_i).ZC(x_j, y_j) < 0 \tag{4.20}$$

where the sampling points (x_i, y_i) and (x_j, y_j) are 8-adjacent, and the derivatives required for the computation of $ZC(x, y)$ are approximated by convolutions with the masks proposed by Beaudet [24]. Notice that (x_i, y_i) and (x_j, y_j) are in the direction of maximal contrast calculated at (x_o, y_o), the center of the 8-adjacent neighborhood. In order to improve the spatial location (mostly with larger operator sizes w), a *local minimum* condition is also used (i.e. $|ZC(x_o, y_o)| < \tau$, $\tau \simeq 0$). With the compound detector, the Gaussian noise can be reduced due to the Gaussian smoothing function. Though this operator improves performance in Gaussian noise, it is still sensitive to impulse noise.

Entropy operator. The entropy operator is employed for both monochrome and color images. It yields a small value when the color chromaticity in the local region is uniform and a large value when there are drastic change in the color chromaticity. The entropy in a processing window (i.e., 3x3) centered on vector $\mathbf{v}_o = (r_o, g_o, b_o)$ is defined as:

$$H = q_R H_R + q_G H_G + q_B H_B \tag{4.21}$$

where H_R, H_G, H_B denote the entropies in the R, G, B directions, respectively, and:

$$q_R = \frac{r_o}{r_o + g_o + b_o}, \quad q_G = \frac{g_o}{r_o + g_o + b_o}, \quad q_B = \frac{b_o}{r_o + g_o + b_o} \tag{4.22}$$

Let $X_o, X_1, \ldots, X_N, (X = R, G, B)$ denote the values in each corresponding channel inside the processing window then H_X is defined as:

$$H_X = -\frac{\sum_{i=1}^{N} p_{X_i} log(p_{X_i})}{log(N)} \tag{4.23}$$

where,

$$p_{X_i} = \frac{X_i}{\sum_{j=1}^{N} X_j} \tag{4.24}$$

Edges can be extracted by detecting the change of entropy H in a window region. Since the presence of noise can disturb the local chromaticity in an image, the entropy operator is sensitive to noise[17].

Second derivative operators. A more sophisticated approach which involves a second derivative operator is suggested by Cumani. Given a vector field $\mathbf{f}(x, y)$ for a color image, the squared local contrast of $\mathbf{f}$ at point $P = (x, y)$ in the direction of the unit vector $\boldsymbol{n}(n_1, n_2)$ is defined as:

$$S(P, \boldsymbol{n}) = En_1^2 + Fn_1n_2 + Gn_2^2 \tag{4.25}$$

where

$$E = \frac{\partial \mathbf{f}}{\partial x} * \frac{\partial \mathbf{f}}{\partial x} = \frac{\partial R}{\partial x}\frac{\partial R}{\partial x} + \frac{\partial G}{\partial x}\frac{\partial G}{\partial x} + \frac{\partial B}{\partial x}\frac{\partial B}{\partial x} \tag{4.26}$$

$$F = \frac{\partial \mathbf{f}}{\partial x} * \frac{\partial \mathbf{f}}{\partial y} = \frac{\partial R}{\partial x}\frac{\partial R}{\partial y} + \frac{\partial G}{\partial x}\frac{\partial G}{\partial y} + \frac{\partial B}{\partial x}\frac{\partial B}{\partial y} \tag{4.27}$$

$$E = \frac{\partial \mathbf{f}}{\partial y} * \frac{\partial \mathbf{f}}{\partial y} = \frac{\partial R}{\partial y}\frac{\partial R}{\partial y} + \frac{\partial G}{\partial y}\frac{\partial G}{\partial y} + \frac{\partial B}{\partial y}\frac{\partial B}{\partial y} \tag{4.28}$$

The eigenvalues of the 2x2 matrix $\begin{pmatrix} E & F \\ F & G \end{pmatrix}$ coincide with the extreme values of S(P,$\boldsymbol{n}$) and are attained when $\boldsymbol{n}$ is the corresponding eigenvector. The extreme values are:

$$\lambda_{\pm} = \frac{E + G \pm \sqrt{(E-G)^2 + 4F^2}}{2} \tag{4.29}$$

and the two corresponding eigenvectors $\boldsymbol{n}_+$ and $\boldsymbol{n}_-$ are given as

$$\boldsymbol{n}_{\pm} = (cos\theta_{\pm}, sin\theta_{\pm}) \tag{4.30}$$

$$\theta_{+} = \begin{cases} \frac{\pi}{4} & \text{if (E-G)=0 and F>0} \\ -\frac{\pi}{4} & \text{if (E-G)=0 and F<0} \\ \text{undefined} & \text{if E=F=G=0} \\ \frac{1}{2}\arctan\frac{2F}{E-G} + k\pi & \text{otherwise} \end{cases} \tag{4.31}$$

$$\theta_{-} = \theta_{+} \pm \frac{\pi}{2} \tag{4.32}$$

Possible edge point are considered as point P where the first directional derivative $D_s(P, \boldsymbol{n})$ of maximal squared contrast $\lambda_+(P)$ is zero in the direction of maximal contrast $\boldsymbol{n}_+(P)$. The directional derivative is defined as:

$$\begin{aligned} D_s(P, \boldsymbol{n}) &= \nabla\lambda_+ \cdot \boldsymbol{n}_+ \\ &= \frac{\partial\lambda_+}{\partial x}n_1 + \frac{\partial\lambda_+}{\partial y}n_2 \\ &= E_x n_1^3 + (2F_x + E_y)n_1^2 n_2 + (G_x + 2F_y)n_1 n_2^2 + G_y n_2^3 \end{aligned} \tag{4.33}$$

The edge points are determined by computing zero-crossings of $D_s(P, \boldsymbol{n})$. Since the local directional contrast needs to be a maximum or minimum, the sign of D_s along a curve tangent at P in the direction of $\boldsymbol{n}_+$ is checked and the edge point is located if it is found to be a maximal point.

The ambiguity of the gradient direction in the above method causes some difficulties in locating edge points. A subpixel technique with bilinear interpolation can be employed to solve the problem. A modification in solving the ambiguities by estimating the eigenvector $\boldsymbol{n}_+$, which can avoid the computational costly subpixel approximation was suggested in [30]. Other techniques [3] have also been proposed to improve the performance of the Cumani operator and reduce its complexity. The proposed operator utilizes different sized convolution masks based on the derivatives of the Gaussian distribution in the computation process instead of the set of fixed-sized 3×3 masks. It was argued in [3] that a considerable increase in the quality of the results can be obtained when the Gaussian masks were employed. Similar to the vector gradient operator, the second-order derivative operator is very sensitive to texture variations and impulsive noise, but it produces thinner edges. The regularizing filter applied in this operator causes a certain amount of blurness in the edge map.

4.3 Vector Order Statistic Edge Operators

As seen in Chap. 2, order statistics based operators play an important role in image processing [31], [32]. Order statistics operators had been used extensively in monochrome (gray scale) as well as color image edge detection. This approach is inspired by the morphological edge detectors [33], [34] that have been proposed for monochrome images. This class of color edge detectors is characterized by linear combinations of the sorted vector samples. Different sets of coefficients of the linear combination give rise to different edge detectors that vary in performance and efficiency. The primary step in order statistics is to arrange a set of random variables in ascending order according to certain criteria. However, as described in Chap. 2, there is no universal way to order the color vectors in the different color spaces. The different ordering schemes discussed in this book can be used to define order statistic edge operators.

Let the color image vectors in a window W denote $\mathbf{x}_i$, $i = 1, 2, ..., n$ and $D(\mathbf{x}_i, \mathbf{x}_j)$ be a measure of distance (or similarity) between the color vectors $\mathbf{x}_i$ and $\mathbf{x}_j$. The vector range edge detector (VR) is the simplest color edge detector based on order statistics. It expresses the deviation of the vector outlier in the highest rank from the vector median in W as follows:

$$VR = D(\mathbf{x}_{(n)}, \mathbf{x}_{(1)}) \tag{4.34}$$

VR expresses in a quantitative way the deviation of the vector outlier in the highest rank from the vector median $\mathbf{x}_{(1)}$. Consequently, VR is small in a uniform area where all vectors are closed together, and it returns a large output when discontinuities exist. Its response on an edge will be large since $\mathbf{x}_{(n)}$ will be selected among the vectors from the one side of the edge while $\mathbf{x}_{(1)}$ will be selected among the vectors from the other side of the edge. The actual color edges can be obtained by thresholding the VR outputs.

The VR detector, though simple and efficient, is sensitive to noise, especially to impulse noise. It will respond to a noise pixel at the center of W with (n) pixels. To alleviate this drawback, dispersion measures which are known as more robust estimates in the presence of noise should be considered. To this end, a class of edge detectors can be defined as a linear combination of the ordered image vectors. This class of operators expresses a measure of the dispersion of the ordered vectors. Therefore, the vector dispersion edge detector (VDED), can be defined as follows:

$$VDED = \|\sum_{i=1}^{n} \alpha_i \mathbf{x}_{(i)}\| \tag{4.35}$$

where $\|\cdot\|$ denotes the appropriate norm.

Note that VR is a special case of VDED with $\alpha_1 = -1, \alpha_n = 1$, and $\alpha_i = 0, i = 2, \ldots, n-1$. The above equation can be further generalized by employing k sets of coefficients and combining the resulting vector magnitudes in an appropriate way. The combination proposed in [17] employs a minimum operator which attenuates the effect of the noise. According to the definition the general class of vector dispersion edge detectors can be written as:

$$GVDED = min_j\{\|\sum_{i=1}^{n} \alpha_{ij} \mathbf{x}_{(i)}\|\}, j = 1, 2, \cdots, k \tag{4.36}$$

Specific color edge detectors can be obtained from (4.36) by selecting the set of coefficients α_{ij}. One member of the GVDED family is the minimum vector dispersion detector (MVD), and it's defined as:

$$MVD = min_j\{D(\mathbf{x}_{(n-j+1)}, \sum_{i=1}^{l} \frac{\mathbf{x}_{(i)}}{l})\}, j = 1, 2, \ldots, k, k, l < n \qquad (4.37)$$

The choice of k and l depend on n, the size of W. These two parameters control the trade-off between complexity and noise attenuation. This more computationally involved operator can improve edge detection performance in the presence of both impulse and Gaussian noise. It can eliminate up to $k-1$ impulse noise pixels in W [17]. Let there be $k-1$ impulse noise pixels in a window of n pixels. By their nature, impulse noise differ from the rest of the pixels by a large amount. Therefore, after ordering, the impulse noise pixels have the highest ranks: $\mathbf{x}_{(n-k+2)}, \mathbf{x}_{(n-k+3)}, \ldots, \mathbf{x}_{(n)}$. Since the distance between these noise pixels and the rest of the pixels are large, (4.37) can be reduced to:

$$MVD = D(\mathbf{x}_{(n-k+1)}, \sum_{i=1}^{l} \frac{\mathbf{x}_{(i)}}{l}) \qquad (4.38)$$

Notice that none of the noise pixels appears at this equation, and thus would not affect the edge detection process. MVD has improved noise performance since it is robust to the presence of heavy tailed noise, due to the minimum operation, and short tailed noise due to the averaging operation.

A statistical analysis of MVD must be carried out in order to determine the error probability of the edge detector. The analysis is confined to the case of additive, multivariate normal (Gaussian) distribution. An ideal edge model will be considered in this analysis. According to the model, the sample vectors $\mathbf{X}_i$ are on the one side of the edge as instances of a random variable $\mathbf{X}$ which follows a multivariate Gaussian distribution with known mean μ_x and unit covariance. Similarly, the sample vectors $\mathbf{Y}_i$ on the other side are instances of the random variable $\mathbf{Y}$ which follows a multivariate Gaussian distribution with known mean μ_y and unit covariance. Then, the error probability is given as:

$$P_E = P_e P_M + P_n P_F \qquad (4.39)$$

where P_e and P_n denote the prior probabilities of 'edge' and 'no edge', respectively, and P_M and P_F are the probabilities of missing an edge and false edge alarm, respectively.

If $\hat{\mathbf{X}}$ is the mean of the vectors $\mathbf{X}_i$, then P_M can be calculated as:

$$P_M = Pr\{min\|\mathbf{Y}_i - \hat{\mathbf{X}}\| < t\|\|\mu_y - \mu_x\|.t\} \qquad (4.40)$$

Denoting with $d_{(i)}$ the sorted distance $\|\mathbf{Y}_i - \hat{\mathbf{X}}\|$, it can be claimed that $d_{(1)} = min\|\mathbf{Y}_i - \hat{\mathbf{X}}\|$. Furthermore, defining $\|\mu_y - \mu_x\| = \tau$ (4.40) can be rewritten as:

$$P_M = Pr\{d_{(1)} - \tau < t' | t' < 0\}, \quad t' = t - \tau$$

$$P_M = \frac{Pr\{d\tau_{(1)} < t', t' < 0\}}{Pr\{t' < 0\}}$$

$$P_M = \frac{F_{d\tau_{(1)}}(t')}{Pr\{t' < 0\}}, \quad t' < 0 \tag{4.41}$$

where $d\tau_{(1)} = d_{(1)} - \tau$. Carrying out similar computations, P_F is given as:

$$P_F = 1 - \frac{F_{d\tau_{(1)}}(t')}{Pr\{t' < 0\}}, \quad t' \geq 0 \tag{4.42}$$

It can easily be seen that $P_e = Pr\{t' < 0\}$, $P_n = Pr\{t' \geq 0\}$ and consequently:

$$P_E = F_{d\tau_{(1)}}(t')u(-t') + Pr\{t' \geq 0\} - F_{d\tau_{(1)}}(t')u(t') \tag{4.43}$$

where $u(x)$ is the unit step and $F_{d\tau_{(1)}}(t') = F_{d\tau_{(1)}}(t)$ since $F_{d\tau_{(1)}}(t') = Pr\{d\tau_{(1)} \leq t'\} = Pr\{d_{(1)} \leq t\}$ with

$$F_{d_{(1)}}(x) = 1 - [1 - F_d(x)]^p \tag{4.44}$$

obtained from F_d, the distribution function of d, and p the number of samples distances.

In order to complete the calculations, the distribution function should be determined. Assuming Euclidean distances, d^2 follows a non central chi-square distribution with m degrees of freedom and non centrality parameter $s = (\mu_y - \mu_x)^T(\mu_y - \mu_x)$ [17]. The cumulative distribution function of the non-central chi-square distribution when $z = \frac{m}{2}$ is an integer can be expressed in terms of the generalized Q function as $F_{d^2}(y) = 1 - Q_z(s, \sqrt{y})$ [41]. Since the Euclidean distances are non-negative, F_d can be obtained by a simple change in variables:

$$F_d(y) = Pr\{d \leq y\} = Pr\{d^2 \leq y^2\} = F_{d^2}(y^2) = 1 - Q_z(s, y) \tag{4.45}$$

From (4.45), (4.44) and (4.43) can be computed provided that $Pr\{t' \geq 0\}$ is known. For the model assumed in this analysis, $t' = t - \tau$ where t is the detector's threshold and $\tau = ||\mu_y - \mu_x||$. Given that t is a deterministic quantity and τ is a constant, t' is also a deterministic quantity and $Pr\{t' \geq 0\}$ is unit or zero for $t' \geq 0$ or $t' \leq 0$, respectively.

An alternative design of the GVDED operators utilizes the adaptive nearest-neighbor filter [38], [39]. The coefficients are chosen to adapt to local image characteristics. Instead of constants, the coefficients are determined by an adaptive weight function for each window W. The operator is defined as the distance between the outlier and the weighted sum of all the ranked vectors:

$$NNVR = D(\mathbf{x}_{(n)}, \sum_{i=1}^{n} w_i \mathbf{x}_{(i)}) \tag{4.46}$$

The weight function w_i is determined adaptively using transformations of a distance criterion at each image location and it is not uniquely defined. There are two constraints on the weight function:

- Each weight coefficient is positive, $w_i \geq 0$
- The weight function is normalized, $\sum_{i=1}^{n} w_i = 1$

Since the operator should also attenuate noise, it is important to assign a small weight to the pixels with high ranks (i.e., outliers). A possible weight function can be defined as follows:

$$w_i = \frac{d_{(n)} - d_{(i)}}{n \cdot d_{(n)} - \sum_{j=1}^{n} d_{(j)}} \tag{4.47}$$

where $d_{(i)}$ is the aggregated distance associated with vector $\mathbf{x}_i$ inside the processing window W.

One special case for this weight function occurs in highly uniform area where all pixels have the same distance. The above weight function cannot be used since the denominator is zero. Since no edge exists in this area, the difference measure NNVR is set to zero.

The MVD operator can also be incorporated with the NNVR operator to further improve its performance in the presence of impulse noise as follows:

$$NNMVD = min_j\{D(\mathbf{x}_{(n-j+1)} - \sum_{i=1}^{n} w_i \mathbf{x}_{(i)})\}, \tag{4.48}$$

$$j = 1, 2, \ldots, k, k < n$$

A final annotation on the class of vector order statistic operators concerns the distance measure $D(\mathbf{x}_i, \mathbf{x}_j)$. By convention, the Euclidean distance measure (L_2 norm) is adopted. The use of L_1 norm is also considered because it reduces the computational complexity by computing the absolute values instead of squares and square root, without any notable deviation in performance. A few other distance measures are also considered in the attempt to locate an optimal measure, namely: the Canberra Metric implementation; the Czekanowski coefficient; and the angular distance measure. Their performances will be addressed later. The Canberra Metric implementation [36] is defined as:

$$D(\mathbf{x}_i, \mathbf{x}_j) = \sum_{k=1}^{m} \frac{|\mathbf{x}_{i,k} - \mathbf{x}_{j,k}|}{\mathbf{x}_{i,k} + \mathbf{x}_{j,k}} \tag{4.49}$$

The Czekanowski coefficient [36] is defined as:

$$D(\mathbf{x}_i, \mathbf{x}_j) = 1 - \frac{2\sum_{k=1}^{m} min(\mathbf{x}_{i,k} + \mathbf{x}_{j,k})}{\sum_{k=1}^{m} (\mathbf{x}_{i,k} + \mathbf{x}_{j,k})} \tag{4.50}$$

where the m is the number of vector components, in the case of a color image, $m = 3$, corresponding to the three channels (R, G, B). In addition, the angular distance measure of [16] can be used:

$$D(\mathbf{x}_i, \mathbf{x}_j) = \arccos(\frac{\mathbf{x}_i \cdot \mathbf{x}_j}{||\mathbf{x}_i|| ||\mathbf{x}_j||}) \tag{4.51}$$

where $|| \cdot ||$ denotes the magnitude of the color vector $\mathbf{x}_i$.

Based on these three distance measures, a variety of color edge detectors can be established.

4.4 Difference Vector Operators

The class of difference vector operators (DV) can be viewed as first derivative-like operators. This group of operators is extremely effective from the point of view of the computational aspects of the human visual system. In this approach, each pixel represents a vector in the RGB color space, and a gradient is obtained in each of the four possible directions $(0^o, 45^o, 90^o, 135^o)$ by applying convolution kernels to the pixel window. Then, threshold can be applied to the maximum gradient vector to locate edges. The gradients are defined as:

$$|\nabla f|_{0^\circ} = ||\mathbf{y}_{0^\circ} - \mathbf{x}_{0^\circ}|| \tag{4.52}$$

$$|\nabla f|_{90^\circ} = ||\mathbf{y}_{90^\circ} - \mathbf{x}_{90^\circ}|| \tag{4.53}$$

$$|\nabla f|_{45^\circ} = ||\mathbf{y}_{45^\circ} - \mathbf{x}_{45^\circ}|| \tag{4.54}$$

$$|\nabla f|_{135^\circ} = ||\mathbf{y}_{135^\circ} - \mathbf{x}_{135^\circ}|| \tag{4.55}$$

$$DV = max\{|\nabla f|_{0^\circ}, |\nabla f|_{90^\circ}, |\nabla f|_{45^\circ}, |\nabla f|_{135^\circ}\} \tag{4.56}$$

where $|| \cdot ||$ denotes L_2 norm, and $\mathbf{x}$ and $\mathbf{y}$ are three dimensional vectors used as convolution kernels. The variation in the definitions of these convolution kernels give rise to a number of operators. Fig. 4.2 shows the partition of the pixel window into two sub-windows within which each convolution kernel is calculated in all four directions.

The basic operator of this group employs a (3×3) window involving a center pixel and eight neighboring pixels. Let $\mathbf{v}(x, y)$ denote a pixel. The convolution kernels for the center pixel $\mathbf{v}(x_0, y_0)$ in all four directions are defined as:

$$\boldsymbol{X}_{0^\circ} = \mathbf{v}(x_{-1}, y_0), \ \boldsymbol{Y}_{0^\circ} = \mathbf{v}(x_1, y_0) \tag{4.57}$$

$$\boldsymbol{X}_{45^\circ} = \mathbf{v}(x_{-1}, y_1), \ \boldsymbol{Y}_{45^\circ} = \mathbf{v}(x_1, y_{-1}) \tag{4.58}$$

$$\boldsymbol{X}_{90^\circ} = \mathbf{v}(x_0, y_{-1}), \ \boldsymbol{Y}_{90^\circ} = \mathbf{v}(x_0, y_1) \tag{4.59}$$

$$\boldsymbol{X}_{135^\circ} = \mathbf{v}(x_1, y_1), \ \boldsymbol{Y}_{135^\circ} = \mathbf{v}(x_{-1}, y_{-1}) \tag{4.60}$$

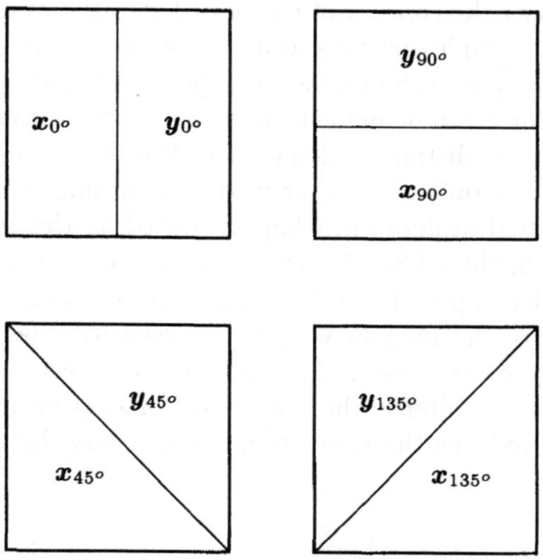

Fig. 4.2. Sub-window Configurations

This operator requires the least amount of computation among the edge detectors considered so far. However, as with the VR operator, DV is also sensitive to impulsive and Gaussian noise [25].

As a result, more complex operators with sub-filtering are designed. A larger window size is needed in this case to allow more data for processing. Although there is no upper limit on the size of the window, usually a (5x5) window is preferred since the computational complexity is directly linked to the size of the window. In addition, when the window becomes too large it can no longer represent the characteristics of the local region. For a $(n \times n)$ window $(n = 2k+1, k = 2, 3, \ldots)$, the number of pixels in each of the sub-windows illustrated in Fig. 4.2 is $N = \frac{n^2-1}{2}$. A filter function can be applied to these N pixels in each sub-window to obtain the respective convolution kernels:

$$\boldsymbol{X}_{d^o} = f(\mathbf{v}^{sub_1}_{d^o,1}, \mathbf{v}^{sub_1}_{d^o,2}, \ldots, \mathbf{v}^{sub_1}_{d^o,N}) \tag{4.61}$$

$$\mathbf{Y}_{d^o} = f(\mathbf{v}^{sub_2}_{d^o,1}, \mathbf{v}^{sub_2}_{d^o,2}, \ldots, \mathbf{v}^{sub_2}_{d^o,N}) \tag{4.62}$$

where $d = 0, 45, 90, 135$.

Depending on the type of noise one wishes to attenuate, different filters can be utilized. Four types of non-linear image filters based on order statistics are employed in our work.

The first type of color edge detector incorporates the vector median filter [37]. This filter outputs the vector median of the N vector samples in the sub-window by using the concept of vector order statistics introduced above, where the N vector samples are arranged in ascending order using R-ordering, $\mathbf{v}_{(1)} \leq \mathbf{v}_{(2)} \leq \ldots \mathbf{v}_{(N)}$, and the vector with the lowest rank, $\mathbf{v}_{(1)}$, is the vector median. This operator can be made more efficient by locating the vector with the minimum reduced distance calculated in R-ordering instead of ordering all (N) samples since only the vector median is of importance here. Vector median was discussed in detail in Chap. 3. Based on the analysis presented there, the vector median filter is very effective for reducing impulse noise because it can reject up to $(N-1)$ impulse noise pixels in a sub-window. However, since only the median vectors are used for edge detection, some edges may be rejected as noise and not able to be detected.

The second type of filter is the arithmetic (linear) mean filter (hereafter $f_{\overline{VM}}$). This filter reduces the effect of Gaussian noise by averaging all the vector samples:

$$f_{\overline{VM}}(\mathbf{v}_1, \mathbf{v}_2, \ldots, \mathbf{v}_N) = \frac{1}{N} \sum_{i=1}^{N} \mathbf{v}_i \tag{4.63}$$

Due to the simplicity of the averaging operation, the vector mean operator is much more efficient than the vector median operator. The vector mean operator may cause certain false edges since the pixels used for edge detection are no longer the original pixels.

The third type of filter, α-trimmed mean filter, is a compromise between the above two filters. It is defined as:

$$f_{\alpha-trim}(\mathbf{v}_1, \mathbf{v}_2, \ldots, \mathbf{v}_N) = \frac{1}{N(1-2\alpha)} \sum_{i=1}^{N(1-2\alpha)} \mathbf{v}^{(i)} \tag{4.64}$$

where α is in the range [0, 0.5). When α is 0, no vector is rejected and the filter reduces to a vector mean filter. When α is 0.5, all vectors except vector median are rejected and the filter reduces to a vector median filter. For other α values, this operator can reject $200\alpha\%$ of impulse noise pixels and it outputs the average of the remaining vector samples. Therefore the α-trimmed mean filter can improve noise performance in the presence of both Gaussian and impulse noise.

The last type of filter to be addressed is the adaptive nearest-neighbor filter [38]. The output of this filter is a weighted vector sum with a weight function that varies adaptively for each sub-window:

$$f_{adap}(\mathbf{v}_1, \mathbf{v}_2, \ldots, \mathbf{v}_N) = \sum_{i=1}^{N} w_i \mathbf{v}_i \tag{4.65}$$

where the weight function w_i was given in (4.47), and it assigns a higher weight to the vectors with lower ranks and a lower weight to the outliers. This filter is also effective with mixed Gaussian and impulse noise and it bears approximately the same complexity as the α-trimmed mean filter since they both need to perform the R-ordering. Again since edge detection is performed on the outputs of the filter instead of the original pixels, there may be a reduction in resulting edge qualities.

Another group of operators denotes a similar concept as the sub-filtering operators where pre-filtering is used instead. Any one of the above filters can be used to perform pre-filtering on an image with a (3×3) window, and then the DV operator with the same window size is used for edge detection. Unlike the previous group, in this family the pixel window is not divided into sub-windows during filtering, and the filter is applied only once to the whole window. The advantage with this group of operators is that it is considerably more efficient than the previous group since the filtering operation, which accounts for most of the complexity, is performed only once instead of eight times (two for each of the four directions) for each pixel.

One last proposed variation for the difference vector operators considers edge detection in only two directions, horizontal and vertical, instead of four directions:

$$DV\_hv = max\{|\nabla f|_{0^\circ}, |\nabla f|_{90^\circ}\} \tag{4.66}$$

It is anticipated that such a design will be as powerful as the other DV operator due to the facts that:

- human vision is more sensitive to horizontal and vertical edges than to others
- the horizontal and vertical difference vectors are able to detect most of the diagonal edges as well, which in turn can reduce the thickness of these edges by eliminating the redundancy from the diagonal detectors. In addition, the amount of computation involved with this operator is slightly reduced.

4.5 Evaluation Procedures and Results

To investigate further the performance of the vector order statistic operators and the difference vector operators, it is necessary to determine how these two classes of operator compare to each other and how the individual edge detector in each class ranks among themselves. Both quantitative and qualitative measures are used to evaluate the performance of the edge detectors in terms of accuracy in edge detection and robustness to noise. The quantitative performance measures can be grouped into two types, probabilistic measures and distance measures. The first type is based on the statistic of correct edge detection and false edge rejection. The second type is based on edge deviation from true edges. The first type of measure can be adopted to evaluate the accuracy of edge detection by measuring the percentage of correctly and falsely detected edges. Since a pre-defined edge map (ground truth) is needed, synthetic images are used for this experiment. The second type of measure can be adopted to evaluate the noise performance by measuring the deviation of edges caused by noise from the true edges [42], [43]. Since numerical measures are not sufficient to model the complexity of human visual systems, qualitative evaluation using subjective tests is necessary in most image processing applications. Also, evaluation based on synthetic images has limited value because they cannot be extrapolated to real images easily [40]. As a result, real images are also used in the evaluation process. All the images used for evaluation are defined in the RGB color space.

A total of 24 edge detectors from the class of the vector order statistic operators and the difference vector operators are implemented and their performance are evaluated along with the Sobel edge detector (see Tables 4.1 and 4.2).

Table 4.1. Vector Order Statistic Operators

VR0	Vector Range operator (W: 3x3) (with L_1 norm)
VR1	Vector Range operator (W: 3x3) with Canberra Metric implementation
VR2	Vector Range operator (W: 3x3) with Czekanowski coefficient
VR3	Vector Range operator (W: 3x3) with angular distance measure
MVD_3	MVD operator (W: 3x3) with k=3, l=4
MVD_5a	MVD operator (W: 5x5) with k=3, l=4
MVD_5b	MVD operator (W: 5x5) with k=6, l=9
NNVR_3	NNVR operator (W: 3x3)
NNVR_5	NNVR operator (W: 5x5)
NNMVD_3	NNMVD operator (W: 3x3) with k=3
NNMVD_5a	NNMVD operator (W: 5x5) with k=3
NNMVD_5b	NNMVD operator (W: 5x5) with k=6

4.5.1 Probabilistic Evaluation

Several artificial images with pre-specified edges are created for accessing the probabilistic performance of selected edge detectors. In order to analyze the responses of the edge detectors to different types of edges, these images contain: vertical, horizontal, and diagonal edges; round and sharp edges; edges caused by variation in only one, only two or all three components; isoluminant and non-isoluminant areas. In this experiment, noise is not added to the images. The resulting edge maps from each detector are compared with the pre-defined edge maps, and the number of correct and false edges detected are computed and are represented as hit and fault ratio as shown in Table 4.2 [39]. The hit ratio is defined as the percentage of correctly detected edges and the fault ratio is the ratio between the number of false edges detected and

Table 4.2. Difference Vector Operators

DV	DV operator(W: 3x3) in four directions
DV_hv	DV operator (W: 3x3) (in only horizontal and vertical directions)
DVadap	DV operator (W: 5x5) with adaptive subfilter
DVadap_hv	same as DVadap except in only two directions
DVαtrim	DV operator (W: 5x5) (with αtrim subfilter)
DVmean	DV operator (W: 5x5) (with vector mean subfilter)
DVmedian	DV operator (W: 5x5) (with vector median subfilter)
fDVadap	DV operator (W: 3x3) (with adaptive pre-filter on entire window)
fDVadap_hv	same as fDVadap except in only two directions
fDV*alpha*trim	DV operator (W: 3x3) (with αtrim pre-filter on entire window)
fDVmean	DV operator(W: 3x3) (with vector mean pre-filter on entire window)
fDVmedian	DV operator(W: 3x3) (with vector median pre-filter on entire window)

the number of true edges in the pre-defined edge map. These two parameters are selected for this evaluation because they characterize the accuracy of an edge detector.

Table 4.3. Numerical Evaluation with Synthetic Images

Edge Detector	% Hit	Fault Ratio
Sobel	97.9%	1.21
VR0	99.4%	1.55
VR1	93.9%	1.49
VR2	92.9%	1.48
VR3	91.3%	1.46
MVD_3	88.7%	0.95
MVD_5a	99.2%	3.33
MVD_5b	98.3%	1.53
NNVR_3	99.4%	1.55
NNVR_5	99.6%	4.01
NNMVD_3	87.5%	0.95
NNMVD_5a	94.4%	3.3
NNMVD_5b	93.6%	1.51
DV	99.4%	1.55
DV_hv	99.1%	1.14
DVadap	4.6%	0.06
DVαtrim	60.5%	0.65
fDVadap	98.4%	2
fDVadap_hv	97.7%	1.58
fDVαtrim	98.4%	1.99

From the results in Table 4.3, a few conclusions can be drawn:

1. Detectors such as the Sobel operator, VR operator with L_1 norm, and DV operator without any filtering all give good performance for images free of noise contamination.
2. MVD with 3×3 window size has a lower hit ratio, but it also gives less false edges. The MVD operators with larger window size (e.g. 5×5) are able to provide high hit ratio.
3. The NNVR operators also show good performance but the NNMVD operators give a slightly lower hit ratio than those achieved by the MVD operators.
4. The L_1 norm used in VR operators shows superior performance than other distance measures.
5. For the difference vector operators, the detectors with only horizontal and vertical direction detection have almost the same hit ratio as the DV operator with all four directions, but they detect considerably less false edges.
6. The DV operator with adaptive and α-trimmed subfiltering show very poor results. It is worth pointing out that this is not the case with real images, as will be seen later. The sub-filtering seems to have undesirable effects on synthetic images. When pre-filtering is performed (fDVadap, fDVα-trim), this undesirable effect does not exist and these operators show good performances.

4.5.2 Noise Performance

Real images with corrupted mixed noise are used for this experiment. The mixed noise contain 4% impulsive noise and Gaussian noise with standard deviation ($\sigma = 30$). The edge maps of the images corrupted with noise are compared with the edge maps of the original image for each edge detector. The noise performance is measured in terms of the PSNR values, and the results are given in Table 4.4. The PSNR is an easily quantifiable measure of image quality, although it only provides a rough evaluation of the actual visual quality the eye may perceive in an edge map.

A few observation can be made from the results:

1. The simple operators such as Sobel, VR and DV are sensitive to both impulsive and Gaussian noise. The noise performance can be improved with added complexity.
2. In the case of vector order statistic operators, the MVD and NNMVD operators show more robustness in the presence of noise. It can also be confirmed that the noise performance improves with the increase complexity of the operators, which are controlled by the two parameters k and l.

Table 4.4. Noise Performance

Edge Detector	PSNR
Sobel	30.9 dB
VR0	24.4 dB
DV	29.4 dB
MVD_3	26.3 dB
MVD_5a	33.6 dB
MVD_5b	35.4 dB
NNVR_3	23.2 dB
NNVR_5	28.6 dB
NNMVD_3	25.9 dB
NNMVD_5a	33.5 dB
NNMVD_5b	35.2 dB
DVadap	52.4 dB
DVadap_hv	52.2 dB
DVα-trimmed	45.5 dB
fDVadap	62.6 dB
fDVadap_hv	62.3 dB
fDVα-trimmed	59.6 dB

3. For the class of difference vector operators, the added filtering improve the performance drastically. Since mixed noise are present, adaptive and α-trimmed filters are used for this experiment. The use of adaptive filters as pre-filters on the whole window demonstrates the best performance in noise suppression. Hence it can be concluded that the adaptive filter outperforms the α-trimmed filter and the pre-filtering method is better than the sub-filtering method in terms of noise suppression. Operators in only the horizontal and vertical directions show very slight deviation in PSNR values from the ones in all four directions.

4.5.3 Subjective Evaluation

Since subjective evaluation is very important in image processing, the forth-mentioned operators have been applied to a collection of real and artificial images ranging from face features to outdoor scenes. The subjective evaluation allows for further investigation of the characteristics of the obtained edge maps through the involvement of human factors. The operators are rated in terms of several criterion: (i) ease at organizing objects; (ii) continuity of edges; (iii) thinness of edges; (iv) performance in suppressing noise. The results obtained are in good agreement in all cases with the selected criterion [43].

After examining a large quantities of edge maps produced by each edge detector, the following conclusion can be drawn:

1. As suggested in the quantitative tests, the performance of Sobel, VR, and DV operators are very similar in that they all produce good edge maps for noiseless images.

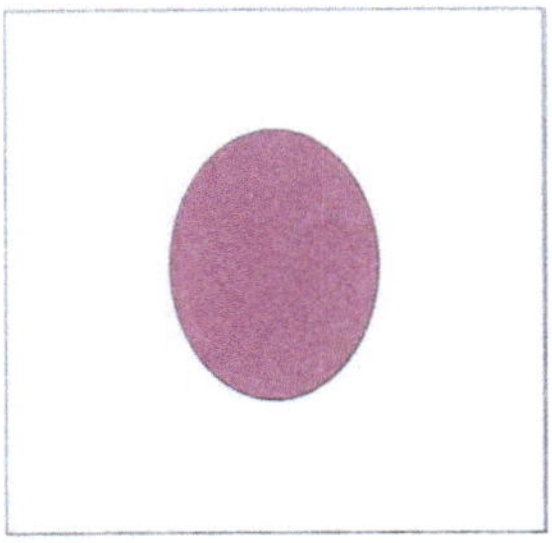

Fig. 4.3. Test color image 'ellipse'

Fig. 4.4. Test color image 'flower'

Fig. 4.5. Test color image 'Lenna'

2. The MVD and NNMVD operators produce thinner edges and are less sensitive to small texture variations because of the averaging operation which smooth out small variations. Also as expected, these two groups of operators are able to extract edges even in noise corrupted images.
3. The two groups of difference vector operators with sub-filtering and pre-filtering all demonstrate excellent performance for noise corrupted images. The vector mean operator performs best in impulsive noise, vector median operator performs best in Gaussian noise, and adaptive and α-trimmed operators perform best in mixed noise. The sub-filtering operator with adaptive filter is able to produce fair edge maps for real images despite its unsuccessful attempts with synthetic images during the numerical evaluation. However, the visual assessments are in agreement with the numerical tests in that the group of pre-filtering operators outperform the group of sub-filtering operators of the same filter.
4. One last note on the difference vector operators is that the operators with only horizontal and vertical directions produce thinner diagonal edges than those in all four directions.

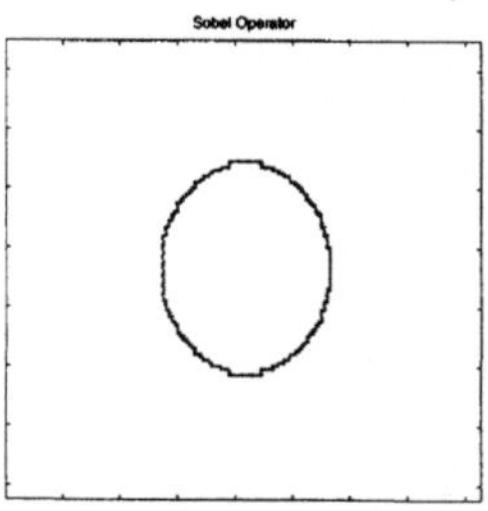

Fig. 4.6. Edge map of 'ellipse': Sobel detector

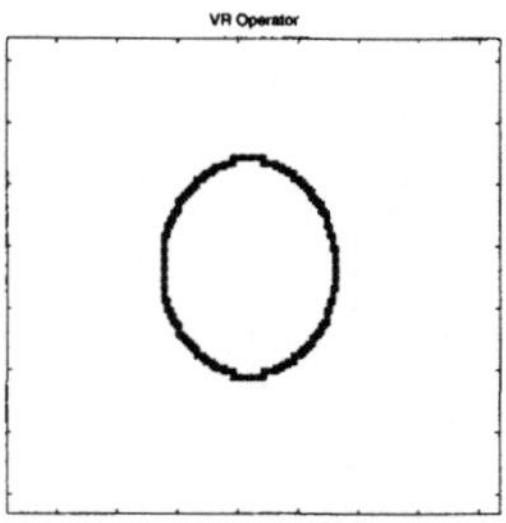

Fig. 4.7. Edge map of 'ellipse': VR detector

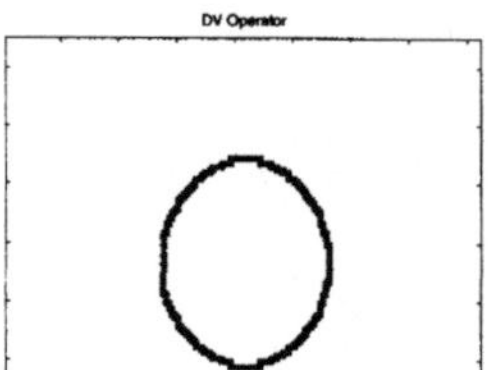

Fig. 4.8. Edge map of 'ellipse': DV detector

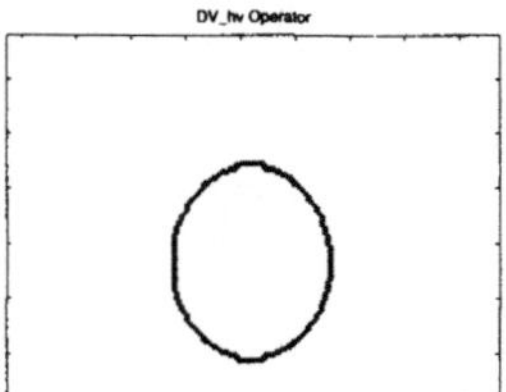

Fig. 4.9. Edge map of 'ellipse': DV_hv detector

The color images 'ellipse', 'flower' and 'Lenna' used in the experiments are shown in Fig. 4.3-4.5. The last image is corrupted by 4% of impulse noise and 30% of Gaussian noise. Edge maps of the synthetic image 'ellipse' is shown in Fig. 4.6-4.9. The figures in Fig. 4.10-4.17 provides the edge maps produced by four selected operators for the test images 'flowers' and 'Lenna'.

4.6 Conclusion

Accurate detection of the edges is of primary importance for the later steps in image analysis, such as segmentation and object recognition. Many effective methods for color edge detection have been proposed for the past few years and a comparative study of some of the representative edge detectors has been presented in this chapter. Two classes of operators, vector order statistic operators and vector difference operators have been studied in detail because both of them are effective with multivariate data and are computationally efficient. Several variations have been introduced to these two

Fig. 4.10. Edge map of 'flower': Sobel detector

Fig. 4.11. Edge map of 'flower': VR detector

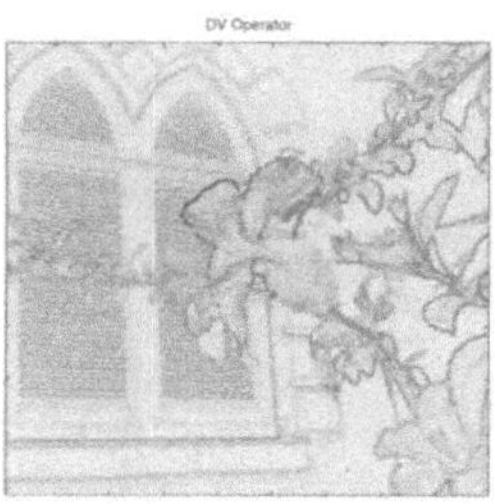

Fig. 4.12. Edge map of 'flower': DV detector

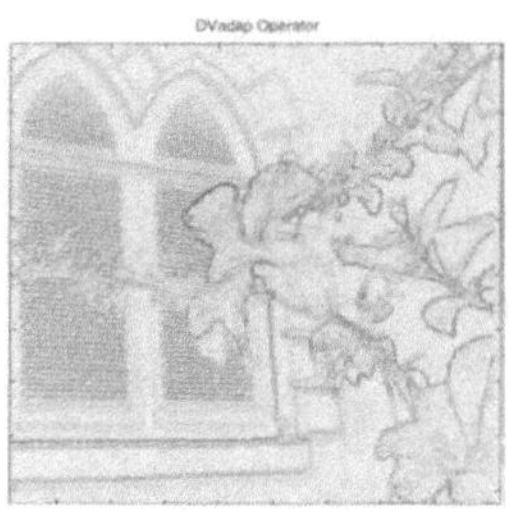

Fig. 4.13. Edge map of 'flower': DVadap detector

classes of operators for the purpose of better noise suppression and higher efficiencies. It has been discovered that both classes offer a mean of improving noise performance at the cost of increasing complexity. The performance of all edge detectors has been evaluated both numerically and subjectively. The results presented demonstrate a superiority of the difference vector operator with adaptive pre-filtering over other detectors. This operator scores high points in numerical tests and the edge maps it produces are perceived favorably by human eyes. It should be noted that different applications have different requirements on the edge detectors, and though some of the general characteristics of various edge detectors have been addressed, it is still better to select edge detectors that are optimum for the particular application.

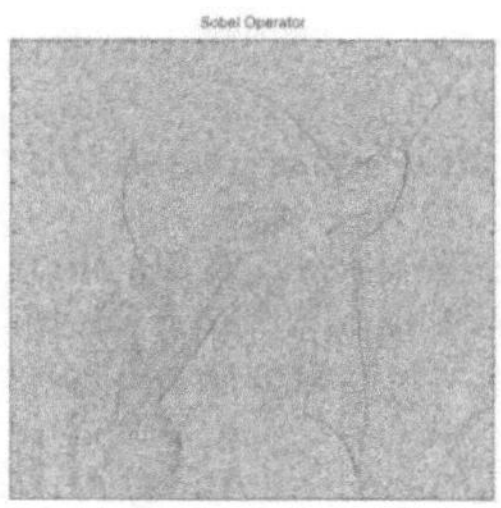

Fig. 4.14. Edge map of 'Lenna': Sobel detector

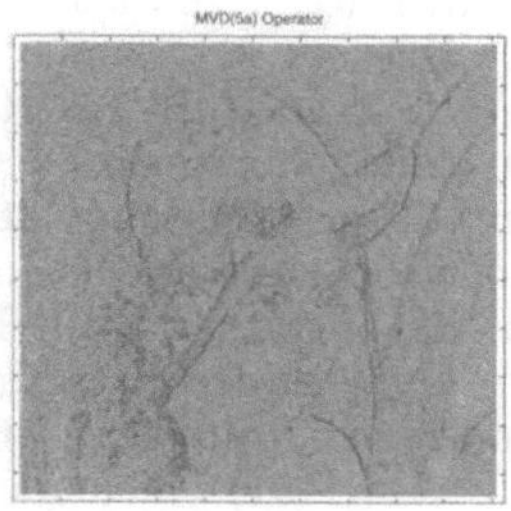

Fig. 4.15. Edge map of 'Lenna': VR detector

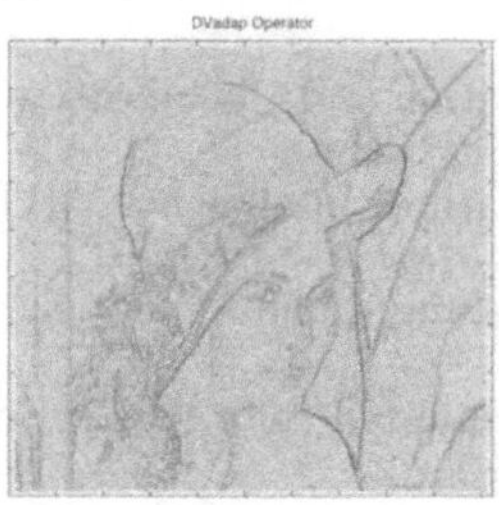

Fig. 4.16. Edge map of 'Lenna': DV detector

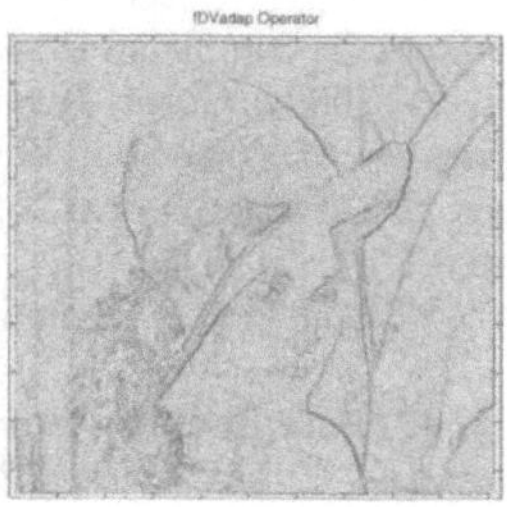

Fig. 4.17. Edge map of 'Lenna': DVadap detector

References

1. Treisman, A., Gelade, G. (1980), A feature integration theory of attention, Cogn. Psych. , **12**, 97–136.
2. Treisman, A. (1986): Features and objects in visual processing, Scientific America, **25**, 114B–125.
3. A. Koschan, A. (1995): A comparative study on color edge detection, Proc. 2nd Asian Conf. on Computer Vision ACCV'95, **III**, 574–578.
4. Gonzales, R.C., Wood, R. E. (1992): Digital Image Processing. Addison-Wesley, Massachusetts.
5. Pratt, W.K. (1991): Digital Image Processing. Wiley, New York, N.Y.
6. Androutsos, P., Androutsos, D., Plataniotis, K.N., Venetsanopoulos, A.N. (1997): Color edge detectors: an overview. Proceedings, Canadian Conference on Electrical and Computer Engineering, **2**, 827-831.
7. Cinque, L., Guerra, C., Levialdi, C. (1994): Reply: On the Paper by R.M. Haralick, CVGIP: Image Understanding, **60**(2), 250–252.
8. Heath, M., Sarkar, S., Sanocki, T., Bowyer, K. (1998): Comparison of Edge Detectors, Computer Vision and Image Understanding, **69**(1), 38–54.

9. Heath, M., Sarkar, S., Sanocki, T., Bowyer, K. (1997): A robust visual method for assessing the relative performance of edge-detection algorithms, IEEE Trans. Pattern Analysis and Machine Intelligence, **19**(12), 1338–1359.
10. Sobel, L.E. (1970): Camera Models and Machine Perception, Ph. D dissertation, Standford University, California.
11. D. Marr, D., Hildreth, E. (1980): Theory of Edge Detection, Proceedings of the Royal Society of London, **B-207**, 187–217.
12. Zenzo, S.D. (1986): A note on the Gradient of a multi-image, Computer Vision Graphics and Image Processing, **36**, 1–9.
13. Scharcanski, J., Venetsanopoulos, A.N. (1997): Edge detection of colour images using directional operators, IEEE Transactions on Circuits and Systems, **xx**, –.
14. Shiozaki, A. (1986): Edge extraction using entropy operator, Computer Vision Graphics and Image Processing, **36**, 116–126.
15. A. Cumani, A. (1991): Edge detection in multispectral images," CVGIP: Graphical Models and Image Processing, **53**, 40–51.
16. Tranhanias, P.E., Venetsanopoulos, A.N. (1993): Color edge detection using vector order statistics, IEEE Transaction on Image Processing, **2**(2), 259–264..
17. Tranhanias, P.E., Venetsanopoulos, A.N. (1996): Vector order statistics operators as color edge detectors, IEEE Transaction on Systems Man and Cybernetics-Part B, **26**(1), 135–143.
18. Scharcanski, J. (1993): Color Texture Representation and Classification, Ph.D. Thesis, University of Waterloo, Waterloo, Ontario, Canada.
19. S. Grossberg, S. (1988): Neural Networks and Natural Intelligence, MIT Press, Massachussets.
20. W.K. Pratt, W.K. (1991) Digital Image Processing, Jone Wiley, N.Y., New York.
21. Healey, G. (1992): Segmenting images using normalized color, IEEE Trans. on Systems, man and Cybernetics, **22**(1), 64–73.
22. Poggio, T., Torre, V., Koch, C. (1995): Computational vision and regularization theory, Nature, **317**.
23. Witkin, A. (1983): Scale-space filtering, Proceedings of the 8th Int. Joint Conf. on Artificial Intelligence, **2**, 1019–1022.
24. P. Beaudet, *Rotationally Invariant Image Operators* in Int. Joint Conf. on Pattern Recog., pp. 579 - 583, 1987.
25. Y. Yang, Y. (1992): Color edge detection and segmentation using vector analysis, M.A.Sc. Thesis, University of Toronto, Toronto, Ontario, Canada.
26. Rosenfeld, A., Kak, A.C. (1982): Digital Picture Processing, Second Edition, Academic Press, N.Y., New York.
27. Nevatia, R. (1977): A color edge detector and its use in scene segmentation, IEEE Trans. on Systems, Man cand Cybernetics, **7**(11), 820–825.
28. Robinson, G.S. (1977): Color edge detection, Optical Engineering, **16**(5), 479–484.
29. R. Machuca, R., Phillips, K. (1983): Applications of vector fields to image processing, IEEE Trans. on Pattern Analalysis and Machine Intelligence, **5**(3), 316–329.
30. Alshatti, W., Lambert, P. (1993): Using eigenvectors of a vector field for deriving a second directional derivative operator for color images", Proceedings of the 5th International Conference, **CAIP'93**, 149–156.
31. David, H.A. (1980): Order Statistics, Wiley, N.Y., New York.
32. Barnett, V. (1976): The ordering of multivariate data, J. Royal Statist. Soc. A, **139**(3), 318–343.
33. Feechs, R.J., Arce, G.R. (1987): Multidimensional morphologic edge detection, Proceedings SPIE Conf. Visual Comm. and Image Processing, **845**, 285–292.

34. Lee, J.S.J., Haralick, R.M., Shapiro, L.G. (1987): Morphologic edge detection, IEEE Journal of Robotic Automation, **RA-3**(2), 142–156.
35. Pitas, I., Venetsanopoulos, A.N. (1990): Nonlinear Digital Filters: Principles and Applications, Kluwer Academic Publishers.
36. K. Krzanowski, K., (1994): Multivariate Analysis I: Distributions, ordination and inference, Halsted Press, N.Y., New York.
37. Astola, J., Haavisto, P., Neuvo, Y. (1990): Vector median filters, Proceedings of the IEEE, **78**(4), 678–689.
38. Plataniotis, K.N., Androutsos, D., Venetsanopoulos, A.N. (1997): Color image filters: The vector directional appoach. Optical Engineering, **36**(9), 2375–2383.
39. Zhu, Shu-Yu, Plataniotis, K.N., Venetsanopoulos, A.N. (1999): A comprehensive analysis of edge detection in color image processing. Optical Engineering, **38**(4), 612-625.
40. Zhou, Y.T., Venkateshwar, T., Chellappa, R. (1989): Edge detection and linear feature extraction using a 2D random field model, IEEE Trans. on Pattern Analysis and Machine Intelligence, **11**(1), 84–95.
41. Proakis, J. G. (1984): Digital Communications. McGraw Hill, New York, N.Y.
42. Androutsos, P., Androutsos, D., Plataniotis, K.N., Venetsanopoulos, A.N. (1997): Subjective analysis of edge detectors in color image processing. Image Analysis and Processing, Lecture Notes in Computer Science, **1310**, 119-126, Springer, Berlin, Germany.
43. Androutsos, P., Androutsos, D., Plataniotis, K.N., Venetsanopoulos, A.N. (1998): Color edge detectors: a subjective analysis. Proceedings Nonlinear Image Processing IX, **3304**, 260-267.

5. Color Image Enhancement and Restoration

5.1 Introduction

Enhancement techniques can be used to process an image so that the final result is more suitable than the original image for a specific application. Most of the image enhancement techniques are problem oriented. Image enhancement techniques fall into two broad categories: spatial domain techniques and frequency domain methodologies. The spatial domain refers to the image itself, and spatial domain approaches are based on the direct manipulation of pixels in the image. On the other hand, frequency domain techniques are based on modifying the Fourier transform of the image. Only spatial domain techniques are discussed in this chapter.

The histogram of a monochrome image presents the relative frequency of occurance of the various levels of the image. Histogram equalization has been proposed as an efficient technique for the enhancement of monochrome images. This technique modifies an image so that the histogram of the resulting image is uniform. Variations of the technique known as histogram modification and histogram specification, which result in a histogram having a desired shape, have also been proposed.

The extension of this histogram equalization to color images is not trivial due to the multidimensional nature of the histogram in color images. Pixels in a monochrome image are defined only by gray-level values, so monochrome histogram equalization is a scalar process. On the contrary, pixels in color images are defined by three primary values, which implies that color equalization is a vector process. If histogram equalization is applied to each of the primary colors independently, changes in the relative percentage of primaries for each pixel may occur. This can lead to color artifacts. For this reason, various methods have been proposed for color histogram equalization which spread the histogram either along the principal component axes of the original histogram or spread repeatedly the three two dimensional histograms. Other enhancement methods have been proposed recently which operate mainly on the luminance component of the original image.

This chapter focuses on the direct three-dimensional histogram equalization approaches. The first method presented is actually a histogram specification method. The specified histogram is a uniform 3-D histogram and consequently, histogram equalization is achieved. The theoretical background of the method is first presented and then issues concerning the computational implementation are discussed in detail. Finally, experimental results from the application of this method are presented. A method of enhancing color images by applying histogram equalization to the hue and saturation component in the HSI color space is also presented.

5.2 Histogram Equalization

In histogram equalization the objective is to obtain a uniform histogram for the output image. The theoretical foundation, underlying histogram equalization, can be found in probability theory. If an absolute continuous random variable (monochrome) X, $a < X < b$, with cumulative probability density function $F_X(x) = Pr(X \leq x)$ is considered, then the random variable $Y = F_X(x)$ will be uniformly distributed over $(0, 1)$. In the discrete case, the assumption of continuity of the variable X is not satisfied and therefore Y will be uniformly distributed only approximately. However, despite the approximate uniform distribution of Y, histogram equalization effectively spreads the monochrome values resulting in a powerful techniques in image enhancement.

For a three dimensional RGB color space you can proceed in an analogous manner. Consider three continuous variables R, G and B with a joint probability density function $f_{R,G,B}(r, g, b)$ and joint probability distribution function $F_{R,G,B} = Pr(R \leq r, G \leq g, B \leq b)$. As in the scalar case three new variables R_s, G_s and B_s are defined as:

$$
\begin{aligned}
R_s &= F_R(R) \\
G_s &= F_G(G) \\
B_s &= F_B(B)
\end{aligned}
\tag{5.1}
$$

The joint probability distribution function of R_s, G_s and B_s is given as:

$$
\begin{aligned}
F_{R_s,G_s,B_s}(r_s, g_s, b_s) &= Pr(R_s \leq r_s, G_s \leq g_s, B_s \leq b_s) \\
&= Pr(F_R(R) \leq r_s, F_G(G) \leq g_s, F_B(B) \leq b_s) \\
&= Pr(R \leq F_R^{-1}(r_s), G \leq F_G^{-1}(g_s), B \leq F_B^{-1}(b_s)) \\
&= F_{R,G,B}(F_R^{-1}(r_s), F_G^{-1}(g_s), F_B^{-1}(b_s))
\end{aligned}
\tag{5.2}
$$

If independence of the R, G and B components is assumed the last equation can be further decomposed as a product of the probability distribution functions of the three primaries:

$$
\begin{aligned}
F_{R_s,G_s,B_s}(r_s, g_s, b_s) &= F_R(F_R^{-1}(r_s)) F_G(F_G^{-1}(g_s)) F_B(F_B^{-1}(b_s)) \\
&= r_s g_s b_s
\end{aligned}
\tag{5.3}
$$

From (5.3):

$$f_{r_s,g_s,b_s} == F_{R_s,G_s,B_s}(r_s, g_s, b_s) = 1 \tag{5.4}$$

From the above result it is concluded that the uniform distribution of the histograph in the R_s, G_s, B_s space is only guaranteed in the case of independent R_s, G_s, and B_s components. However, it is known that the three primaries at the RGB color space are correlated and thus, this assumption is not valid. Many methods have been proposed to overcome this difficulty. Most of them spread the histogram along the principal component axes of the original histogram or spread repeatedly the three two dimensional histograms.

However, since the aim is a uniform 3-D histogram, the problem can be viewed as a histogram specification one. In other words the 3-D uniform histogram is specified as the output histogram and therefore, histogram equalization is achieved.

In the scalar case, the method of histogram specification works as follows. Assume that X and Y are the input and output variables that take the values x_{i_x} and y_{i_y}, $x_{i_x}, y_{i_y} = 0, 1, \cdots, L-1$, with L the number of discrete gray-levels, with probabilities $p_X(x_{i_x})$ and $p_Y(y_{i_y})$, respectively. Then the following auxiliary parameters can be defined:

$$C_{I_x} = \sum_{i_x=0}^{I_x} p_X(x_{i_x}), I_x = 0, \cdots, L-1 \tag{5.5}$$

$$\hat{C}_{I_y} = \sum_{i_y=0}^{I_y} p_Y(y_{i_y}), I_y = 0, \cdots, L-1 \tag{5.6}$$

For the 3-D case, the following method is proposed for uniform histogram specification.

Let $\mathbf{X}$ and $\mathbf{Y}$ be the input and output vector variables which assume as value triplets $(x_{r_x}, x_{g_x}, x_{b_x})$ and $(y_{r_y}, y_{g_y}, y_{b_y})$ with $r_x, g_x, b_x, r_y, g_y, b_y = 0, 1, \cdots, L-1$ and probabilities $p_{\mathbf{X}}(x_{r_x}, x_{g_x}, x_{b_x})$ and $p_{\mathbf{Y}}(y_{r_y}, y_{g_y}, y_{b_y})$

The probabilities $p_{\mathbf{X}}$ are computed from the original color image histogram. The probabilities $p_{\mathbf{Y}}$ are all set to $\frac{1}{L^3}$ since there are L^3 histogram entries and the same uniform distribution is wanted.

The 3-D equivalents of the variables defined in (5.3)-(5.5), $C_{R_xG_xB_x}$ and $\hat{C}_{R_yG_yB_y}$, are computed from $p_{\mathbf{X}}$ and $p_{\mathbf{Y}}$ as follows:

$$C_{R_xG_xB_x} = \sum_{r_x=0}^{R_x} \sum_{g_x=0}^{G_x} \sum_{b_x=0}^{B_x} p_{\mathbf{X}}(x_{r_x}, x_{g_x}, x_{b_x}) \tag{5.7}$$

$$\hat{C}_{R_yG_yB_y} = \sum_{r_y=0}^{R_y} \sum_{g_y=0}^{G_y} \sum_{b_y=0}^{B_y} p_{\mathbf{Y}}(y_{r_y}, y_{g_y}, y_{b_y})$$

$$= \sum_{r_y=0}^{R_y} \sum_{g_y=0}^{G_y} \sum_{b_y=0}^{B_y} \frac{1}{L^3} = \frac{(R_y+1)+(G_y+1)+(B_y+1)}{L^3} \tag{5.8}$$

It can easily be seen that $\hat{C}_{R_y G_y B_y}$ can be computed as a product instead of a triple summation. Following that, the smallest R_y, G_y, B_y for which the inequality

$$\hat{C}_{R_y G_y B_y} - C_{R_x G_x B_x} \geq 0 \tag{5.9}$$

is true can be found. Summarizing, the following three steps constitute the above described method for 3-D histogram specification:

1. Compute the original histogram.
2. Compute $C_{R_x G_x B_x}$ and $\hat{C}_{R_y G_y B_y}$ using (5.7) and (5.8), respectively.
3. For each value (R_x, G_x, B_x) find the smallest (R_y, G_y, B_y) such that (5.9) is satisfied. The set of values (R_y, G_y, B_y) is the output produced.

Computationally, step 1 is implemented in just one pass through the image data. Step 2 can be implemented recursively, reducing drastically the execution time and memory requirements. Dropping out for simplicity the case where either of R_x, G_x and B_x is zero, $C_{R_x G_x B_x}$ can be computed as:

$$\begin{aligned} C_{R_x G_x B_x} &= C_{R_x-1 G_x-1 B_x-1} + C_{R_x-1 G_x B_x} \\ &+ C_{R_x G_x-1 B_x} + C_{R_x G_x B_x-1} - C_{R_x-1 G_x-1 B_x} \\ &- C_{R_x-1 G_x B_x-1} - C_{R_x G_x-1 B_x-1} + \mathbf{px}(R_x, G_x, B_x) \end{aligned} \tag{5.10}$$

Step 3 presents an ambiguity since many solutions for the (R_y, G_y, B_y) exist and satisfy (5.9). This ambiguity is remedied as follows. The computed value of $C_{R_x G_x B_x}$ at (R_x, G_x, B_x) is initially compared to the product $P = \frac{1}{L^3}(R_x+1)+(G_x+1)+(B_x+1)$, the value of $\hat{C}_{R_y G_y B_y}$ at (R_x, G_x, B_x) since for a uniform histogram the value of this product should also be the value of $C_{R_x G_x B_x}$.

In case of equality the input value is not changed. If $C_{R_x G_x B_x}$ is greater (less) than P then the indexes R_x, G_x, and B_x are repeatedly increased (decreased), one at a time, until (5.9) is satisfied. The final values constitute the output produced. The merit of this is twofold: (a) histogram stretching is achieved simultaneously in all three directions, and (b) the computational requirements are reduced since only a few values are checked. Because this method processes all three dimensions at once and maintains the basic ratio between the three primaries, it does not produce the color artifacts related to the independent processing. The overall computational complexity of the algorithm is manifested by step 2 which computes the cumulative histogram $C_{R_x G_x B_x}$ for a total of L^3 entries resulting in a $O(L^3)$ complexity.

Histogram equalization and modification can be applied directly on RGB images. However, such an application causes significant color hue shifts that are usually unacceptable to the human eye. Thus, color image histogram modification that does not change the color hue should be utilized. Such a modification can be performed in coordinate systems where luminance, hue and saturation of color can be described. A practical approach to developing color image enhancement algorithms is to transform the RGB values into another color coordinate representation which can describe luminance, hue and saturation. In a such a system interaction between colors can cause change in all these parameters. For monochrome images, histogram equalization can be frequently utilized to increase high-frequency components of images and thus to enhance images with poor contrast. Experimentation with color images revealed that the high frequency components of the saturation value can be quite different from that of the luminance values.

Currently, the most common method of equalizing color images is to process only the luminance component. Since most high frequency components of a color image are concentrated in the luminance component, histogram equalization is applied to only the luminance component to enhance contrast of color images, in color spaces, such as YIQ, YC_BC_R or the hue, luminance and saturation space discussed in [4]. However, there is still correlation between the luminance value and the chrominance values in these color spaces. Therefore, histogram equalization of the luminace component also changes chromatic information resulting in color artifacts.

To alleviate the problem, an algorithm for saturation equalization was developed recently [5]. In this approach, histogram equalization is also applied to the saturation component obtained from the two chromatic channels. In all different approaches, after processing the new coordinates, the enhanced image coordinates are inverse transformed to the RGB components for display.

One such system on which the modification can be performed is the HSI color space. Modification of the I or S components does not change the hue. In other words, a color characterized yellow, remains yellow when the algorithm changes its intensity and/or saturation, although a different yellow variant is produced. This observation suggests the application of histogram equalization or modification only to the I or S components. The cone shape of the HSI color space suggests non-uniform densities for I and S if an overall uniform density is desired for the entire RGB cube. If this fact is not taken into account when image intensity is equalized, many color points are concentrated near the points $I = 0$ and $I = 1$. The limited color space provided near these points causes distortion of the color image. Using geometrical concepts it is possible to define the probability density functions that will fill the HSI color space uniformly as follows:

$$\begin{aligned} f_I(I) &= 12I^2 0 \leq I \leq 0.5 \\ &= 12(1-I)^2 0.5 \leq I \leq 1 \end{aligned} \tag{5.11}$$

$$f_S(S) = 6S - 6S^2 0 \leq I \leq 1 \tag{5.12}$$

$$\begin{aligned} f_{IS}(I,S) &= 6S S \leq 2I,\ I \in [0, 0.5] \\ &= 0 S > 0,\ I \in [0, 0.5] \\ &= 6S S \leq 2(1-I),\ I \in [0.5, 1] \\ &= 0 S > 2(1-I),\ I \in [0.5, 1] \end{aligned} \tag{5.13}$$

It can easily be shown that the marginal distributions of the joint pdf $f_{IS}(I,S)$ are $f_I(I)$ and $f_S(s)$, respectively. If $\mathbf{X} = (x_1, x_2, \cdots, x_m)$ needs to be transformed to $\mathbf{Y} = (y_1, y_2, \cdots, y_m)$ where $\mathbf{Y}$ must have a certain joint pdf $f_{\mathbf{Y}}(\mathbf{Y})$, first derive the transformations $\mathbf{Z} = T_1(\mathbf{X})$ and $\mathbf{S} = T_2(\mathbf{Y})$ that equalize the random vectors $\mathbf{X}$ and $\mathbf{Y}$ and then combine the two transforms in one by means of:

$$\mathbf{Y} = T_2{}^{-1}[T_1(\mathbf{X})] \tag{5.14}$$

Transformations of the form of (5.14) can be used for the modification of I, S either jointly or separately. Intensity modification of the form (5.14) and (5.11)-(5.13) have been proven by simulations to provide better results than the straightforward quantization of I. Modifications of saturation S must be done with care. Many natural scenes posses very low saturation values. Sometimes the preferred color for image reproduction is a more saturated version of the original color. Transformed color images by using (5.12) or (5.13), tend to have highly saturated colors. In certain cases, the result of this modification may be too saturated and the colors may not appear natural. In such a case, a conservative saturation equalization is preferable. The mathematical formulation of such an equalization is difficult, because the subjective perception of acceptable saturation may differ from one image to another. For example, highly saturated colors seem to be appropriate in pseudo-coloring because they can enhance the visual representation of an image.

To illustrate the receding discussion, Fig. 5.1 and 5.2 show an original image and the resulting image after the application of the histogram modification technique discussed in this section.

5.3 Color Image Restoration

In the last decade, restoration of multichannel data has become increasingly important in a wide variety of research areas mainly due to the development of powerful digital electronics and computers, and the wide-spread applicability of color images processing techniques.

These areas include the processing of color images, high definition TV and video, remote sensing, environmental studies, astronomy, industrial inspection and biomedical applications [8]-[12]. In video applications the need for

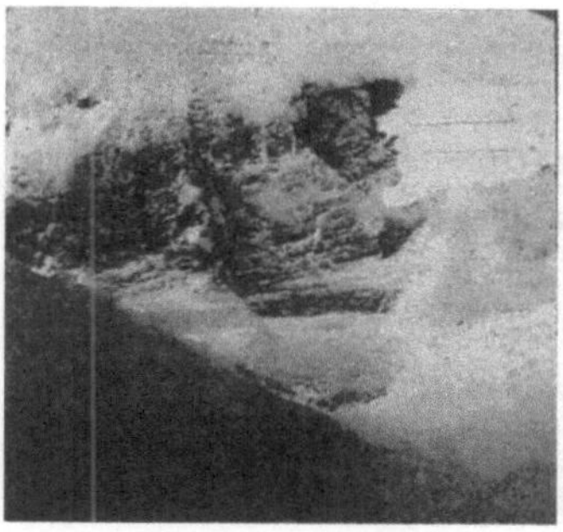

Fig. 5.1. The original color image 'mountain'

Fig. 5.2. The histogram equalized color output

high quality images urges for increased bandwidth. Due to the huge amount of information encoded in color, multichannel processing and compression become crucial in the effective transmission and storage of color images. In the field of industrial inspection multichannel processing is used to obtain quality products and to isolate damaged parts. In the field of robot guidance several video images are processed to acquire information about the environment and the position of the robot within this environment in order to guide the robot autonomously.

A modern field of image restoration applications concerns the decomposition of images into several subbands and/or resolution levels. Subband signal processing has evoked significant attention because of its potential to separate frequency bands and operate on them separately. Similarly, multiresolution signal processing employs wavelet transforms to decompose and represent an image at different levels of detail [13], [14]. The multiband and multiresolution signals can be considered as subclasses of multichannel images.

In these and other related applications the data set is collected from a variety of sensors that capture different properties of the signal. In high definition TV and video applications the data set is composed of high resolution color images obtained at different time instances. In multisensor robot guidance, various sensors receive information from different spatial locations. In multispectral satellite remote sensing signal information is distributed in different frequency bands that cover the visible and/or the invisible wavelength spectrum. Moreover, in satellite imaging the images are characterized by different levels of resolution. In the areas of environmental studies and astronomy the data is collected from different sources at different times instances and various frequency bands. In the area of biomedical applications many modalities exist, with possibly different bands in each modality. In multiresolution image processing that involves subband decomposition, wavelet and other orthogonal transforms, the image is characterized by several levels of resolution. In general, the processing of multiple image frames will be referred to as multichannel image processing.

In this chapter, the problem of image restoration that refers in general to the inversion of processes that degrade the acquired image will be considered. From the multiple encoding of information, it becomes evident that color images convey much more information than single channel images. In most applications, the composite images are highly correlated. However, the corresponding channels are affected differently by physical phenomena. Information lost in one channel, through atmospheric absorption, for instance, may be present in the other channels.

Restoring color or, in general, multichannel images becomes much more involved than the processing of single-channel images not only because of the increased dimensionality, but also due to the need for identification and exchange of information among all different channels. As an example, consider the case of multiframe coherent imagery, such as sequences of synthetic aperture radar (SAR) images. Optimal single-channel algorithms usually fail to restore the original image, especially in cases of low signal-to-noise ratio. Alternatively, temporal processing, or even a simple temporal averaging of multiple registered frames (with motion compensation), is efficient in suppressing the dominant speckle noise and in recovering the signal under consideration [15], [16].

When formulating algorithmic procedures for image restoration, there must be careful consideration of two factors: (i) the image formation process, and (ii) the stochastic models of the image and the noise process. It was argued in Chap. 2 that due to limitations in sensor technology, the observed image is often a degraded version of the original image. If the luminance of the observed scene is large, the sensor is saturated, operating in its nonlinear region. The captured image is also contaminated by noise that reflects either thermal oscillations in electronic devices or intensity quantization due to absorbency and other material limitations in regular film cameras. The noise corruption is often signal dependent and multiplicative in nature. Furthermore, a fast movement of an object or the existence of an out of focus object introduce blurring and noise in the observed image. Overall, degradation processes are encountered in the data, such as non-linear transformation, bandwidth reduction and noise contamination. The degradation process usually varies from pixel to pixel along the area of the observed image, and its effect is space-variant. This means that the restoration process that inverts the degradation should also be space variant. In addition, a realistic image model capturing the statistical properties of the observed scene is highly non-stationary. Image characteristics, such as sharp intensity and color transmissions, edges and texture differences imply varying statistical properties within the image area. Consequently, the stochastic model describing the image process should be non-stationary. These aspects are considered along with algorithms developed for color image restoration. An overview of these algorithms is provided along with conclusions and open problems for further research.

5.4 Restoration Algorithms

Image restoration is practically an ill-posed problem. Through linear degradations that cause bandwidth reduction, significant amount of visual information is lost in the acquisition process. Thus, linear degradation kernels possess small or even zero eigenvalues, whose inversion results in excessive noise amplification and renders the process of image restoration unstable. Most image restoration approaches can be interpreted as attempts to regularize this ill posed problem. Such approaches are based on specific assumptions and reflect their limitations as artifacts in the restored estimate. Due to the coupling of information among the channels, such artifacts are much more significant in multichannel images. Computational errors due to simplifying approximations tend to accumulate in multichannel images, since errors in one channel commute and affect the processing of all other channels. Thus, the requirements for accurate modeling are much more strict in the restoration of color images than in the processing of monochrome images.

Initiating research in multichannel images, several aspects of color image restoration were addressed in [17]. The Karhunen-Loeve (KL) transformation was introduced as a tool in de-correlating the color image set, so that each channel can be processed independently of the others [18], [19]. The de-correlation approach and the subsequent independent restoration of channels is valid only under restrictive assumptions in the imaging model [20].

Moreover the KL transform does not help in de-correlating the cross channel degradation effects or the channel correlated noise process. A more natural approach to color image processing calls for multidimensional image processing techniques. Initial attempts in this direction provide extremely promising results. The minimum-mean-square-error (MMSE) restoration scheme and the Wiener filtering approach were extended to the case of multichannel images [21]. The Wiener filter approach has been also used in the multiframe restoration of sequences of images [22]. Even though the limitations of the Wiener filter are well established, the results obtained from its multichannel extension are very encouraging. Relaxing the requirement for the stationarity assumption within each channel, which is essential in the efficient implementation of the the Wiener filter, a three-dimensional (3-D) Kalman filtering approach in multichannel image restoration was developed in [18]. A related autoregressive approach is proposed in [22] using an extended 2-D Kalman filter, whose coefficients capture the relationship among channels. The improvement, however, achieved through this technique over the stationary MMSE implementation is insignificant [18]. This performance is directly attributed to the limitations of the quadratic error function associated with the linear MMSE estimation approach. Typical regularized approaches, such as the constrained least-squares (CLS) and the Tikhonov-Miller formulations, attempt to compensate for the ill-posedness of the pseudo-inverse solution by utilizing smoothness information in the restoration process [23], [24].

In view of this inherent characteristic, the regularization approaches can be related to the stochastic maximum a-posteriori probability (MAP) estimation of the original image. The inversion of the degradation processes is achieved by estimating the original image from the data assuming specific forms of the prior and posterior distribution functions.

Motivated by the success of regularized approaches in monochrome images the CLS estimation scheme was applied in the restoration of multichannel images in [25].

Even though the CLS approach is a powerful technique that enables the incorporation of a priori information by means of constraints, it defines and operates on quadratic metrics. Suffering from the nature of its penalizing functions, it does not encourage the reconstruction of sharp edges and it often produces noise and ringing artifacts. The maximum likelihood approach to the multichannel inversion problem has been also considered. In particular, the iterative expectation maximization (EM) algorithm has been extended to the multichannel domain and has been used for blur identification and restoration [26]. Despite its increased computational complexity, this iterative technique has comparative advantages over other techniques in cases where the blur operator, and the signal and noise power spectra are unknown. When the degradation operator is known, however, it reduces to an iterative Wiener algorithm.

Several properties of the signal and the noise processes can be expressed through mathematical constraints defined on a set theoretic environment. Motivated by the success of projection algorithms in single-channel restoration, multichannel constraints were introduced in [27] leading to the extension of the projection-onto-convex-sets (POCS) approach to the color case. These multichannel constraints capture the available information regarding the original image in the general structure of prototype estimates. The POCS approach is closely related to the concept of regularization of ill-posed problems through set-theoretic considerations. In essence, the set-theoretic approach constrains the estimation space by means of convex ellipsoids, defined through quadratic constraints [25]. A general framework for the restoration of multichannel images in the frequency domain is presented in [28]. The regularization approaches that can be modeled in this form reflect deterministic constraints on the estimate, which aim to restrain the amplification of noise. However, such a universal deterministic constraint suppresses the high-frequency components of the estimate with no respect to the stochastic structure of the image under consideration. Thus, such approaches do not encourage the reconstruction of sharp edges and often produce noise and signal artifacts.

A structured regularized approach referred to as the constrained mean-square-error (CMSE) estimation scheme is developed in [15]. The smoothing functional in the CMSE restoration approach reflects the structure of the multichannel MMSE estimate, which is essentially utilized as a prototype constraint image. In contrast to the POCS algorithm, the MMSE estimate is not used to define hard constraints on the estimate space, but is rather employed as a means of influencing the structure of the solution. Thus, the CMSE approach always derives a meaningful estimate, which is conceptually located between the MMSE estimate and the pseudo-inverse solution. This approach enables the suppression of streak and ringing artifacts, but does not account for the representation of sharp edges. In order to preserve discontinuities (edges) the multichannel data are modeled as a Markov random field using a Gibbs prior [29]. This approach incorporates non-stationary intra-channel and inter-channel correlations allowing the preservation and reconstruction at sharp edges.

The previous approaches consider variations of quadratic objective functions that reflect Gaussian stochastic models for both the signal and the noise statistics. Quadratic functions are attractive, because they enable the analytic derivation of the corresponding estimators and provide cost-efficient implementation. The Gaussian model, however, does not cover most realistic noise sources, which are characterized by Poisson, Laplacian, binomial, or even signal dependent statistics. Furthermore, the Gaussian distribution cannot characterize the vast majority of images, whose histograms are not even unimodal. Following the channel interactive nature of color processing, it is expected that the limitations of linear algorithms are even more restrictive in the case of multichannel images. Most multichannel applications, including color images as well as satellite remote sensing, SAR and infrared time-varying imagery involve noise models whose distributions possess long tails, such as Cauchy or Laplacian. Moreover, transmission errors in digital channels, as well as abrupt natural phenomena, manifest themselves through the creation of impulses in the image, which can be interpreted as outliers in the actual or assumed noise distribution. Restoration algorithms must reflect the long-tailed characteristics of such noise processes. In addition, they must incorporate good models for the signal statistics that permit the representation of sharp edges. There is growing demand for robust multichannel algorithms that can effectively handle the large variation of signal and noise statistics encountered in practice [30], [31].

The concept of robust estimation has been extensively utilized in the case of monochrome images. Robust approaches have been developed in stochastic environments where the noise statistics are approximated by models that allow uncertainty in the form of noise outliers. Informally stated, an estimator is called robust if it is almost optimal at the nominal noise distribution, and it is only slightly affected by the presence of outliers in the noise distribution, which reflect the uncertainty in the noise model [31], [32]. For example, the class of M-estimators, is obtained from the generalized maximum likelihood approach. In this approach, the objective function deviates from the quadratic structure at large errors, as to restrain their contribution to the overall criterion. The optimal properties of the M-estimators are derived on the basis of the minimax formulation, which minimizes the worst (maximum) estimation error within a specific class of noise processes [31], [33].

The framework for the extension of robust approaches to the multichannel case by means of a generalized MAP formulation is developed in [34]. Nonlinear estimators derived through robust probabilistic measures have been employed in the restoration of images corrupted by mixed-noise processes, with impressive success [30]-[34]. Mixed noise processes involve both medium-tailed noise, such as Gaussian noise, and processes with long-tailed statistics, such as exponential and impulsive noise.

The non-quadratic objective functions can be also utilized in the representation of the prior signal statistics. Such functions have been motivated in maximum a posteriori (MAP) formulations under Markov random fields with non-quadratic Gibbs distributions [29]. These formulations have been proved useful in applications, such as emission tomography [35] and reconstruction of three-dimensional (3-D) surfaces from range information [36]. A novel view to the modeling of the detailed structure, in which the existence of sharp edges manifests uncertainty regarding the distribution of the signal was presented in [34]. In essence, sharp edges can be considered as outliers applied to an assumed medium-tailed (possibly Gaussian) signal distribution. Thus, to account for uncertainty in the overall restoration process, robust functionals are motivated in the representation of not only the noise, but also the signal statistics. In the next section the framework for the development of multichannel (color) restoration algorithms is provided.

5.5 Algorithm Formulation

5.5.1 Definitions

Consider the formation of p channels. For the kth channel, let f_k and g_k denote the original and the degraded image, n_k denote the noise vector, and H_{kk} represent the channel point-spread function (PSF), all in vector ordering. The image and the noise vectors are considered of dimensionality $\{N \times 1\}$ with $\{N = M^2\}$, where $\{M \times M\}$ is the dimensionality of the 2-D problem, and the PSF operator H_{kk} is an $\{N \times N\}$ matrix. Through the lexicographic notation, the original image vector is written as:

$$\mathbf{f} = [\mathbf{f}_1^t \dots \mathbf{f}_k^t]^{\mathbf{T}} \tag{5.15}$$

where T denotes the transpose operation. The vector notation results from the multichannel representation by arranging rows above columns within each channel, and then arranging channels on top of each other. In general, the degradation matrix for the k-th channel couples all the components of the original image. In the case of a color image ($p = 3$), the overall degradation matrix is written in block form as:

$$\mathbf{H} = \begin{bmatrix} H_{11} & H_{12} & H_{13} \\ H_{21} & H_{22} & H_{23} \\ H_{31} & H_{32} & H_{33} \end{bmatrix} \tag{5.16}$$

with dimensionality equal to ($3N \times 3N$). This formulation is general enough to cover a variety of multichannel image processing applications where linear degradation operators are involved. For instance the block elements H_{ij} can encode channel limitations in multispectral and color imagery. Moreover, they encode geometric affine and scale transformations in multisensor robotic applications, in registration or alignment of image data sets, in video and time-varying sequences. In these cases, the component block images f_i represent sequential frames and/or images obtained from a number of displaced cameras. The block elements H_{ij} may implement L-D projection operators as in biomedical applications, or non-invertible data reduction transforms as in lossy image compression. These elements can also incorporate decimation effects in the modeling of sub-band, wavelet and other orthogonal decomposition and/or multiresolution approaches. This representation facilitates the analysis and the optimal design of sub-band and multiresolution filters and is particularly useful in image interpolation and image reconstruction from limited sub-band information [37], [38].

The diagonal block elements H_{ij} represent intra-channel, and band or frame effects and degradations. The off-diagonal block elements H_{ij}, $i \neq j$, enable the consideration of channel interference in the image formation process, representing channel leakage in multispectral imagery, or registration errors in time-varying sequences, for instance [15], [25]. In addition, these elements can be utilized for the simultaneous registration and restoration of time-varying images.

Let $\mathbf{H_k}$ represents the k-th block row of the PSF matrix:

$$\mathbf{H_k} = [\mathbf{H_{k1}} \dots \mathbf{H_{kK}}] \tag{5.17}$$

Following these definitions, the formation of the k-th data channel is written as:

$$g_k = \mathbf{H_k f} + \mathbf{n_k} \tag{5.18}$$

Equivalently, the overall image formation model of p channels is given by:

$$\mathbf{g} = \mathbf{Hf} + \mathbf{n} \tag{5.19}$$

where the data vector $\mathbf{g}$ and the noise vector $\mathbf{n}$ are defined similar to (5.15). This model involves linear degradation or bandwidth reduction and additive noise. Even though it does not exhaust the degradation factors that may affect a multichannel image it covers many useful data formation processes and has been extensively used due to its simplicity and its potential in deriving effective inversion operators.

At this point some fundamental differences between the monochrome and the color formulations will be discussed. In the case of monochrome images, the assumptions of wide-sense stationarity and space invariance are often used. These assumptions lead to block-circulant matrix forms, whose eigenvalue decomposition is easily performed in the 2-D discrete Fourier transform (DFT) domain. Consequently, the invertibility of combined matrix operators involved in the computation of the restored image estimate is also verified in the 2-D DFT domain. In particular, the regularizing operator can be easily selected in relation with the PSF matrix $\mathbf{H}$. In the usual case of a low-pass operator $\mathbf{H}$, it suffices to select a Laplacian high-pass regularizing filter that stabilizes all small eigenvalues of $\mathbf{H}$, in order to derive a well-posed inversion process and guarantee the uniqueness (and stability) of the corresponding solution.

The stationarity assumption is unrealistic in the overall characterization of the multichannel image, because each channel captures different features of the image. Moreover, overall space invariance in the imaging model is unjustifiable [39]-[43].

Each pair of two specific frames embodies the relationship between specific characteristics of the image; wavelength relationship in multispectral imagery, or temporal association in time-varying sequences. A reasonable consequence of this coupling is the assumption of stationary interference only within pairs of specific frames. Thus, for the multichannel model stationarity and space invariance are assumed only within each pair of channels. This assumption results in the henceforth called partially block-circulant structure, whose composite block matrices are in block-circulant forms and can be diagonalized through the 2-D DFT operator [18]. Nevertheless, these blocks may not be related and can be arranged in any structure within the partially block-circulant matrix. This assumption has been employed in the implementation of multichannel algorithms with particular success [8], [17], [18] since it provides considerable reduction of the computational complexity and reasonable characterization of multichannel interaction.

Typical operations with partially block-circulant matrices are efficiently implemented in the so-called multichannel DFT domain. The transformation of a multichannel vector $\mathbf{x}$ in this domain is performed through a multiplication with the block matrix $\overline{\mathbf{W}}$ [18]:

$$\overline{\mathbf{W}}\mathbf{x} \triangleq \begin{bmatrix} \mathbf{W} & 0 & \dots & 0 \\ 0 & \mathbf{W} & \dots & 0 \\ \vdots & \vdots & \vdots & \vdots \\ 0 & 0 & \dots & \mathbf{W} \end{bmatrix} \mathbf{x} \tag{5.20}$$

where $\mathbf{W}$ denotes the 2-D DFT matrix.

The transformation of a partially block-circulant matrix **A** in the multichannel DFT domain is also expressed as matrix multiplication:

$$\overline{\mathbf{A}} = \overline{\mathbf{W}}\mathbf{A}\overline{\mathbf{W}}^{\mathbf{T}} \tag{5.21}$$

The resulting matrix $\overline{\mathbf{A}}$ is composed of diagonal blocks. This particular matrix type is referred to as partially block-diagonal matrix. Operations with such matrices preserve the partially block-diagonal structure. The multiplication of a multichannel image vector with a partially block-diagonal matrix can be decomposed into single-channel multiplications in the 2-D DFT domain. Moreover, the inversion of partially block-diagonal matrices can be efficiently performed in two ways. The first one is a recursive technique [18], while the second method is based on the inversion of small matrices [17]. Thus, the computational complexity for computing the restored image estimate is small, despite the large dimensionality of the problem.

In the sequel two classes of algorithms, direct and robust, are considered. These two classes involve well established approaches and provide the framework for the development of novel algorithms with specific properties that can be used in specialized applications.

5.5.2 Direct Algorithms

In this class, algorithms that derive their estimates in one step are considered. Most of them are derived from variations of either the MAP formulation or the regularization theory applied to the ill-conditioned restoration problem. Their primary goal is to provide an analysis of the ill-posed problem through the analysis of an associated well-posed problem, whose solution will yield meaningful answers and approximations to the ill-posed problem [34]. In broad perspective, these two approaches can be related in terms of the constraints imposed on the ill-posed problem. The MAP estimate is computed by maximizing the log-likelihood function:

$$\hat{\mathbf{f}} = \arg\mathbf{max_f}\left\{\log\mathbf{Pr}(\mathbf{f}|\mathbf{g})\right\} \tag{5.22}$$

where $Pr(\mathbf{f}|\mathbf{g})$ is the posterior density, or equivalently:

$$\hat{\mathbf{f}} = \arg\mathbf{max_f}\left\{\log\mathbf{Pr}(\mathbf{f}|\mathbf{g}) + \mathbf{logPr}(\mathbf{f})\right\} \tag{5.23}$$

Introducing the data formation model and considering the noise $\mathbf{n}$ process uncorrelated from the image f, the optimization problem reduces to:

$$\hat{\mathbf{f}} = \arg\mathbf{max_f}\left\{\log\mathbf{Pr}(\mathbf{g}-\mathbf{Hf}|\mathbf{f}) + \mathbf{logPr}(\mathbf{f})\right\} \tag{5.24}$$

Assuming general exponential distributions, the problem is equivalently expressed as:

$$\begin{aligned}\hat{\mathbf{f}}(\alpha) &= \arg\left\{\mathbf{min_f Q}(\alpha,\mathbf{f})\right\}\\ &= \arg\left\{min_f\left\{R_n(\mathbf{g}-\mathbf{Hf}) + \alpha\mathbf{R_f}(\mathbf{Cf})\right\}\right\}\end{aligned} \tag{5.25}$$

where α is referred to as the *regularization parameter*. In the MAP formulation presented, this parameter depends on the variances of the signal and the noise processes. The functionals $R_n(.)$ and $R_f(.)$ are referred to as $R_n(.)$ and $R_f(.)$ as the residual and the stabilizing terms, respectively. For independently distributed Gaussian noise and Gaussian prior distribution, the functionals $R_n(.)$ and $R_f(.)$ reduce to quadratic norms on weighted Hilbert spaces, defined as:

$$R_n(\mathbf{x}) = \| \mathbf{x}\|^{\mathbf{2}}_{\mathbf{L_n}} \triangleq \mathbf{x_t L_n x} \tag{5.26}$$

and

$$R_f(\mathbf{x}) = \| \mathbf{x}\|^{\mathbf{2}}_{\mathbf{L_f}} \triangleq \mathbf{x^t L_f x} \tag{5.27}$$

where $\mathbf{L_n}$ and $\mathbf{L_f}$ are diagonal weight matrices characterizing the corresponding spaces. The general form of the optimization problem in (5.25) can also be obtained from the Tikhonov-Miller approach, which regularizes the ill-posed restoration problem through the stabilizing functional $R_f(.)$ [25]. In the last formulation, the regularization parameter is set to the ratio (ϵ/E) of two bounds ϵ and E. The first bound represents the fidelity of the solution to the data, whereas the second bound indicates a smoothness requirement on the solution. Similarly, the CLS approach minimizes the functional $R_n(\mathbf{g} - \mathbf{Hf})$ under the constraint that $R_f(\mathbf{Cf})$ remains bounded. The multichannel CLS criterion is solved in [25] geometrically, by finding the center of one of the ellipsoids that bound the intersection of the two individual ellipsoids defined by:

$$\| \mathbf{g} - \mathbf{Hf}\|^{\mathbf{2}} \leq \epsilon^{\mathbf{2}} \tag{5.28}$$

and

$$\| \mathbf{Cf}\|^{\mathbf{2}} \leq \mathbf{E^2} \tag{5.29}$$

respectively. In fact, several regularized approaches have appeared in the literature for the formation of similar optimization problems utilizing quadratic functionals. These approaches derive the same form of estimator $\hat{\mathbf{f}}(\alpha)$ and differ only on the definition of the regularization parameter α [40].

This parameter controls the effect of the stabilizing term on the robust least-squares term and, consequently, the quality of the final estimate. A cross validation approach for the selection of this parameter is extended to the multichannel case in [41].

With these weighted norms, the solution of the MAP criterion can be obtained analytically as:

$$\hat{\mathbf{f}}(\alpha) = \left[\mathbf{H^t L_n H} + \alpha \mathbf{C^t L_f C}\right]^{-1} \mathbf{H^t L_n g} \tag{5.30}$$

This solution represents the estimates of the MAP approach with Gaussian distributions, the CLS formulation, the Tikhonov-Miller formulation, and the set theoretic formulation, extended to the multichannel representation [25], [29], [41], [42]. The operator $\mathbf{C}$ represents a linear, typically highpass operator in the form of (5.16).

It can take the form of the 3-D Laplacian filter, or the 2-D Laplacian filter applied independently on the different channels of the image [25]. In addition, a channel adaptive form $\mathbf{C}$ is proposed in [25], where one channel affects another channel according to the similarity of overall brightness in these channels. Other forms of $\mathbf{C}$ to reflect multichannel correlation are proposed in [15], [29], [41].

A simplified form is obtained if the weight matrices are equal to the unit matrix $\mathbf{I}$. Similar to (5.30), the corresponding estimate is expressed as:

$$\hat{\mathbf{f}}(\alpha) = \left[\mathbf{H^t H} + \alpha \mathbf{C^t C}\right]^{-1} \mathbf{H^t g} \tag{5.31}$$

It is also interesting to note that other estimates can be brought to the form of (5.30). Consider for instance the Wiener estimate expressed as [18]:

$$\hat{\mathbf{f}} = \mathbf{R_{ff} H^t} \left[\mathbf{H R_{ff} H^t} + \mathbf{R_{nn}}\right]^{-1} \mathbf{g} \tag{5.32}$$

where $\mathbf{R_{ff}}$ and $\mathbf{R_{nn}}$ represent the autocorrelation matrices of the multichannel signal and noise processes, respectively. For invertible matrices $\mathbf{A}$ and $\mathbf{B}$, the following property is easily proved:

$$\mathbf{H^t} \left[\mathbf{H A H^t} + \mathbf{B}\right]^{-1} = \left[\mathbf{H^t B^{-1} H A} + \mathbf{I}\right]^{-1} \mathbf{H^t B^{-1}} \tag{5.33}$$

Using this property, the Wiener estimate can be expressed as:

$$\begin{aligned} \hat{\mathbf{f}} &= \mathbf{R_{ff}} \left[\mathbf{H^t R_{nn}^{-1} H R_{ff}} + \mathbf{I}\right]^{-1} \mathbf{H^t R_{nn}^{-1} g} \\ &= \left[\mathbf{H^t R_{nn}^{-1} H} + \mathbf{R_{ff}^{-1}}\right]^{-1} \mathbf{H^t R_{nn}^{-1} g} \end{aligned} \tag{5.34}$$

This is the exact form of the estimate in (5.30) with weight matrices $\{\alpha \mathbf{L_f} = \mathbf{R_{ff}^{-1}}\}$, $\{\mathbf{L_n} = \mathbf{R_{nn}^{-1}}\}$ and $\{\mathbf{C} = \mathbf{I}\}$.

If $\mathbf{H^t}$ commutes with $\mathbf{R_{nn}^{-1}}$ then the Wiener estimate can be written as:

$$\hat{\mathbf{f}} = \left[\mathbf{H^t H} + \mathbf{R_{nn} R_{ff}^{-1}}\right]^{-1} \mathbf{H^t g} \tag{5.35}$$

which is similar to (5.31) with $\{\alpha \mathbf{C^t C} = \mathbf{R_{nn} R_{ff}^{-1}}\}$.

In computing the estimate $\hat{\mathbf{f}}$ the multichannel DFT transform $\overline{\mathbf{W}}$ is used that requires the inversion of partially block diagonal matrices. Even though there exist methods for testing the invertibility of such matrices, there exist no straightforward procedure to select the operator $\mathbf{C}$ for given $\mathbf{H}$, such as to guarantee the invertibility of the matrix $\{\mathbf{H^t H} + \alpha \mathbf{C^t C}\}$, in general. This difficulty motivates the use of iterative algorithms for the multichannel image restoration problem, even in the case of quadratic functionals that allow the analytic derivation of the estimate. Nevertheless, whenever an inversion of a partially block-diagonal matrix is required, the use of the algorithm which decomposes the entire inversion process into inversions of small $(p{\times}p)$ matrices is recommended [25].

This algorithm allows independent regularization of the individual inversions, thus resulting in stable implementation schemes. The recursive algorithm in [18] suffers from singularities caused by numerical computations. This algorithm is extremely sensitive, especially when applied to operators involving correlation matrices that often reflect a large condition number in the inversion.

In the previous estimates, the regularizing operator $\mathbf{C}$ is uniformly applied on the estimate to ensure global smoothness. It has a partially block circular structure with each block representing a high-pass filter kernel. This operator is defined independent of the structure of the data and thus, it cannot account for non-stationary formations that locally appear in the ideal image $\mathbf{f}$. As a side effect of this inefficiency, the restored image does not recover sharp edges and/or suffers from the creation of artifacts.

To alleviate this problem, the CMSE approach defines the regularizing term:

$$R_f(\mathbf{Cf}) = \mathbf{E}\{\| \mathbf{C}(\mathbf{f} - \hat{\mathbf{f}})\|^2\} \tag{5.36}$$

This approach incorporates the MMSE (Wiener) estimate as prior knowledge in the restoration process and influences the restored image towards the structure of the Wiener estimate. The CMSE approach can be interpreted as a regularized optimization scheme, which utilizes the prototype Wiener structure in smoothing its estimate. This approach enables the effective suppression of streak artifacts created in the restored image, especially for regularization parameters that lead to restoration of sharp edges.

Another attempt to influence the MAP estimate through local prior information in order to maintain spatial discontinuities is presented in [29]. It defines a multispectral image model through a Gibbs prior over a Markov random field containing spatial and spectral clique functions . The estimate is obtained as:

$$\hat{\mathbf{f}} = \arg\mathbf{min}_{\mathbf{f}}\{\sum_{c\epsilon C} \mathbf{V_c}(\mathbf{f}) + \| \mathbf{g} - \mathbf{Hf}\|^2_{\mathbf{L}}\} \tag{5.37}$$

where $V_c(.)$ is a function defined on a local group of points c, called cliques, and C is the set of cliques within the image. The cliques are defined in both the spatial and the spectral domains.

In the spatial domain, the local cliques are defined separately on each channel and provide a measure of the local signal activity in each channel. The spatial clique functions have the form of high pass filters. The individual function for each clique operates similar to $\mathbf{Cf}$ in (5.29). These functions must favor smooth prior distributions but they must not penalize sharp signal deviations too severely in order to allow for the restoration of sharp edges. To ensure these properties robust metrics on the result of linear high pass filters, one defined for each clique, were used in [29]. The aspects of robust functions is considered extensively in the next section.

In the spectral domain, clique functions are used only along the image edges to incorporate spectral information and align object edges between frequency planes. The application of each clique functions is performed locally, following the result of spatial edge detectors. The alignment of edges in multichannel image restoration is important, since it can eliminate false colors that can result when frequency planes are processed independently [29].

5.5.3 Robust Algorithms

The MAP restoration approach derives linear estimates under the assumption that both the signal and the noise are samples from Gaussian fields. Several limitations of this approach arise from the underlying stochastic assumptions. In image restoration applications, not only the noise statistics, but also the signal statistics are determined under uncertainty. The Gaussian distribution characterizes the noise process in only a narrow range of applications. It is worth mentioning the need for filtering speckle noise in SAR images and Poisson distributed film-grain noise in chest X-rays, mamograms, and digital angeographic images [43]-[45]. In addition, the Gaussian assumption induces severe smoothing in the representation and the restoration of the detailed structure of the original signal [30], [34]. Artifacts created by linear algorithms are even more pronounced in the case of multichannel image processing, due to the coupling of information among the channels and the propagation of errors. This section reviews the framework for the development of robust regularized approaches that address the accurate representation of both the noise and the signal statistics. In order to account for and tolerate stochastic uncertainty in the restoration process, the concept of robust functionals globally in the noise and the signal distributions is considered.

The robust approach for the multichannel problem, has been interpreted as a generalized MAP approach in [33]. A non-quadratic kernel function $r_n(.)$ is applied on the entries of the residual-error vector $\{\mathbf{g} - \mathbf{Hf}\}$ constructing the functional:

$$R_n(\mathbf{g} - \mathbf{Hf}) \triangleq \sum_{\mathbf{m=1}}^{\mathbf{N}} \mathbf{r_n} \left(\mathbf{g[m]} - \sum_{\mathbf{j=1}}^{\mathbf{N}} \mathbf{H[mj]f[j]} \right) \tag{5.38}$$

where $H[mj]$ denotes the mj-th scalar element of the matrix $\mathbf{H}$, and $f[m], g[m]$ denote the m-th scalar elements of the vectors $\mathbf{f}$ and $\mathbf{g}$, respectively. According to the generalized MAP formulation, the robust functional $R_n(.)$ induces a non-Gaussian noise distribution Pr_n which, computed at the residual, reflects the following conditional distribution of $\mathbf{g}$ given $\mathbf{f}$:

$$Pr_{g|f} = Pr_n|_{n=g-Hf} = K_n exp\{-\sigma_n R_n(\mathbf{g} - \mathbf{Hf})\} \tag{5.39}$$

with K_n and σ_n representing the normalizing constants of this distribution. Since large deviations from zero are penalized lighter by the robust metric than by a quadratic metric, this distribution can assign significant probability to large values and supports the existence of long tails. The distribution in (5.39) with an absolute-value functional represents the Laplacian distribution, while it can still reflect the Gaussian distribution with a quadratic functional. Moreover, the Huber measure in (5.39) enforces robust performance in the presence of outliers and derives asymptotically efficient estimates [31].

Alternatively, the robust metric on the signal space can be selected as to reflect long tails in the signal distribution and allow the accurate representation of the detailed structure. This term defines a robust functional $R_f(.)$ on the signal space, based on a non-quadratic kernel function $r_f(.)$, as:

$$R_f(\mathbf{Cf}) \triangleq \sum_{\mathbf{m}=1}^{\mathbf{N}} \mathbf{r_f}\left(\sum_{\mathbf{j}=1}^{\mathbf{N}} \mathbf{C[mj]f[j]}\right) \tag{5.40}$$

where $C[mj]$ denotes the mj-th scalar element of the matrix $\mathbf{C}$. The prior distribution induced by the signal functional in (5.40) is given by:

$$Pr_f = K_f exp\{-\sigma_f R_f(\mathbf{Cf})\} \tag{5.41}$$

where K_f and σ_f are the normalizing constants of the signal distribution.

The operator $\mathbf{C}$ is defined again as a high pass operator, possibly having the adaptive form in [25] or the combined clique form in [29]. The generalized distribution Pr_f essentially characterizes the highpass content of the image $\mathbf{f}$.

The quadratic stabilizing function utilized in conventional regularized approaches causes a smoothing influence on sharp edges, degrading the detailed structure of the estimate. In contrast, a robust function allows the existence of sharper transitions in the estimate, since it penalizes such deviations lighter than the quadratic scheme.

The robust measures $R_f(.)$ and $R_n(.)$ on the domains of the signal and the noise represent functionals which pertain robust characteristics, so that an uncertainty related to either the noise or the signal distribution does not degrade significantly the quality of the estimate. The signal kernel function $r_f(.)$ and the noise kernel function $r_n(.)$ are defined in terms of their derivatives $\phi_f(.)$ and $\phi_n(.)$, respectively, which in a robust estimation environment are referred to as the influence functions. Overall, the noise function accounts for efficient representation of the noise statistics and provides robustness with respect to noise outliers, whereas the signal function accounts for efficient representation of the signal statistics and for effective reconstruction of sharp edges in the estimate. The gradient descent derivation of the robust multichannel algorithm updates the estimate on the basis of the gradient. More specifically:

$$\mathbf{f_{i+1}} \triangleq \mathbf{f_i} - \mathbf{l}\nabla_\mathbf{f}^\mathbf{t}\mathbf{Q}(\alpha, \mathbf{f_i}) = \mathbf{f_i} + \mathbf{lH^t}\phi_\mathbf{n}(\mathbf{g} - \mathbf{Hf_i}) - \mathbf{l}\alpha\mathbf{C^t}\phi_\mathbf{f}(\mathbf{Cf_i}) \tag{5.42}$$

with l representing the iteration parameter. This algorithm is efficiently implemented in a mixed multichannel DFT and image space domain. In the former domain vector/matrix multiplications are computed wheres in the latter the point operations of the influence functions are performed. The convergence of such gradient descent algorithms has been extensively studied in the case of gray-level images. A sufficient condition for convergence requires that the influence functions $\phi_n(.)$ and $\phi_f(.)$ be continuous almost everywhere and be non-decreasing functions [46].

In this case, the mapping defined in (5.42) is non-expansive. Moreover, in image restoration applications the estimate of an iterative algorithm can be restricted to a closed, bounded, and convex subset of R^N through the use of a non-expansive constraint operating on the gray-levels of the estimate [47]. The robust algorithm in (5.42) utilizing such a constraint defines a non-expansive mapping on a closed, bounded, and convex set. This algorithm is guaranteed to converge to one of its fixed points, for **l** appropriately selected [47].

The soft limiter as an influence function describes the Huber error measure [29], [31], [46] which is extensively employed within the noise kernel function $r_n(.)$. This measure has been successfully used in applications with noise outliers [33]. Its operation in the residual term requires the specification of a structural parameter that can be derived using typical stochastic information regarding the noise process, such as the noise variance. In general, the soft limiter associated with the residual term is considered. The Huber measure also has been used along with the stabilizing term [29]. To maintain spatial discontinuities for this term, however, functions are needed whose robust performance is independent of structural parameters, since the specification of such parameters is difficult due to the lack of information regarding the signal process. Two classes of robust functions devoid of structural parameters have been proposed. The first class involves the l_p-norms, $1 < p < 2$ [48], whereas the second class involves entropic functions that operate in accordance to the human visual system [46]. In particular, the $l_{1.2}$-norm [48] and the absolute entropy function in [46] is recommended to be used as kernel functions $r_f(.)$ in (5.25).

5.6 Conclusions

Color image restoration finds important applications in a variety of fields. However, the inversion of degradation processes in multichannel data is involved not because of the increased dimensionality of the problem, but rather due to the peculiarities of the factors that affect multichannel images. The need for accurate modeling and identification of the intra- and the inter-channel correlation characteristics of the image is of critical importance. Moreover, the restoration algorithm must deal efficiently with information exchange among all different channels. The critical issues that determine the success of a particular restoration approach are:

1. accurate blur identification;
2. efficient modeling and identification of color prior and posterior distributions;
3. appropriate modeling of the constraint operators employed by the restoration algorithm.
4. appropriate use of a priori information.

The study of all the issues has been restricted to the frame work imposed by the partially block circulant structure of multichannels operators. The concept of color blur identification has been treated only within the assumption of space invariance within each channel and between each pair of channels. The EM algorithm has shown good potential in the computation of the particular block circulant components of such an operator structure [26]. Moreover, from the processing of gray scale images the neural networks emerge as promising tools for blur identification [47].

The stochastic form of the multichannel prior and the posterior distributions is the issue that seems to receive the most significant attention. Nevertheless, the structure of log likehood functions preserve partial block-circularity, mainly due to the computational efficiency of resulting algorithms. This specific structure implies wide-sense stationarity within channels and pairs of channels. Only a few approaches deviate from this assumption by incorporating either local-edge information from the data or apriori information by means of a prototype constrained image. In addition, only a few approaches break away from the Gaussian model. The development of robust multichannel algorithms presents an important challenge for accurate modeling of the distributions of the signal and the noise processes, at least locally.

The multichannel operators employed in restoration have been only considered heuristically. Their effects on the restored estimate have not been carefully analyzed nor thoroughly understood. Furthermore, the concept of the prior information has not been utilized effectively. It appears that such useful information could be used in all aspects of image restoration, from identification of the degrading operator, to the modeling of signal and/or noise statistics, to the structure of the restoration algorithm and its constrained operators.

Towards the study of the four issues in the multichannel restoration mentioned above, the wavelet analysis of the multichannel problem can play a determinant role. To justify this argument it is worth mentioning some results from the gray scale processing that trace the utility of wavelet analysis in studying and designing restoration algorithms.

Consider an image $\mathbf{f}$ in its vector form of dimensionality $N \times 1$. The multiresolution analysis utilizes an orthonormal wavelet basis and decomposes the original signal into its projection to a lower resolution space and the detail signals [48]. Because of a dyadic increase in the duration of each basis function in the new space, this transformation implies decimation of the composite images by a factor of two in each direction. The original image can be exactly reconstructed from the multiresolution image. Each sub-image can be equivalently obtained through a filtering operation followed by dyadic decimation. The last approach leads to the subband decomposition of the image which, under specific assumptions, becomes equivalent to the multiresolution decomposition.

The first level of multiresolution decomposition of an image $\mathbf{f}$ defines four filters $\mathbf{T_i}$, $i = 1, \ldots, 4$, which are represented in the same lexicographic form as $\mathbf{f}$. Each filter is essentially a separable operator that defines either a lowpass or a highpass filter on each image direction. The decimation in each direction is represented by the operator D. Thus, the decimation of an image vector in both directions is represented by the Kronecker product $\{D \otimes D\}$ of dimensionality $N/4 \times N$.

According to the previous convention, the image-vector $\mathbf{f}$ is decomposed through the 2-D wavelet transform into four filtered and decimated $(N/4 \times 1)$ signals as [46]:

$$\tilde{f}_i \triangleq (D \otimes D)\mathbf{T_i f} \quad \mathbf{i = 1, \ldots, 4} \tag{5.43}$$

The overall signal in the wavelet domain is formulated as:

$$\tilde{\mathbf{f}} = [\tilde{f}_1^t \ \tilde{f}_2^2 \ \tilde{f}_3^t \ \tilde{f}_4^t] \tag{5.44}$$

Thus, the image decomposed in the wavelet domain can be equivalently expressed in the form of a multichannel image composed of four channels. The wavelet transform can be repeatedly applied to any subband, resulting in higher orders of multiresolution decomposition. Accordingly, the multichannel representation in (5.44) can be readily expanded to any resolution level. Define the K-dimensional $(K = 4)$ unit vector $\{e_k, k = 1, \ldots, K\}$. The multiresolution signal can now be expressed in the compact form:

$$\tilde{\mathbf{f}} = \sum_{\mathbf{k=1}}^{\mathbf{K}} \mathbf{e_k} \otimes \tilde{\mathbf{f}}_\mathbf{k} = \sum_{\mathbf{k=1}}^{\mathbf{K}} \mathbf{e_k} \otimes (\mathbf{D} \otimes \mathbf{D})\mathbf{T_k f} \triangleq \mathbf{Tf} \tag{5.45}$$

Because of the orthonormality of the wavelet basis, the transform matrix $\mathbf{T}$ is an orthonormal operator $(\mathbf{T^t T = I})$.

The wavelet transform on a matrix operator is similarly defined. More specifically, a matrix $\mathbf{A}$ is represented in the multiresolution domain by a block matrix [49]:

$$\tilde{\mathbf{A}} = \mathbf{TAT^t} \tag{5.46}$$

whose mj-th block element is given by:

$$\tilde{A}_{mj} = (D \otimes D)\mathbf{T_m A T_j^t}(\mathbf{D^t} \otimes \mathbf{D^t}) \tag{5.47}$$

Following the vector and matrix decompositions in the wavelet transform, it is readily proved that a multiplication $\mathbf{Af}$ in the image domain is equivalent to the multiplication $\tilde{\mathbf{A}}\tilde{\mathbf{f}}$ in the wavelet domain.

The representation of images in the wavelet domain provides new insight into commonly used operators. One very important attribute of image processing operators is that of block circularity. This structure is derived under the assumptions of wide-sense stationarity of the image and noise, and the consideration of space-invariant operations. It is well known that a linear restoration operator in the image domain with block-circular (space-invariant) structure is transformed into a wavelet-domain operator with partially block circulant structure. Such a transformed operator functions almost independently in the different bands of the wavelet transform. Similarly, a block-circulant correlation matrix preserves little information regarding the cross-bands of the wavelet transform. Thus, the wide-sense stationarity assumption in the image space leads to loss of cross-band correlation in the wavelet domain.

The formulation of the signal in the wavelet domain elucidates the implications of typically used assumptions in the signal domain regarding wide-sense stationarity processes and space-invariant operators. In addition, it provides the framework for the development of novel implementation schemes that relax unrealistic assumptions [46]. Towards this direction, partially block-circulant operators can be utilized and the implementation of conventional algorithms can be developed directly in the wavelet domain. Such an implementation scheme has two advantages compared with the 2-D DFT implementation. First it replaces the wide-sense stationarity assumption in the image domain with a weaker assumption in the wavelet domain. The new assumption, namely of wide-sense stationarity within each band and each pair of bands, implies non-stationary image process in general. It provides better representation of the image's detailed structure and results in the reconstruction of sharper edges in the estimate. Second, it can implement non-stationary, signal-dependent, and space-variant operators that take under consideration the localized space-frequency characteristics of the image. In essence, the design of partially block-circulant operators in the wavelet domain relaxes unnecessary assumptions and can sustain additional information regarding the statistics of the signal and the structure of the degradation, as compared with conventional designs in the 2-D DFT domain.

The implementation of single-channel algorithms designed in the wavelet domain is directly associated with the implementation of multichannel algorithms. The multichannel DFT operator $\overline{\mathbf{W}}$ in (5.20) diagonalizes the blocks of a partially block-circulant operator such as (5.47). Thus, computations with such operators in the wavelet domain can be efficiently performed in the multichannel DFT domain. It becomes evident that the extension of these issues to the design of multichannel restoration algorithms will provide powerful tools in relaxing assumptions used up to now and in exploiting non-stationary and space varying multichannel correlation structures that are needed for effective modeling and efficient algorithmic design and implementation. The study of multichannel algorithms in the wavelet domain is an area that is expected to receive important attention in the near future.

References

1. Woods, R. E., Gonzalez, R. C. (1981): Real-time digital image enhancement. Proceedings of the IEEE, **69**, 634-654.
2. Bockstein, I. M. (1986): Color equalization method and its application to color image processing. Journal Optical Society of America, **3**(5), 735-737.
3. Trahanias, P. E., Venetsanopoulos, A. N. (1992): Color image enhancement through 3-D histogram equalization. Proceedings of the 15th IARP International Conference on Pattern Recognition, **1**, 545-548.
4. Faugeras, O. D. (1979): Digitl color image processing within the framework of a human visual model. IEEE Transaction on Acoustics, Speech and Signal Processing, **27**(4), 380-393.
5. Weeks, A. R., Haque, G. E., Myler, H. R. (1995): Histogram equalization of 24-bit color images in color difference (C-Y) color space. Journal of Electronic Imaging, **4**(1), 15-22.
6. Strickland, R., Kim, C., McDonell, W. (1987): Digital color image enhancement based on the saturation component. Optical Engineering, **26**, 609-616.
7. Trahanias, P.E., Pitas, I., Venetsanopoulos, A.N. (1994): Color Image Processing. (Advances In 2D and 3D Digital Processing: Techniques and Applications, edited by C.T. Leondes), Academic Press, N.Y..
8. Jain, A.K. (1989): Fundamentals of Digital Image Processing. Prentice Hall, Englewood Cliffs, New Jersey.
9. Kuan, D., Phipps, G., Hsueh, A.C. (1998): Autonomous robotic vehicle road following. IEEE Transaction on Pattern Analysis and Machine Intelligence **10**(5): 648-658.
10. Holyer, R.J., Peckinpaugh, S.H. (1989): Edge detection applied to satellite imagery of the oceans. IEEE Transaction on Geoscience and Remote Sensing **27**(1): 46-56.
11. Rignot, E., Chellappa, R. (1992): Segmentation of polarimetric synthetic aperture radar data. IEEE Transaction on Image Processing, **1**(3): 281-299.
12. Robb, R.A. (ed) (1985): Three-Dimensional Biomedical Imaging. CRC Press, Boca Raton FL.
13. Mallat, S. G. (1989): A theory for multiresolution signal decomposition: The wavelet representation. IEEE Transaction on Pattern Analysis and Machine Intelligence, **11**(7): 674-693.
14. Mallat, S. G. (1989): Multifrequency channel decompositions of images and wavelet models. IEEE Transaction on Acoustics, Speech, and Signal Processing, **37**(12): 2091-2110.
15. Zervakis, M. E. (1992): Optimal restoration of multichannel images based on constrained mean-square estimation. Journal Of Visual Communication and Image Representation, **3**(4): 392-411.
16. Sadjadi, F. A. (1990): Perspective on techniques for enhancing speckled imagery. Optical Engineering, **29**(1): 25-31.
17. Hunt, B. R., Kubler, O. (1984): Karhunen-Loeve multispectral image restoration, part I: Theory. IEEE Transaction on Acoustics, Speech, and Signal Processing, **32**(3): 592-600.

18. Galatsanos, N. P., Chin, R. T. (1991): Restoration of color images by multichannel Kalman filtering. IEEE Transaction on Signal Processing, **39**(10): 2237-2252.
19. Angwin, D., Kaufman, H. (1987): Adaptive restoration of color images. Proceedings of the 26th Conference on Decision and Control, Los Angeles, CA.
20. Galatsanos, N. P., Chin, R. T. (1989): Digital restoration of multichannel images. IEEE Transaction on Acoustics, Speech, and Signal Processing, **37**(3): 415-422.
21. Ozkan, M. K., Erdem, A. T., Sezan, M. I., Tekalp, A. M. (1992): Efficient multiframe Wiener restoration of blurred and noisy image sequences. IEEE Transaction on Image Processing **1**: 453-476.
22. Tekalp, A. M., Pavlovic, G. (1990): Multichannel image modeling and Kalman filtering for multispectral image restoration. Signal Processing **19**: 221-232.
23. Hunt, B. R. (1973): The application of constrained least squares estimation to image restoration by digital computer. IEEE Transaction on Computers, **C-22**(9).
24. Galatsanos, N. P., Katsaggelos, A.K. (1992): Methods for choosing the regularizing parameter and estimating the noise variance in image restoration and their relation. IEEE Transaction on Image Processing **1**(3): 322-336.
25. Galatsanos, N. P., Kataggelos, A.K., Chin, R. T., Hillery, A. D. (1991): Least squares restoration of multichannel images. IEEE Transaction on **39**(10): 2222-2236.
26. Tom, B.C.S., Lay, K. T., Katsaggelos, A. K. (1996): Multi channel image identification and restoration using the Expectation-Maximization algorithm. Optical Engineering, **35**: 241-254.
27. Sezan, M. I., Trussel, H. J. (1991): Prototype image constraints for set-theoretic image restoration. IEEE Transaction on Signal Processing, **39**(10): 2227-2285.
28. Katsaggelos, A. K., Lay, K. T., Galatsanos, N. P. (1993): A general framework for frequency domain multichannel signal processing. IEEE Transaction on Image Processing, **2**: 417-420.
29. Schultz, R., Stevenson, R. (1995): Stochastic modeling and estimation of multispectral image data. IEEE Transaction on Image Processing, **4**(8): 1109-1119.
30. Zervakis, M. E., Kwon, T. M. (1992): Robust estimation techniques in regularized image restoration. Optical Enginnering, **31**(10).
31. Kassam, S. A., Poor, H. V. (1985): Robust techniques for signal processing: A survey. Proceedings of the IEEE, **73**(3): 433-481.
32. Herbrt, T., Leahy, R. (1989): A generalized EM algorithm for 3-d Bayesian reconstruction from Poisson data using Gibbs priors. IEEE Transaction on Medical Imaging, **8**(2).
33. Zervakis, M. E., Venetsanopoulos, A. N. (1990): M-Estimators in robust nonlinear image restoration. Optical Engineering, **29**(5): 455-470.
34. Zervakis, M. E. (1996): Generalized maximum a posteriori processing of multichannel images and applications. Circuits Systems Signal Processing, **15**(2): 233-260.
35. Green, P. J. (1990): Bayesian reconstruction from emission tomography data using the modified EM algorithm. IEEE Transaction on Medical Imaging, **9**(1).
36. Stevenson, R., Delp, E. (1990): Fitting curves with discontinuities. Proceedings of the First International Workshop on Robust Computer Vision, Seattle, WA.
37. Tirakis,A., Delopoulos, A., Kollias, S. (1995): 2-D filter bank design of optimal reconstruction using limited subband information. IEEE Transaction on Image Processing, **4**(8): 1160-1165.

38. Delopoulos, A., Kollias, S. (1996): Optimal filterbanks for reconstruction from noisy subband components. IEEE Transaction on Signal Processing, **44**(2): 212-224.
39. Katsaggelos, A. K. (ed.): (1991): Digital Image Restoration. Springer Verlag, New York, N. Y.
40. Zhu, W., Galatsanos, N. P., Katsaggelos, A.K. (1995): Regularized multichannel restoration using cross-validation. Graphical Models and Image Processing, **57**: pp.38-54.
41. Chan, C.L., Katsaggelos, A. K., Sahakian, A. V. (1993): Image sequence filtering in quantum-limited noise with applications to low-dose fluoroscopy. IEEE Transaction on Medical Imaging, **12**: 610-621.
42. Han, Y. S., Herrington, D. H., Snyder, W. E. (1992): Quantitative angiography using mean field annealing. Proceedings of Computers in Cardiology 1992, **1**: 119-122.
43. Slump, C. H. (1992): Real time image restoration in diagnostic X-Ray imaging: The effects on quantum noise. 11th IAPR International Conference on Pattern Recognition, **2**: 693-696.
44. Zervakis, M. E., Katsaggelos, A. K., Kwon, T. M. (1995): A class of robust entropic functionals for image restoration. IEEE Transaction on Image Processing, **4**: 752-773.
45. Schafer, R. W., Mersereau, R. M., Richards, M. A. (1981): Constrained iterative restoration algorithms. Proceedings of the IEEE, **69**(4): 432-451.
46. Bouman, C., Sauer, K. (1993): A generalized Gaussian image model for the edge-preserving MAP estimation. IEEE Transaction on Image Processing, **2**(3): 296-310.
47. Figueiredo, M. A., Leitao, J. M.M. (1994): Sequential and parallel image restoration: Neural networks implementations. IEEE Transaction on Image Processing **3**: 789-801.
48. Daubechies, I. (1992): Ten Lectures on Wavelets. SIAM, Philadelphia, PA.
49. Zervakis, M. E., Kwuon, T. W., Yang, J-S. (1995): Multiresolution image restoration in the wavelet domain. IEEE Transaction on Circuits and Systems II, **42**(9): 578-591.

6. Color Image Segmentation

6.1 Introduction

Image segmentation refers to partitioning an image into different regions that are homogeneous with respect to some image feature. Image segmentation is an important aspect of the human visual perception. Humans use their visual sense to effortlessly partition their surrounding environment into different objects to help recognize them, guide their movements, and for almost every other task in their lives. It is a complex process that includes many interacting components that are involved with the analysis of color, shape, motion, and texture of objects in images. However, for the human visual system, the segmentation of images is a spontaneous, natural activity. Unfortunately it is not easy to create artificial algorithms whose performance is comparable to that of the human visual system. One of the major obstacles to the successful development of theories of segmentation has been a tendency to underestimate the complexity of the problem exactly because the human performance is mediated by methods which are largely subconscious. Because of this, segmentation of images is weakened by various types of uncertainty making most simple segmentation techniques ineffective [1].

Image segmentation is usually the first task of any image analysis process. All subsequent tasks, such as feature extraction and object recognition rely heavily on the quality of the segmentation. Without a good segmentation algorithm an object may never be recognizable. Over-segmenting an image will split an object into different regions while under-segmenting it will group various objects into one region. In this way, the segmentation step determines the eventual success or failure of the analysis. For this reason, considerable care is taken to improve the probability of successful segmentation.

Emerging applications, such as multimedia databases, digital photography and web-based visual data processing generated a renewed interest on image segmentation, so that the field has become an active area of research not only in engineering and computer science but also in other academic disciplines, such as geography, medical imaging, criminal justice, and remote sensing. Image segmentation has taken a central place in numerous applications, including, but not limited to, multimedia databases, color image and video transmission over the Internet, digital broadcasting, interactive TV, video-on-demand, computer-based training, distance education, video-conferencing and tele-medicine, and with the development of the hardware and communications infrastructure to support visual applications.

Many reasons can be cited for the success of the field. There is a strong underlying analytical framework based on mathematics, statistics, and physics. Thus, well-founded, robust algorithms that eventually lead to consumer applications can be designed. The field has also been helped tremendously by the advances in computer and memory technology, enabling faster processing of images, as well as in scanning and display.

Most attention on image segmentation has been focused on gray scale (monochrome) image segmentation. A common problem in segmentation of a gray scale image occurs when an image has a background of varying gray-level such as gradually changing shades, or when regions assume some broad range of gray-levels. This problem is inherent since intensity is the only available information from monochrome images. It is known that the human eye can detect only in the neighborhood of one or two dozen intensity levels at any point in a complex image due to brightness adaptation, but can differentiate thousands of color shades and intensities [2].

This paper surveys the existing techniques (past and present) of color image segmentation. The techniques are reviewed in six major classes: pixel-based, edge-based, region-based, model-based, physics-based, and hybrid-based techniques. There have been a number of image segmentation survey papers published [3–5], and [6], but either they don't consider color image segmentation techniques or they don't survey any modern segmentation techniques.

Because of the uncertainty problems encountered while trying to model the human visual system, there are currently a large number of image segmentation techniques available. However, no general methods have been found that perform adequately across a varied set of images. The early attempts at gray scale image segmentation are based on three techniques:

- Pixel-based techniques
- Region-based techniques
- Edge-based techniques

Even though these techniques were introduced three decades ago, they still find great attention in color image segmentation research today. Three of the major techniques that have been recently introduced include motion-based, physics-based, and model-based color image segmentation techniques. The following sections will survey the various techniques of color image segmentation starting with pixel-based techniques, edge-based, region-based, model-based, physics-based, and the last section surveying hybrid-based techniques. The final section describes the applicability of a specific region-based color image segmentation technique.

6.2 Pixel-based Techniques

Pixel-based techniques do not consider the spatial context but decide solely on the basis of the color features at individual pixels. This attribute has its advantages and disadvantages. Simplicity of the algorithms are an advantage to pixel-based techniques while lack of spatial constraints make them susceptible to noise in the images. Model-based techniques which utilize spatial interaction models to model images are used to further improve pixel-based techniques.

6.2.1 Histogram Thresholding

The simplest technique of pixel-based segmentation is histogram thresholding . It is one of the oldest and most popular techniques. If an image is composed of distinct regions, the color histogram of the image usually shows different peaks, each corresponding to one region and adjacent peaks are likely to be separated by a valley. For example, if the image has a distinct object on a background, the color histogram is likely to be bimodal with a deep valley. In this case, the bottom of the valley is taken as the threshold so that pixels that belong above and below this value on the histogram are grouped into different regions. This is called bi-level thresholding [3]. For multi-thresholding the image is composed of a set of distinct regions. In this case, the histogram has one or more deep valleys and the selection of the thresholds becomes easy because it becomes a problem of detecting valleys. However, normally, detection of the valleys is not a trivial job.

A method of color image segmentation which incorporated the histograms of nine color features from the three color spaces RGB, HSI, and YIQ: three from each color space was proposed in [9]. The most dominant peak of the nine histograms determines the intervals of the subregion. Pixels falling in this interval create one region and pixels falling out of it other ones. The dominant peak selection is driven by a priority list of seven.

The general segmentation algorithm consists of the following steps:

1. Put the image domain into the initially empty region list.
2. Determine the nine histograms of the region being considered. The entire image is the original region.
3. Locate all peaks in the set of histograms.
4. Select the best peak in the list of peaks using the priority list of seven. If none, then output this uniform region and go to step 2.
5. Determine and apply threshold.
6. Regions produced are then added to the list of region.
7. Go to step 2.

Additional processing, such as removal of small regions and addition of textural features, e.g. density of edges, can be used to improve the performance of the basic method.

Another attempt to derive a set of effective color features by systematic experiments of region segmentation is presented in [10]. This method is based on the fact that the color feature which has the deep valleys on its histogram and has the largest discriminant power to separate the clusters in a given region need not be the R, G, and B color features. Since a feature is said to have large discriminant power if its variance is large, color features with large discriminant power were derived by the Karhunen-Loeve (KL) transformation. At every step of segmenting a region, calculation of the new color features is done for the pixels in that region by the KL transform of R, G, and B data. Based on extensive experiments, it was found in [10] that the following three color features constitute an effective set of features for segmentation:

$$I1 = \frac{(R+G+B)}{3} \tag{6.1}$$

$$I2 = (R-B) \tag{6.2}$$

$$I3 = \frac{(2G-R-B)}{2} \tag{6.3}$$

The proposed color features were compared with the RGB, XYZ, YIQ, $L^*a^*b^*$ and HSI color primaries. Results reported in [10] indicated that the I1I2I3 color space has only a slight advantage over the other seven color spaces. However, according to [10] the I1I2I3 space should be selected because of the simplicity of transforming to this space form the RGB color space.

The opponent color space representation of I1I2I3 has also been used for segmentation purposes in [11]. A model of the human visual system is introduced in [11] and is used as a preprocessor for scene analysis. The proposed human visual system yields a pair of opponent colors as a two-dimensional feature for further scene analysis. A rough segmentation can be performed only upon the base of the 2-D histogram of the opponent colors. The procedure starts by transforming the RGB values to the opponent color pairs red-green (RG), yellow-blue (YB), and the intensity feature (I). Then the three channels are smoothed by applying band-pass filters. The center frequencies of the filters dispose of the proportion $I : RG : YB = 4 : 2 : 1$ so that the intensity channel shows the strongest high-pass character which puts emphasis on the edges of the image. Then peaks and flat levels in the 2-D $RG - YB$ histogram are searched for. These peaks and flat points determine the areas in the $RG - YB$ plane. Pixels falling into one of these areas create one region and pixels falling into another area create another region. Although this method leaves some non-attachable pixels in the image, it was argued in [11] that the proposed technique is superior to the methodologies suggested in [9] and [10].

The opponent color based methodology can be improved further by merging pixels that are not attached to a region [12]. Spatial neighborhood relations are used for the merging criterion. The improvements consists of an additional refinement process that is employed to the segmentation results obtained in [11]. If one or more pixels of the eight neighbors of a non-assigned pixels are assigned to the same region, the non-assigned pixel is marked for assignment to this region. Nothing is done if none of the neighborhood pixels are assigned or if several pixels in the neighborhood belong to different regions. After the entire image is scanned the marked pixels are assigned to the corresponding regions. This procedure is applied five times to the intermediate results. According to [12] while 30% to 80% of the pixels in an image are assigned to regions when employing the approach of [11], less than 10% are not assigned to regions when using the modified algorithm.

Another segmentation technique based on histogram thresholding is the one suggested in [13]. This technique attempts to detect the peaks of the three histograms in the hue, value, and chroma (HVC) components of the Munsell color space. Since no analytical formula exists for the transformation from the CIE standard system to the Munsell system, conversion is based on a table [8].

The segmentation algorithm of [13] consists of the following steps:

1. The histogram, of the region under consideration, is computed for each of the color features (HVC). Initially the entire image is regarded as the region. The histograms are smoothed by an average operator.
2. The most dominant peak in either of the three histograms is found. The peak selection is based on the shape analysis of each peak under consideration. First, some clear peaks are selected. Next, the following criterion function is calculated for each candidate peak:

$$f = \frac{S_p}{T_a}\frac{100}{F_p} \tag{6.4}$$

   where S_p denotes a peak area between two valleys, F_p is the full-width at half the maximum of the peak, and T_a is the total number of pixels in the specified region; the area of the histogram.
3. The two thresholds, one on each side, of the most dominant peak of the three histograms are found. Applying the thresholds, partitions the region into two sets of subregions: one consists of subregions corresponding to the color attributes within the threshold limits, and the other is a set of subregions with the remaining attribute values.
4. The threshold process is repeated for the extracted subregions. If all the histograms become mono-model, a suitable label is assigned to the latest extracted subregions.
5. Steps 1 through 4 are repeated for the remaining regions. The segmentation is terminated when the areas of the regions are sufficiently small in comparison the original image size or no histogram has significant peaks.

The remaining pixels which have not been assigned to a region are merged into the neighboring regions of similar color.

In summary, histogram thresholding is one of the simplest methods of image segmentation. This attribute lends it great consideration in current segmentation research when a rough segmentation of the image is needed. Many of the current image and video database systems [14] employ histogram thresholding for image segmentation in image and video retrieval. A common drawback of histogram thresholding is that it often produces unsatisfactory segmentation results on color images of natural scenes.

6.2.2 Clustering

Clustering is another pixel-based technique that is extensively used for image segmentation. The rationale of the clustering technique is that, typically, the colors in an image tend to form clusters in the histogram, one for each object in the image. In the clustering-based technique, a histogram is first obtained by the color values at all pixels and the shape of each cluster is found. Then, each pixel in the image is assigned to the cluster that is closest to the pixel color. Many different clustering algorithms are in existence today [15], [16]. Among these, the K-means and the fuzzy K-means algorithms have received extensive attention [21–23,27,28], and [29].

Clustering techniques can be combined with histogram thresholding approaches. In [17] the histogram thresholding method of [13] was extended to include clustering. The method consists of two steps. The first step is a modification of the algorithm introduced in [13] and reviewed in the previous section (Section 6.2.1). The modification consists of computing the principal components axes in the CIE L*a*b* color space for every region to be segmented. In other words, the color features have been transformed onto the principal component axes. Peaks and valleys are searched for in the three 1-D histograms of the three coordinate axis. The second step is a reclassification of the pixels on a color distance measure. Suppose a set of K representative colors $\{\mathbf{m_1}, \mathbf{m_2}, \ldots, \mathbf{m_k}\}$ are extracted from the image. The first cluster center $\mathbf{a_1}$ in the color space is chosen as $\mathbf{a_1} = \mathbf{m_1}$. Next, the color difference from $\mathbf{m_2}$ to $\mathbf{a_1}$ is computed. If this difference exceeds a given threshold T, a new cluster center $\mathbf{a_2}$ is created as $\mathbf{a_2} = \mathbf{m_2}$. Otherwise $\mathbf{m_2}$ is assigned to the domain of the class $\mathbf{a_1}$. In a similar fashion, the color difference from each representative color $(\mathbf{m_3}, \mathbf{m_4}, \ldots)$ to every established cluster center is computed and thresholded. A new cluster is created if all of these distances exceed T, otherwise the color is assigned to the class to which it is closest.

A method of detecting clusters by fitting to them some circular-cylindrical decision elements in the CIE L*a*b* uniform color coordinate system was proposed in [18] and [19]. This estimates the clusters' color distributions without imposing any constraints on their forms. Boundaries of the decision elements are formed with constant lightness and constant chromaticity loci. Each boundary is obtained using only 1-D histograms of the L*H°C* cylindrical coordinates of the image data. The Fisher linear discriminant method [68] is then used to simultaneously project the detected color clusters onto a line. For two clusters w_1 and w_2 the Fisher line W is given by:

$$W = (K_1 + K_2)^{-1}(M_1 - M_2) \tag{6.5}$$

where (K_1, K_2) and (M_1, M_2) are the covariance matrices and the mean vectors, respectively, of the two clusters. The color vectors of the image points, which are the elements of clusters w_1 and w_2, are then projected onto this line using the equation $d(C) = W^T C$, where C is a color vector in one of the clusters and the linear discriminant function. The 1-D histogram is calculated for the projected data points and thresholds are determined by the peaks and valleys in the histogram. Projecting the estimated color clusters onto a line permits utilization of all the property values of clusters for segmentation and inherently recognizes their respective cross correlation. This way, the region acceptance is not limited to the information available from one color component. Which gives the method an advantage over the multidimensional histogram thresholding techniques presented in Section 6.2.1.

Recently a color segmentation algorithm that uses the watershed algorithm to segment the 3-D color histogram of an image was proposed in [69]. An explanation of the morphological watershed transform can be found in [67]. The L*u*v* color space is utilized for the development of the algorithm. The non-linearity of the transformation from the RGB space to the L*u*v* space transforms the homogeneous noise in the RGB space to inhomogeneous noise. Even if the RGB data is smoothed prior to the transformation, due to the non-linearity of the transform, any small residual amount of noise may be significantly amplified. To this end, an adaptive filter is employed. The filter removes noise from a 3-D color histogram in the L*u*v* color space with subsequent perceptual coarsening.

The algorithm is as follows:

1. Calculate the color histogram of the image.
2. Filter it for noise reduction.
3. Perform perceptual coarsening.
4. Perform clustering using the watershed algorithm in the L*u*v* color space.

A new segmentation algorithm for color images based on mathematical morphology has been recently presented in [71]. The algorithm employs the scheme of thresholding the difference of two Gaussian smoothed 3-D histograms, that differ only in the standard deviation used, to get the initial seeds for clustering, and then uses a closing operation and adaptive dilation to extract the number of clusters and their representative values, and to include the suppressed bins during Gaussian smoothing, without a priori knowledge of the image. Through experimentation on various color spaces, such as the RGB, XYZ, YIQ, and I1I2I3, it was concluded that the proposed algorithm yields almost identical segmentation results in any color space. In other words, the algorithm works independently of the choice of color space.

Among the most popular clustering algorithms in existence today [15], [16], the K-means and the fuzzy K-means algorithms have received extensive attention [21–23,27,28], and [29]. A survey of segmentation techniques that utilize these clustering algorithms will be presented next.

K-means algorithm. The K-means algorithm for cluster-seeking is based on the minimization of a performance index which is defined as the sum of the squared distances from all points in a cluster domain to the cluster center. This algorithm consists of the following steps [16]:

1. Determine or choose K initial cluster centers $\mathbf{c_1(1)}, \mathbf{c_2(1)}, \ldots, \mathbf{c_K(1)}$. Here $\mathbf{c_1(1)}$ is the color features of the first cluster center during the first iteration.
2. At the kth iteration each pixel $\mathbf{a}$ is assigned to one of the K clusters $C_1(k), \ldots, C_K(k)$, where $C_j(k)$ denotes the set of pixels whose cluster center is $\mathbf{c_j(1)}$. $\mathbf{a}$ is assigned to cluster $C_j(k)$ if:

$$\|\mathbf{a} - \mathbf{c_j(k)}\| \le \|\mathbf{a} - \mathbf{c_i(k)}\| \tag{6.6}$$

$$\forall i, j = 1, 2, \ldots, K, \quad i \neq j \tag{6.7}$$

3. From the results of step 2, compute the new cluster centers $\mathbf{c_j(k+1), j = 1, 2, \ldots, K}$, such that the sum of the squared distances from all points in $C_j(k)$ to the new cluster center is minimized. In other words, the new cluster center $\mathbf{c_j(k+1)}$ is computed so that the performance index

$$J_j = \sum_{\mathbf{a} \in \mathbf{C_j(k)}} \|\mathbf{a} - \mathbf{c_j(k+1)}\|^2, \quad \mathbf{j = 1, 2, \ldots, K} \tag{6.8}$$

is minimized. The new cluster center which minimizes this is the sample mean of $C_j(k)$. Therefore, the new cluster center is given by:

$$\mathbf{c_j(k+1)} = \frac{\mathbf{1}}{\mathbf{N_j}} \sum_{\mathbf{a} \in \mathbf{C_j(k)}} \mathbf{a}, \quad \mathbf{j = 1, 2, \ldots, K} \tag{6.9}$$

where N_j is the number of pixels of cluster $C_j(k)$.
4. If $\mathbf{c_j(k+1)} = \mathbf{c_j(k)}$ for $j = 1, 2, \ldots, K$, the algorithm has converged and the procedure is terminated. Otherwise, go to step 2.

The determination of the initial cluster centers plays a crucial part because the better the initial partition is, the faster the algorithm will converge.

A comparison between the K-means clustering algorithm technique of color image segmentation and a region-based technique is given in [21]. The two algorithms were compared in color spaces, such as the RGB, XYZ, HSI, L*a*b*, and the I1I2I3 color space of [10]. In the clustering algorithm the initial cluster centers of the image were determined by first generating the m-dimensional histogram of the image and then determining the dominant peaks in the histogram. The K initial cluster centers correspond to K dominant peaks in the histogram. Their results showed that the K-means clustering method didn't perform as well as the region-based technique.

Similarly, a segmentation method which uses the K-means algorithm to locate clusters within the HSI color space was proposed in [22]. The hue, saturation and luminance components of the image are determined and used to form a three-dimensional vector that represents the color of any pixel within the image. The algorithm treats each color within the image simply as a three-dimensional vector. K initial cluster centers are initially chosen at random. K-means clustering is then implemented with the Euclidean distance, in the three-dimensional vector space, as a metric to distribute the pixels (Step 2 above). A modified algorithm can be obtained by separately segmenting the hue feature followed by segmentation of the two-dimensional saturation and luminance features. This approach biases the segmentation process towards the hue color value. However, the selection of the hue component is based on the fact that hue corresponds well to human visual perception. A parallel K-means clustering algorithm to track the centroids of clusters formed from moving objects in a sequence of colored images was proposed in [23]. The resulting tracking algorithm is robust with respect to shape variations and partial occlusions of the objects.

Fuzzy K-means algorithm. The K-means algorithm discussed in the previous section can be extended to include fuzzy inference rules. The so-called fuzzy K-means algorithm , which is also referred to as the fuzzy c-means algorithm, was first generalized in [25], [26]. The algorithm uses an iterative optimization of an objective function based on a weighted similarity measure between the pixels in the image and each of the K cluster centers. A local extremum of this objective function indicates an 'optimal' clustering of the input data. The objective function that is minimized is given by:

$$J_m(U, \mathbf{v}) = \sum_{k=1}^{n} \sum_{i=1}^{K} (\mu_{ik})^m (d_{ik})^2 \tag{6.10}$$

where μ_{ik} is the fuzzy membership value of pixel k in cluster center i, d_{ik} is any inner product induced norm metric (i.e. the Euclidean norm), m varies the nature of clustering with hard clustering at $m = 1$ and increasingly fuzzier clustering at higher values of m, $\mathbf{v}$ is the set of K cluster centers and U is the fuzzy K-partition of the image. The algorithm relies on the appropriate choices of U and $\mathbf{v}$ to minimize the objective function given above. The minimization of the objective function can also be done in an iterative fashion [27].

For the given set of data points $\mathbf{x}_1, \mathbf{x}_2, \cdots, \mathbf{x}_n$:

1. Fix the number of clusters K, $2 \leq K < n$, where n is the number of pixels. Fix $m, 1 \leq m < \infty$. Choose any inner product induced norm metric $\| * \|$.
2. Initialize the fuzzy K-partition, $U^{(b)} \in$ all possible fuzzy partitions, with $b = 0$ initially.

3. Calculate the K cluster centers $\{\mathbf{v_i^b}\}$ with $U^{(b)}$ and:

$$\mathbf{v_i} = \frac{\sum_{\mathbf{k=1}}^{\mathbf{n}} (\mu_{\mathbf{ik}})^{\mathbf{m}} \mathbf{x_k}}{\sum_{\mathbf{k=1}}^{\mathbf{n}} (\mu_{\mathbf{ik}})^{\mathbf{m}}}, \mathbf{i} = \mathbf{1}, \ldots, \mathbf{c}. \tag{6.11}$$

4. Update $U^{(b)}$. Let $d_{ik} = ||\mathbf{x_k} - \mathbf{v_i}||$: if $d_{ik} \neq 0$,

$$\mu_{ik} = \frac{1}{\left[\sum_{j=1}^{K} \left(\frac{d_{ik}}{d_{jk}}\right)^{2/(m-1)}\right]} \tag{6.12}$$

else, $\mu_{ik} = 0$.
5. Compare $U^{(b)}$ and $U^{(b+1)}$ in a matrix norm: if $||U^{(b)} - U^{(b+1)}|| \leq \varepsilon$, stop; otherwise, set $b = b + 1$ and return to step 3.

There are a number of parameters that need to be set in the system before the algorithm can be used. These are: $K, m, \varepsilon, U^{(0)}$, the inner product induced norm metric, and the number of items in the data set n. Due to the large amount of data items n being processed at any one time, a randomly chosen training subset of pixels taken form the input picture can be initially clustered [27] . An arbitrary number of initial clusters can also be used in the beginning of the segmentation process. The cluster center of the training set are used to calculate membership functions for all of the pixels in the image using (6.12) above.

These membership functions are examined and any pixel with a membership above a pre-defined threshold, called an α-cut, is assigned to the feature space cluster of that membership function. All of the pixels that remain are put back into the algorithm and the process is repeated until either all or a pre-determined amount of the pixels are identified as belonging to the clusters that were found during each iteration. Experiments were done in both the RGB and the I1I2I3 color spaces. It was suggested in [27] that the difference in results between the two is minimal. This type of algorithm will produce spherical or ellipsoidal shaped clusters in the feature space parallelizing the human visual color matching for constant chromaticity that has been shown to follow the spherical or ellipsoidal shaped cluster pattern [27].

In [28] a segmentation algorithm for aerial images that utilizes the fuzzy clustering principle was proposed. The method employs region growing concepts and pyramidal data structure for hierarchical analysis. Segmentation of the image at a particular processing level is done by the fuzzy K-means algorithm. Four values are replaced by their mean value to construct a higher level in the pyramid. Starting from the highest level, regions are created by pixels that have their fuzzy membership value above α-cut. If the homogeneity test fails, regions are split to form the next level regions which are again subjected to the fuzzy K-means algorithm. This algorithm is a region splitting algorithm.

A color image segmentation algorithm based upon histogram thresholding and fuzzy K-means techniques was proposed in [29]. The segmentation technique can be considered as a kind of coarse to fine technique. The strategy was adopted to reduce the computational complexity required for the fuzzy K-means algorithm. The coarse segmentation stage attempts to segment coarsely by using histogram scale space analysis [30], [31]. This analysis enables reliable detection of dominant peaks in the given histogram and the intervals around those peaks. The bounds of the intervals are found as zero-crossings of the second derivative for a τ-scaled version of the histogram. The τ-scaling of the histogram $h(x)$ is defined by the convolution of h with a Gaussian function which has a mean of zero and the standard deviation equal to τ. The second derivative of the scaled function can be computed by the convolution with the second derivative of the Gaussian function. Those pixels which are not segmented by the coarse segmentation are further segmented using the fuzzy K-means algorithm, proposed by Bezdek [25], [26], in the fine segmentation stage with the pre-determined clusters. In [29] different color spaces, such as the RGB, XYZ, YIQ U*V*W*, and the I1I2I3 were utilized

It is widely recognized that the clustering technique to image segmentation suffer from problems related to: (i) adjacent clusters frequently overlap in color space, causing incorrect pixel classification, and, (ii) clustering is more difficult when the number of clusters is unknown, as is typical for segmentation algorithms [29].

The pixel-based segmentation techniques surveyed in this section do not consider spatial constraints which make them susceptible to noise in the images. The resulting segmentation often contains isolated, small regions that are not present in noise-free images. In the past decade, many researchers have included spatial constraints in their pixel-based segmentation techniques using statistical models. These techniques will be surveyed in the model-based segmentation techniques section (Sect. 6.5).

6.3 Region-based Techniques

Region-based techniques focus on the continuity of a region in the image. Segmenting an image into regions is directly accomplished through region-based segmentation which makes it one of the most popular techniques used today [33]. Unlike the pixel-based techniques, region-based techniques consider both color distribution in color space and spatial constraints. Standard techniques include region growing and split and merge techniques. Region growing is the process of grouping neighboring pixels or a collection of pixels of similar properties into larger regions. The split and merge technique constitutes iteratively splitting the image into smaller and smaller regions and testing to see if adjacent regions need to be merged into one. The process of merging pixels or regions to produce larger regions is usually governed by a homogeneity criterion, such as the distance measures discussed in Chap. 2.

6.3.1 Region Growing

Region growing is the process of grouping neighboring pixels or a collection of pixels of similar properties into larger regions. Testing for similarity is usually achieved through a homogeneity criterion. Quite often after an image is segmented into regions using a region growing algorithm regions are furthered merged for improved results. A region growing algorithm typically starts with a number of seed pixels in an image and from these grows regions by iteratively adding unassigned neighboring pixels that satisfy some homogeneity criterion with the existing region of the seed pixel. That is, an unassigned pixel neighboring a region, that started from a seed pixel, may be assigned to that region if it satisfies some homogeneity criterion. If the pixel is assigned to the region, the pixel set of the region is updated to include this pixel. Region growing techniques differ in choice of homogeneity criterion and choice of seed pixels. Several homogeneity criteria linked to color similarity or spatial similarity can be used to analyze if a pixel belongs to a region. These criteria can be defined from local, regional, or global considerations. The choice of seed pixel can be supervised or un-supervised. In a supervised method the user chooses the seed pixels while in an un-supervised method choice is made by the algorithm.

In [34] a region growing segmentation algorithm was compared against an edge detection algorithm and a split and merge algorithm. The algorithms were tested in the RGB, YIQ, HLS (hue, saturation, and brightness), and L*a*b* color spaces. The region growing algorithm of [34] is a supervised one where the seed pixels and threshold values are chosen by a user. The Euclidean distance in color space was used to determine which pixels in the image satisfy the homogeneity condition. If the color of the seed pixel is given as $S = (s_1, s_2, s_3)$ and the color of a pixel in consideration is $P = (p_1, p_2, p_3)$, all pixels which satisfy

$$(s_1 - p_1)^2 + (s_2 - p_2)^2 + (s_3 - p_3)^2 < T^2 \tag{6.13}$$

would be included in the region. Here T is the threshold value which is chosen by the user. The algorithm can be summarized with the following steps:

1. Choose next seed pixel. This seed pixel is the first pixel of the region.
2. Test to see if the four neighboring pixels (vertical and horizontal neighbors) of the pixel belong to the region with condition (6.13).
3. If any of the four neighboring pixels satisfy the condition, they are assigned to the region and step 2 is repeated and their four neighbors are considered and tested for homogeneity.
4. When the region is grown to its maximum (e.g. no neighbors of the pixels on the edge of the region satisfy (6.13)), go to step 1.

It was found in [34] that the region growing algorithm performed best in the HLS and L*a*b* color spaces. The authors of the study also suggest that instead of comparing the unassigned pixel to the seed pixel, to compare it to the mean color of the set of pixels already assigned to the region. Every time a pixel is assigned to the region the mean value is updated. They never conducted any experiments with this new homogeneity criterion.

Another color segmentation algorithm which combines region growing and region merging processes was recently proposed in [35]. This algorithm starts with the region growing process which is based on three homogeneity criteria that take into account color similarity and spatial proximity. The resulting regions are then merged on the basis of a homogeneity criterion that takes into account only color similarity. The three criteria they used for the region growing approach include:

1. a local homogeneity criterion, which corresponds to a local comparison between adjacent pixels;
2. a first average homogeneity criterion, which corresponds to a local and regional comparison between a pixel and its neighborhood, considering only the region under study;
3. a second average homogeneity criterion, which corresponds to a global and regional comparison between a pixel and the studied region.

For a visual point of view, the authors in [35] consider that regions which present similar color properties belong to the same class, even if they are spatially disconnected. Consequently, these regions are merged using a global homogeneity criterion which corresponds to a global comparison of the average color features representative of the two regions under study. They have also considered that regions which are spatially dispersed in the image, such as details, edges, or high-frequency noise have to be merged to the other regions either locally pixel by pixel, or globally. All color comparisons are accomplished using the Euclidean distance measure in the RGB color space. Threshold values are computed according to an adaptive process relative to the color distribution of the image. Finally, it was suggested in [35] that the algorithm listed there can be extended to other uniform color spaces but new thresholds have to be defined.

A graph-theoretic approach to the problem of color image segmentation was proposed in [36]. The algorithm is based on region growing in the RGB and L*a*b* color spaces using the Euclidean distance metric to measure the color similarity between pixels. The suppression of artificial contouring is formulated as a dual graph-theoretic problem. A hierarchical classification of contours is obtained which facilitates the elimination of the undesirable contours. Regions are represented by vertices in the graph and links between geometrically adjacent regions have weights that are proportional to the color distance between the regions they connect. The link with the smallest weight determines the regions to be merged. At the next iteration of the algorithm the weights of all the links that are connected to a new region are recomputed before the minimum weight link is selected. The links chosen in this way define a spanning tree on the original graph and the order in which links are chosen defines a hierarchy of image representations. Results presented in [36] suggested that no clear advantage was gained through the utilization of the L*a*b* color space.

6.3.2 Split and Merge

As opposed to the region growing technique of segmentation where a region is grown from a seed pixel, the split and merge technique subdivides an image initially into a set of arbitrary, disjointed regions and then merge and/or split the regions in an attempt to satisfy a homogeneity criterion between the regions.

In [2] a split and merge algorithm that iteratively works toward satisfying these constraints was presented. The authors describe the split and merge algorithm initially proposed in [37]. The image is subdivided into smaller and smaller quadrant regions so that for each region a homogeneity criterion holds. That is, if for region R_i the homogeneity criterion does not hold, divide the region into four sub-quadrant regions, and so on. This splitting technique may be represented in the form of a so-called quad-tree. The quad-tree data structure is the most common used data structure in split and merge algorithms because of its simplicity and computational efficiency [38]. A split artificial image and the corresponding quad-tree is shown in Fig. 6.1 and 6.2, respectively. Note that the root of the tree corresponds to the entire image. Merging of adjacent sub-quadrant regions is allowed if they satisfy a homogeneity criterion. The procedure may be summarized as:

1. Split into four disjointed quadrants any region where a homogeneity criterion does not hold.
2. Merge any adjacent regions that satisfy a homogeneity criterion.
3. Stop when no further merging or splitting is possible.

Most split and merge approaches to image segmentation follow this simple procedure with varying approaches coming from different color homogeneity criteria.

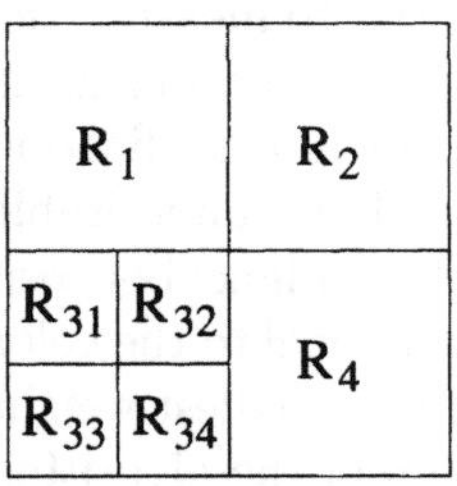

Fig. 6.1. Partitioned image

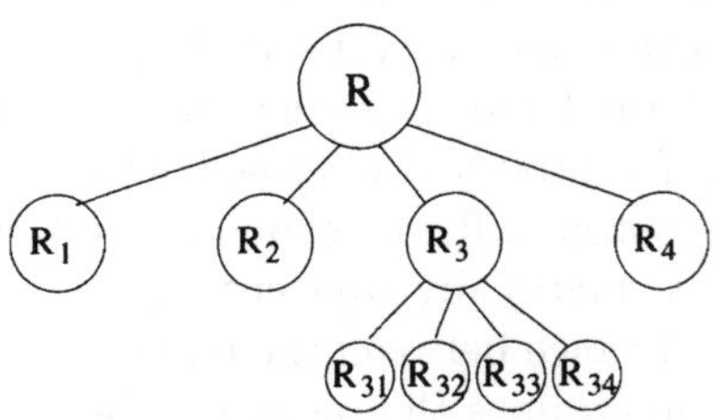

Fig. 6.2. Corresponding quad-tree

As mentioned in the previous section (Sect. 6.3.1), in [34] the authors compared a split and merge segmentation algorithm against an edge detection algorithm and a region growing algorithm. They tested all algorithms in the RGB, YIQ, HLS (hue, saturation, and brightness), and L*a*b* color spaces. They used statistical properties of the image regions to determine when to split and when to merge. They used the trace of the covariance matrix for a region to determine how homogeneous a given region is. If the mean color in a region with n pixels is:

$$M = (m_1, m_2, m_3) = \left(\frac{\sum_i c_{1i}}{n}, \frac{\sum_i c_{2i}}{n}, \frac{\sum_i c_{3i}}{n} \right) \tag{6.14}$$

where (m_1, m_2, m_3) and (c_1, c_2, c_3) representing the three color features of the mean of the region and of a pixel, respectively, then the trace of the covariance matrix is equal to:

$$\begin{aligned} T &= v_{11} + v_{22} + v_{33} \\ &= \left(\sum_i (c_{1i} - m_1)^2 + \sum_i (c_{2i} - m_2)^2 + \sum_i (c_{3i} - m_3)^2 \right) /n. \end{aligned} \tag{6.15}$$

If the trace is above a user specified threshold, the region is recursively split. Otherwise, the rectangular region is added to a list of regions to be subsequently merged.

Two statistical measures for merging regions were employed in [34]. The first is based on the trace of the covariance matrix of the merged region. This value is calculated for the two regions that are being considered. If this value is below the specified threshold, then the two regions are merged. Otherwise, they are not. The second method considers the Euclidean color distance between the means of the two regions to be merged. As with their region growing method, the two regions are merged when this distance is below the specified threshold and not otherwise.

As mentioned in Sect. 6.2.2, the authors in [21] had compared the quad-tree split and merge algorithm to the K-means clustering algorithm. They compared the two algorithms in seven color spaces. They tested the quad-tree split and merge algorithm explained above with two homogeneity criteria: (i) a homogeneity criterion based on functional approximation and (ii) the mean and variance homogeneity criterion.

The functional approximating criterion assumes that the color over a region may either be constant or variable due to intensity changes caused by shadows and surface curvatures. They used low-order bivariate polynomial approximating functions as the set of approximating functions, because these functions detect useful information, such as abrupt changes in the color features, relatively well and ignore misleading information, such as changes in intensity caused by shadows and surface curvature, when the order is not too high. The set of low-order polynomials can be written as:

$$f_m(x,y) = \sum_{i+j\leq m} a_{ij}x^i y^j, \quad m \leq 4 \tag{6.16}$$

Using the above formula, the planar polynomial is obtained by $m = 1$, and the bi-quadratic polynomial by $m = 2$.

The vector **a** is calculated with a least-square solver. The fitting error ϵ for a region R is:

$$\epsilon = \sqrt{\sum_{x,y\in R} \frac{(g(x,y) - f_m(x,y))^2}{n}} \tag{6.17}$$

where n is the number of pixels in R and $g(x,y)$ is the pixel value at coordinates (x,y). The fitting error ϵ is compared to the mean noise variance in the region and is considered homogeneous if it is less than this value. The mean and variance homogeneity criterion assumes that the color of the pixels, discarding noise, over a region is constant and is based on the mean and variance of a region, which is the case for $m = 0$. That is,

$$f_0 = a_{00}.$$

The fitting error ϵ is calculated and compared to the mean noise variance of the region, as before. They found that the split and merge method of image segmentation outperforms the K-means clustering method.

A major drawback of quad-tree-structured split and merge algorithms is their inability to adjust their tessellation to the underlying structure of the image data because of the rigid rectilinear nature of the quad-tree structure.

In [39] an image segmentation algorithm to reduce this drawback was introduced. The proposed split and merge algorithm employs the incremental Delaunay triangulation as a directed region partitioning technique which adjusts the image tessellation to the semantics of the image. A Delaunay triangulation of a set of points is a triangulation in which the circum-circle of any of its triangles does not contain any other point in its interior [40]. The homogeneity criterion is the same used in [21]. Region-based techniques of image segmentation are very common today because of their simplicity and computational simplicity. This lends them great attention when hybrid segmentation techniques are created. Region-based techniques are often mixed with other techniques, such as edge detection. These hybrid techniques will be described in Section 6.7.

6.4 Edge-based Techniques

Edge-based segmentation techniques focus on the discontinuity of a region in the image. The color edge detection techniques discussed in Chap. 4 are being used today for image segmentation purposes. Once the edges within an image have been identified, the image can be segmented into different regions based upon these edges. A disadvantage with edge-based techniques is their sensitivity to noise .

6.5 Model-based Techniques

Recently, much work has been directed toward stochastic model-based segmentation techniques [42,50,52], and [55]. In such techniques, the image regions are modeled as random fields and the segmentation problem is posed as a statistical optimization problem. Compared to previous techniques, the stochastic model-based techniques often provide more precise characterization of the image regions. In fact, various stochastic models can be used to synthesize color textures that closely resemble natural color textures in real-world natural images [43]. This characteristic, along with the optimization formulation, provides better segmentation when the image regions are complex and otherwise difficult to discriminate by simple low-order techniques. Most of the techniques introduced use the spatial interaction models like Markov random field (MRF) or Gibbs random field (GRF) to model digital images. Although interest in MRF models for tackling image processing problems can be traced back to [41], only recently have the applicable mathematical tools for exploitation of the full power of MRF in image segmentation found their way into image processing literature. Research methodologies reported in [43–49,52,53], and [54] all make use of the Gibbs distributions for characterizing MRF.

Stochastic model-based color image segmentation techniques can be either supervised or un-supervised. In a supervised segmentation approach, the model parameters are obtained from training data, whereas in an unsupervised segmentation approach, the model parameters have to be estimated directly from the observed color image. Therefore, the unsupervised segmentation problem can be considered as a model fitting problem where a random field model is fitted to an observed image. The unsupervised approach is often necessary in many practical applications where training data is not available, for example when only one image is available.

In [49], the authors developed an unsupervised segmentation algorithm which uses Markov random filed models for color textures. These models characterize a texture in terms of spatial interaction within each color plane and interaction between different color planes. The algorithm consists of a region splitting phase and an agglomerative clustering phase and is performed in the RGB color space. In the region splitting phase, the image is partitioned into a number of square regions that are recursively split until each region satisfies a homogeneity criterion. The agglomerative clustering phase is divided into a conservative merging process followed by a stepwise optimal merging process. Conservative merging uses color mean and covariance estimates for the efficient processing of local merges. After the conservative merging is the stepwise optimal merging process that at each iteration maximizes a global performance functional based on the conditional pseudo-likelihood of the color image. The stepwise optimal merging process is stopped using a test based on rapid changes in the pseudo-likelihood of the image. In [51] the maximum posteriori (MAP) probability approach to image segmentation was introduced. The main points of the approach are presented in the following sections.

6.5.1 The Maximum A-posteriori Method

The maximum a-posteriori probability (MAP) approach is motivated by the desire to obtain a segmentation that is spatially connected and robust in the presence of noise in the image. The MAP criterion functional consists of two parts, the class conditional probability distribution, which is characterized by a model that relates the segmentation to the data, and prior probability distribution, which expresses the prior expectations about the resulting segmentation .

In [47] the authors have proposed a MAP approach of the segmentation of monochromatic images, and have successfully used GRF's as a-priori probability models for the segmentation of labels. The GRF prior model expresses the expectation about the spatial properties of the segmentation. In order to eliminate isolated regions in the segmentation that arise in the presence of noise, the GRF model can be designed to assign a higher probability for segmentation results that have contiguous, connected regions. Thus, estimation of the segmentation is not only dependent on the image intensity, but also constrained by the expected spatial properties imposed by the GRF model.

The observed monochrome image data is denoted by $\mathbf{y}$. Each individual pixel intensity is denoted by y_s, where s denotes the pixel location. A segmentation field, denoted by the N-dimensional vector $\mathbf{x}$, is obtained by assigning labels to each pixel site in the image. A label $x_s = i, \quad i = 1, \ldots, K$, implies that the site s belongs to the i'th class among the K classes. The desired estimate of the segmentation label field is defined as the one that maximizes the a-posteriori pdf $p(\mathbf{x}|\mathbf{y})$ of the segmentation label field, given the observed image $\mathbf{y}$. Using the Baye's rule:

$$p(\mathbf{x}|\mathbf{y}) \propto p(\mathbf{y}|\mathbf{x})p(\mathbf{x}) \tag{6.18}$$

where $p(\mathbf{y}|\mathbf{x})$ represents the conditional pdf of the data given the segmentation labels, namely the class-conditional probability density function (pdf). The term $p(\mathbf{x})$ is the a-priori probability distribution that can be modeled to impose a spatial connectivity constraint on the segmentation.

A spatial connectivity constraint on the segmentation field can be imposed by modeling it as a discrete-valued GRF. Detailed discussion of GRF models can be found in [42], [44], [47]. The a-priori probability $p(\mathbf{x})$ can be modeled as a Gibbs distribution:

$$p(\mathbf{x}) = \frac{1}{Z} \exp[-U(\mathbf{x})/T] \tag{6.19}$$

where the normalizing constant Z is called the partition function, T is the temperature constant, and $U(\mathbf{x})$, the Gibbs potential (Gibbs energy).

The authors in [47] model the mean intensity of each image region as a constant, denoted by the scalar $\mu_i, i = 1, 2, \ldots, K$. The conditional probability distribution is expressed as:

$$p(\mathbf{y}|\mathbf{x}) \propto \exp\left[-\sum_s (y_s - \mu_{x_s})^2 / 2\sigma^2\right] \tag{6.20}$$

where μ is the mean intensity function. Note that this is the probability distribution used in the case of estimating the segmentation on the basis of the maximum likelihood (ML) criterion. It should be observed that the MAP estimation follows a procedure that is similar to that of the K-means algorithm, namely it starts with an initial estimate of the class means and assign each pixel to one of the K classes by maximizing, then update the class means using these estimated labels, and iterate between these two steps until convergence.

The MAP method presented in [47] can be extended to color images. The three color channels of the image are denoted by a $3N$-dimensional vector $[\mathbf{y_1}, \mathbf{y_2}, \mathbf{y_3}]^t$. A single segmentation field $\mathbf{x}$, which is consistent with all 3 channels of data and is in agreement with the prior knowledge, is desired. By assuming the conditional independence of the channels given the segmentation field, the conditional probability in (6.20) becomes:

$$\begin{aligned} p(\mathbf{y}|\mathbf{x}) &= p(\mathbf{y_1}, \mathbf{y_2}, \mathbf{y_3}|\mathbf{x}) \\ &= p(\mathbf{y_1}|\mathbf{x})p(\mathbf{y_2}|\mathbf{x})p(\mathbf{y_3}|\mathbf{x}) \end{aligned} \tag{6.21}$$

The image is modeled as consisting of K distinct regions, where the ith region has the uniform mean color represented by $(\mu_i^1, \mu_i^2, \mu_i^3)$. The posterior probability distribution can be written as:

$$p(\mathbf{x}|\mathbf{y}) \propto \exp\left\{\sum_s \left[\sum_{j=1}^{3} \frac{1}{2\sigma_j^2}(y_{j,s} - \mu_{x_s}^j)^2\right] + \sum_{C \in \mathcal{C}} V_C(\mathbf{x})\right\} \tag{6.22}$$

where $y_{j,s}$ represents the intensity data in channel j at site s, and σ_j^2 denotes the variance of the combined additive noise for the j'th color channel. Thus, a pixel represented by the color triplet $(y_{1,s}, y_{2,s}, y_{3,s})$ is assigned to the region characterized by the class mean $(\mu_i^1, \mu_i^2, \mu_i^3)$ according to the single segmentation label $x_s = i$.

6.5.2 The Adaptive MAP Method

In [53] an adaptive clustering algorithm for monochrome image segmentation was proposed. The algorithm is based on improving the conditional probability model in (6.20). The uniform region mean intensity μ_{x_s} is not adequate in modeling actual images intensities. The author proposed using a space variant intensity μ_{s,x_s} to model each region as a slowly varying function of the site location s. The modified conditional probability becomes:

$$p(\mathbf{y}|\mathbf{x}) \propto \exp\left[-\sum_s (y_s - \mu_{s,x_s})^2/2\sigma^2\right] \tag{6.23}$$

In [54] the authors extended the results reported in [53] for monochrome image segmentation to color segmentation using an adaptive MAP framework in the L*u*v* and L*a*b* color spaces. They assume that each region i in the color image has a distinct space-variant mean color, denoted by $(\mu^1_{s,i}, \mu^2_{s,i}, \mu^3_{s,i})$ for each site s. The posterior probability distribution for estimating an adaptive color segmentation is now:

$$p(\mathbf{x}|\mathbf{y}) \propto \exp\left\{\sum_s \left[\sum_{j=1}^{3} \frac{1}{2\sigma_j^2}(y_{j,s} - \mu^j_{s,x_s})^2\right] + \sum_{C\in C} V_C(\mathbf{x})\right\} \tag{6.24}$$

Note that a non-adaptive color clustering algorithm, similar to K-means, is obtained when both the spatial dependence of the class means and the prior probability distribution are ignored. Results obtained when using most stochastic model-based techniques for color image segmentation are favorable in most cases, especially with natural scenes. The problems encountered with these techniques is there complexity. These techniques are computationally intensive. There is a trade-off between complexity and segmentation results.

6.6 Physics-based Techniques

Physics-based segmentation techniques use the underlying physical models of the color image formation process in developing color difference metrics. The objective of these techniques is to segment a color image at object boundaries and not at the edges of highlights and shadows in an image [56]. Physics-based techniques allow the segmentation of color images based on physical models of image formation. The basic mathematical methods used by these techniques are often similar to those already discussed in the previous sections. They differ regarding the reflection models employed for segmenting color images. For example, the authors in [57] use region splitting guided by preliminary edge detection to classify regions. Some approaches intend to be applied preliminary to the segmentation process. For example, they try to distinguish material changes from shadow and possibly highlight boundaries.

In [58], [59] the authors investigated the influence of highlights, shading, and camera properties as, for example, color clipping, color balancing, and chromatic lens aberration on the results of color image segmentation. They classify physical events with measured color variation in the image by employing the Dichromatic Reflection Model from dielectrics. The Dichromatic

Reflection Model (DRM) of [60] describes the light, $L(\lambda, i, e, g)$, which is reflected from a point on a dielectric non-uniform material as a mixture of the light $L_s(\lambda, i, e, g)$ reflected at the material surface and the light $L_b(\lambda, i, e, g)$ reflected from the material body. The parameters i, e, g, and λ denote the angle of incident light, the angle of emitted light, the phase angle, and the wavelength, respectively.

The DRM is given by [60]:

$$L(\lambda, i, e, g) = L_s(\lambda, i, e, g) + L_b(\lambda, i, e, g) \tag{6.25}$$

Using this classification, the author in [59] developed a hypothesis-based segmentation algorithm. The algorithm searches for color clusters from local image areas that show the characteristic features of the body and surface reflection processes in a bottom-up manner. When a promising cluster is found in an image area, a hypothesis is generated which describes the object color and highlight color in the image area and the shading and highlight components of every pixel in the area is determined. The new hypothesis is then applied to the image using a region growing approach. This determines the exact extent of the image area to which the hypothesis applies. This step verifies the applicability of the hypothesis. Accurate segmentation results are presented in [59] for images of plastic objects.

There are many rigid assumptions of the DRM, e.g. the illumination conditions, and the type of materials. For most realistic scenes, these assumptions do not hold. Therefore, the DRM can be used to segment color scenes taken only within a controlled environment.

6.7 Hybrid Techniques

A number of hybrid color image segmentation techniques were introduced recently [61], [65]. These techniques combine the benefits of the various techniques mentioned in past sections and masks the disadvantages of others.

In [61] the authors proposed a segmentation scheme that first splits the color image into chromatic and achromatic regions and then employs a histogram thresholding technique to the two regions, separately. The scheme can be summarized into the following steps:

1. Convert RGB color values through the XYZ and L*u*v* space to HSI color values.
2. Define the effective ranges of hue and saturation in the HSI space and determine chromatic and achromatic regions in the image.
3. Use hue, saturation, and/or intensity one-dimensional histogram thresholdings to further segment the image.
4. Detect and recover over-segmentation regions using a region merging technique.

The proposed algorithm is employed in the HSI color space due to its close relation to human color perception. The authors suggest to split up the color image into chromatic and achromatic regions to determine effective ranges of hue and saturation. The criteria for achromatic areas were measured by experimental observation of human eyes and are defined as follows:

1. (intensity$>$95) or (intensity$\leq$25),
2. (81$<$intensity$\leq$95) and (saturation$<$18),
3. (61$<$intensity$\leq$81) and (saturation$<$20),
4. (51$<$intensity$\leq$61) and (saturation$<$30),
5. (41$<$intensity$\leq$51) and (saturation$<$40),
6. (25$<$intensity$\leq$41) and (saturation$<$60),

while intensity is re-scaled from 1 to 100, and saturation is variable with a maximal value of 180. In step 3 chromatic regions are segmented using hue histogram thresholding and achromatic regions are segmented using intensity histogram thresholding. The histogram thresholding they employ in their algorithm is the one proposed in [13]. Over-segmentation regions are recovered using an 8×8 mask. The mask is evenly divided into sixteen 2×2 sub-masks. If there is at least a chromatic and an achromatic pixel in the 2×2 sub-mask, then the sub-mask has a vote to the dispersion of the mask. A special label is assigned the 8×8 region if the mask possesses more than seven votes. After convoluting the mask throughout the image, region merging is used to merge the labeled regions with the segmented region or to form some new regions.

In [62] the authors proposed a segmentation technique that combines edge-based segmentation results with region-based segmentation results. Their algorithm utilizes the RGB color components of the image. The algorithm consists of the gradient operator proposed in [70], for edge detection and the region growing algorithm for region-based segmentation results. They obtain an accurate superposition between the edge pixels supplied by the gradient operator and the contours provided by the region growing pixels approach. This good correlation performs a significant matching for both edges and contour images, improves linkage between dislocated edges, and closes pixel elements of contours. Closing is achieved by a local operation which combines an iterative labeling method associated with a probabilistic relaxation approach.

In [63] two color image segmentation algorithms were proposed. The algorithms employ a fuzzy region growing technique and an edge detection technique in the RGB color space. One of them is used for fine segmentation towards compression and coding of image and the other for coarse segmentation towards other applications, such as object recognition and image understanding. Edge detection and region growing approaches are combined to find large and crisp segments for coarse segmentation. Segments can grow or expand based on two fuzzy criteria. The fuzzy region growing and expanding approaches use histogram tables for fine segmentation.

In [64] the authors propose an image segmentation technique which is based both on edge detection and region extraction. Suitable fuzzy sets representing the color information of a given image are automatically generated, which are then used for the intuitive edge detection and region extraction (pixel classification) approaches. The method is based on doing both in tandem, in order to make up for the disadvantages inherent in applying them singly. In this technique, the HLS (hue, lightness, saturation) color features are used by employing fuzzy sets to quantify them. The HLS color values are obtained from the RGB values using the following formulas:

$$L = 0.3R + 0.59G + 0.11B \tag{6.26}$$
$$H = \tan^{-1}(C_1/C_2) \tag{6.27}$$
$$S = \sqrt{C_1^2 + C_2^2} \tag{6.28}$$
$$C_1 = R - L = 0.7R - 0.59G - 0.11B \tag{6.29}$$
$$C_2 = B - L = -0.3R - 0.59G + 0.89B \tag{6.30}$$

Membership functions are based on histograms of hue and lightness and are generated by the following steps:

1. The histograms of lightness and hue are smoothed. If the pixel's saturation is too low, the hue attribute of that pixel is ignored.
2. A triangular membership function is set up around a mountain on the histogram, with the apex of the triangular membership function corresponding to the peak of the mountain, and the slope according to the frequencies at peaks and valleys. The lightness or hue attribute corresponding to the peak is called the representative color of that fuzzy set. Trapezoidal membership functions are assigned when the frequencies are too large.
3. Among the fuzzy sets generated at step 2, those in which representative color cannot be distinguished are merged into one set.

They proposed a method of combining the results of the two segmentation processes which have been run in parallel, with neither procedure providing the bulk of the segmentation process. The method is based on classifying a pixel as an edge pixel based on four cases which take into account the results of the edge detection and the region extraction processes.

In [65] the authors proposed a new method for color image segmentation that combines edge detection, a split and merge algorithm, and the model-based technique of [54]. The approach rests on the principle that the segmentation map which is an indicator of similarities between pixels must be consistent with the edge map which represents discontinuities between pixels. First, an initial color segmentation map is computed where labels form spatially contiguous regions. Then, region labels are optimized by split and merge procedures to enforce consistency with the edge map. Their method is performed in the luminance-chrominance (YES) color space defined by a linear transformation from the RGB color space [66]. However, they state that there method can be easily applied in any other suitable color space.

6.8 Application

In this section, the applicability of a color based image segmentation scheme is discussed. The proposed scheme utilizes the HSI (hue, saturation, intensity) color space because of its close relation to the human perception of colors. It was mentioned in Chap. 1 that although color receptors in the human eye (cones) absorb light with the greatest sensitivity in the blue, green, and red part of the color spectrum, the signals from the cones are further processed in the visual system [72], it is hard to visualize a color based on the R, G, and B components. It is also impossible to evaluate the perceived differences between colors on the basis of distance in the RGB color space. In terms of segmentation, the RGB color space is usually not prefered because it is psychologically non-intuitive and non-uniform. For all of these reasons the HSI color space was selected for development purposes [73].

The proposed region-based scheme utilizes region growing and region merging techniques. The scheme can be split into four general steps:

1. The pixels in the image are classified as chromatic or achromatic by examining their HSI color values.
2. Seeds pixels are found for the chromatic regions.
3. The region growing algorithm is employed to segment the image into regions starting from the seeds found is step 2.
4. Regions that are similar in color are merged.

6.8.1 Pixel Classification

When comparing the colors of two pixels, a problem is encountered when one or both of the pixels have no or very little chromatic information. That is, a gray scale object can not successfully be compared to an object that has substantial chromatic information. For this reason, all the pixels in the image are classified as either chromatic or achromatic pixels. Pixels that have very little or no chromatic information are reffered to as achromatic pixels. Achromatic pixels are never compared to chromatic pixels.

Of the three HSI color components of a pixel, hue has the greatest discrimination power because it is independent of any intensity attribute. Even though hue is the most useful attribute, there are two problems in using this value: (i) hue is meaningless when the intensity is very low or very high; and (ii) hue is unstable when the saturation is very low [2]. In Fig. 6.3 the HSI cone with the hue problem areas in yellow can be seen. Because of these hue attributes, in hue-based color models, the image is first divided into chromatic and achromatic regions by defining effective ranges of hue, saturation, and intensity values.

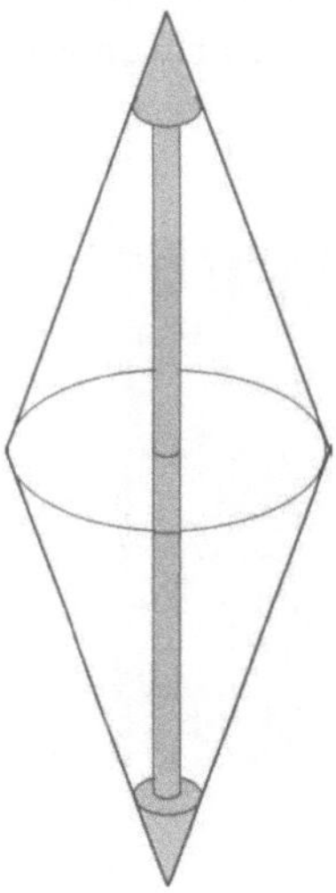

Fig. 6.3. The HSI cone with achromatic region in yellow

Since the hue value of a pixel is meaningless when the intensity is very low or very high the achromatic pixels in the image are defined as the pixels that have low or high intensity values. Pixels can also be categorized as achromatic if their saturation value is very low, since hue is unstable for low saturation values. From the concepts discussed above, the pixels in the image with low saturation, low intensity, or high intensity values are classified as achromatic. These threshold values are defined as: (i) $SATLOW$, (ii) $INTLOW$, and (iii) $INTHIGH$. It was found that achromatic pixels are best defined as follows:

$$\text{intensity (I)} > SATLOW = 90\% \text{ of MAX} \tag{6.31}$$
$$\text{or}$$
$$\text{intensity (I)} < INTLOW = 10\% \text{ of MAX}$$
$$\text{or}$$
$$\text{saturation (S)} < INTHIGH = 10\% \text{ of MAX}$$

where MAX is the maximum possible value. The threshold values were determined by experimental human observation. Pixels that do not fall into the achromatic category are categorized as chromatic pixels. In Fig. 6.4-6.15 images with the chromatic pixels in blue and the achromatic pixels as they are in the original image are depicted. In all the figures, the saturation and intensity values will be given on a scale of 0 to 100. In Fig. 6.5-6.7 the results when only the $SATLOW$ threshold value changes can be seen, while in

Fig. 6.9-6.11 the results obtained when only the $INTLOW$ threshold value changes are depicted. Finally, in Fig. 6.13-6.15 the results obtained when only the $INTHIGH$ threshold value changes are summarized. It can be observed in all three scenarios, that having low threshold values classifies achromatic pixels as chromatic and having high values classifies chromatic pixels as achromatic. It may be noted that most color images do not have many achromatic pixels, as is observed in Fig. 6.16-6.19.

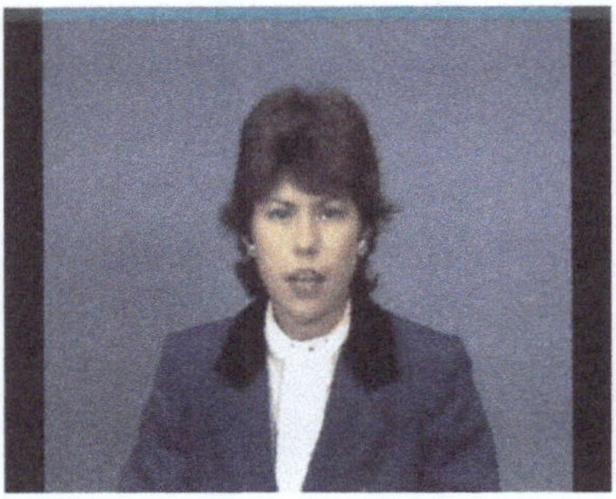

Fig. 6.4. Original image. Achromatic pixels: intensity < 10, > 90

Fig. 6.5. Saturation < 5

Fig. 6.6. Saturation < 10

Fig. 6.7. Saturation< 15

6.8.2 Seed Determination

The region growing algorithm starts with a set of seed pixels and from these grows regions by appending to each seed pixel those neighboring pixels that satisfy a certain homogeneity criterion, which will be described later. An unsupervised algorithm is used to find the best chromatic seed pixels in the

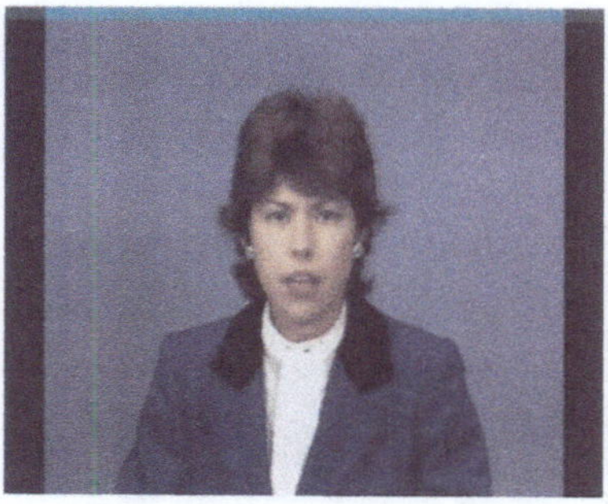

Fig. 6.8. Original image. Achromatic pixels: saturation < 10, intensity> 90

Fig. 6.9. Intensity < 5

Fig. 6.10. Intensity < 10

Fig. 6.11. Intensity < 15

image. These pixels will be the pixels that are in the center of the regions in the image. Usually the pixels in the center of a homogeneous region are the pixels that are dominant in color. The algorithm is used only to determine the seeds of the chromatic regions.

The seed determination algorithm employs variance masks to the image on different levels. Only the hue value of the pixels are considered in this approach because it is the most significant feature that may be used to detect uniform color regions [74]. All the pixels in the image are first considered as level zero seed pixels. At level one, a (3×3) non-overlapping mask is applied to the chromatic pixels in the image. The mask determines the variance, in hue, of the nine level zero pixels. If the variance is less than a certain threshold and the nine level zero pixels in the mask are chromatic pixels then the center pixel of the mask is categorized as a level one seed pixel. The first level seeds represent (3×3) pixel regions in the image. In the second level, the non-overlaping mask is applied to the level one seed pixels in the image.

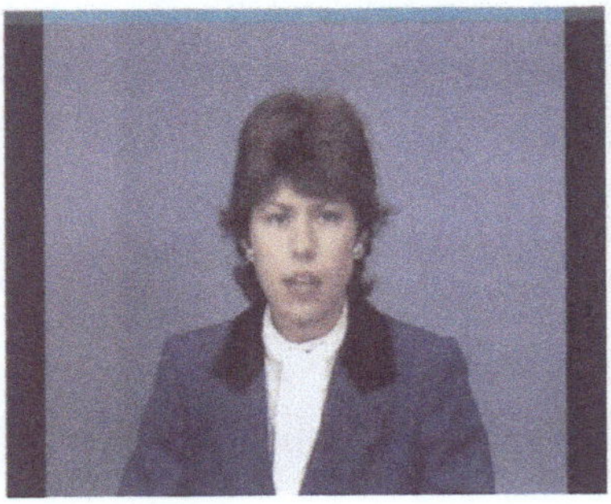

Fig. 6.12. Original image. Achromatic pixels: saturation< 10, intensity< 10

Fig. 6.13. Intensity > 85

Fig. 6.14. Intensity > 90

Fig. 6.15. Intensity > 95

Once again, the mask determines the variance in the average hue values of the nine level one seed pixels. If the variance is less than a certain threshold, the center pixel of the mask is considered as a level two seed pixel and the eight other level one seeds are disregarded as seeds. The second level seeds represent regions of (9×9) pixels. The process is repeated for successive level seed pixels until the seed pixels at the last level represent regions of a size just less than the size of the image. Typically, this is level 5 for an image that is a minimum of $(3^5 \times 3^5)$ in dimension. Fig. 6.20 shows an example of an image with level 1, 2, and 3 seeds. The algorithm is summarized in the following steps, with a representing the level:

1. All chromatic pixels in the image are set as level 0 seed pixels. Set a to 1.
2. Shift the level a mask to the next nine pixels (beginning corner of image if just increased a).

Fig. 6.16. Original image

Fig. 6.17. Pixel classification with chromatic pixels in red and achromatic pixels in the original color

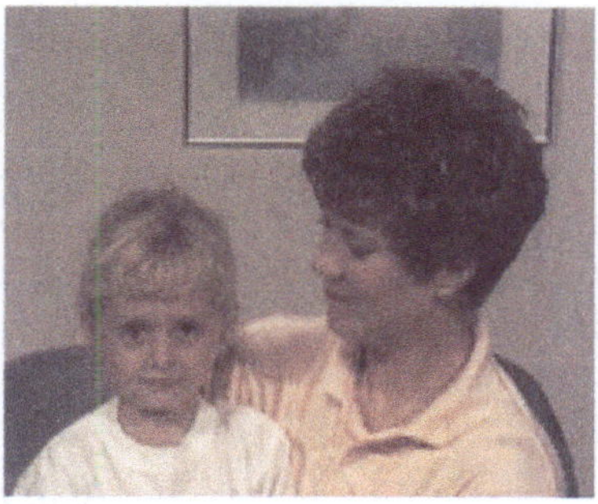

Fig. 6.18. Original image

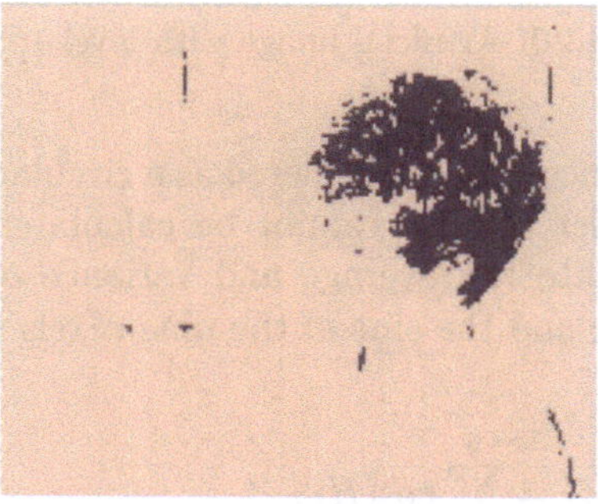

Fig. 6.19. Pixel classification with chromatic pixels in tan and achromatic pixels in the original color

3. If the mask reaches the end of the image increase a and go to step 2.
4. If all the seed pixels in the mask are of level $a-1$, continue. If not, go to step 2.
5. Determine the hue variance of the nine level $a-1$ seed pixels in the (3×3) mask. The variance is computed by considering, if $a=1$, the hue values of the nine pixels. Otherwise, the average hue values of the level $a-1$ seed pixels are considered.
6. If the variance is less than a threshold T_{VAR} then the center level $a-1$ seed pixel is changed to a level a seed pixel and the other eight level $a-1$ seed pixels are no longer considered as seeds.
7. Go to step 2.

Although the image is not altered in the algorithm it can be considered as a crude segmentation of the image.

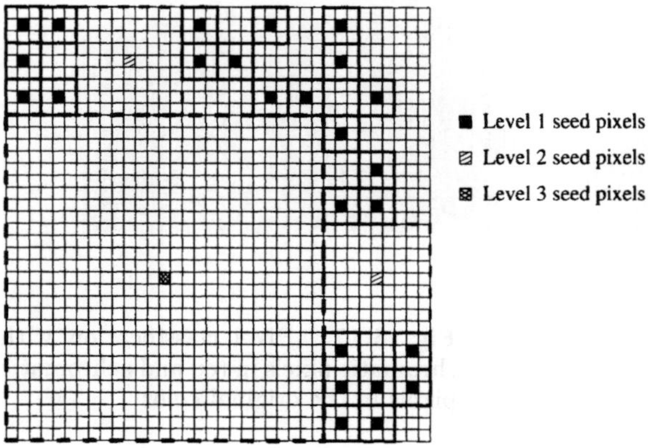

Fig. 6.20. Artificial image with level 1, 2, and 3 seeds.

Since hue is considered as a circular value, the variance and average values of a set of hues cannot be calculated using standard linear equations. To calculate the average and variance of a set of hue values the sum of the cosine and the sine of the nine pixels must first be determined [75]:

$$C = \sum_{k=1}^{9} \cos(H_k) \tag{6.32}$$

$$S = \sum_{k=1}^{9} \sin(H_k) \tag{6.33}$$

where H_k is the hue value of pixel k in the (3×3) mask. The average hue $AVGHUE$ of the nine pixels is then defined as:

$$AVGHUE = \begin{cases} \arctan(S/C) & \text{if} \quad S > 0 \quad \text{and} \quad C > 0 \\ \arctan(S/C) + \pi & \text{if} \quad C < 0 \\ \arctan(S/C) + 2\pi & \text{if} \quad S < 0 \quad \text{and} \quad C > 0 \end{cases} \tag{6.34}$$

The variance $VARHUE$ of the nine pixels is determined as follows:

$$VARHUE = (-2\ln(R))^{\frac{1}{2}} \tag{6.35}$$

where R is the radiance of the hue and is defined as:

$$R = \frac{1}{9}\sqrt{C^2 + S^2} \tag{6.36}$$

If the value of $VARHUE$ is less than the threshold T_{VAR} then the center level $a-1$ seed is changed to a level a seed. The value of T_{VAR} varies depending on the level. The threshold value for each level is determined with the following formula:

$$T_{VAR} = VAR * a \tag{6.37}$$

where a and VAR are the level and an initial variance value, respectively.

6.8.3 Region Growing

The region growing algorithm starts with a set of seed pixels and from these grows regions by appending to each seed pixel those neighboring pixels that satisfy a homogeneity criterion. The general growing algorithm is the same for the chromatic and achromatic regions in the image. The algorithm is summarized in Fig. 6.21. The first seed pixel is compared to its 8-connected neighbors: eight neighbors of the seed pixel. Any of the neighboring pixels that satisfy a homogeneity criterion are assigned to the first region. This neighbor comparison step is repeated for every new pixel assigned to the first region until the region is completely bounded by the edge of the image or by pixels that do not satisfy the criterion. The color of each pixel in the first region is changed to the average color of all the pixels assigned to the region. The process is repeated for the next and each of the remaining seed pixels.

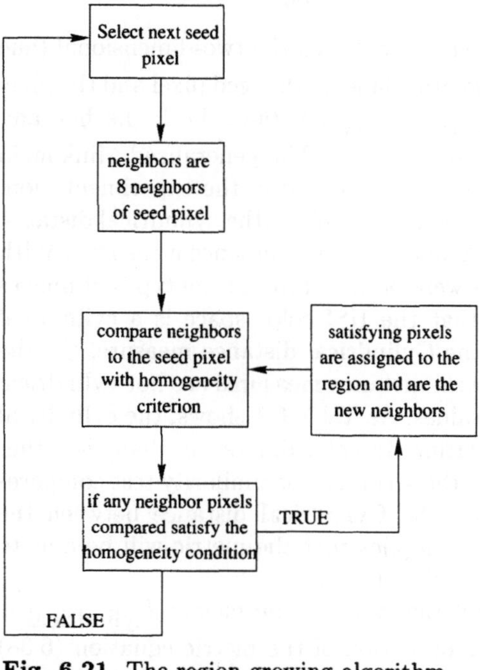

Fig. 6.21. The region growing algorithm.

For the chromatic regions, the algorithm starts with the set of varied level seed pixels. The seed pixels in the highest level are considered first, followed by the next highest level seed pixels, and so on, until level zero seed pixels are considered. The homogeneity criterion used for comparing the seed pixel and the unassigned pixel is that if the value of the distance metric used to compare the unassigned pixel (i) and the seed pixel (s) is less than a threshold value T_{chrom} than the pixel is assigned to the region.

The distance measure used for comparing pixel colors is a cylindrical metric. The cylindrical metric computes the distance between the projections of the pixel points on a chromatic plane. It is defined as follows [61]:

$$d_{\text{cylindrical}}(s,i) = \sqrt{(d_{\text{intensity}})^2 + (d_{\text{chromaticity}})^2} \tag{6.38}$$

with

$$d_{\text{intensity}} = |I_s - I_i| \tag{6.39}$$

and

$$d_{\text{chromaticity}} = \sqrt{(S_s)^2 + (S_i)^2 - 2S_s S_i \cos\theta} \tag{6.40}$$

where

$$\theta = \begin{cases} |H_s - H_i| & \text{if} \quad |H_s - H_i| < 180^\circ \\ 360^\circ - |H_s - H_i| & \text{if} \quad |H_s - H_i| > 180^\circ \end{cases} \tag{6.41}$$

The value of $d_{\text{chromaticity}}$ is the distance between the two-dimensional (hue and saturation) vectors, on the chromatic plane, of the seed pixel and the pixel under consideration. Therefore, $d_{\text{chromaticity}}$ combines both the hue and saturation (chromatic) components of the color. The generalized Minkowski and the Canberra distance measures were also used in the experimentation. But, in [76], it was found that when comparing colors, the cylindrical distance metric is superior over the Minkowski and Canberra distance measures . With the Cylindrical metric, good results were obtained for all the types of images tested. A reason for this may be that the HSI color space is a cylindrical color space which correlates with the Cylindrical distance measure. On the contrary, the Canberra and Minkowski distance measures are not cylindrical and don't compensate for angular values. As Table 6.1 shows, the cylindrical distance measure is more discriminating, in color difference, than the other two distance measures. Even though the second color similarity test compares two colors that are visually similar, the Cylindrical distance between the color is 3.43% of the maximum. This implies that the metric will be able to discriminate two colors that are virtually similar.

A pixel is assigned to a region if the value of the metric $d_{\text{cylindrical}}$ is less than a threshold T_{chrom}. An examination of the metric equation (6.38) shows that it can be considered as a form of the popular Euclidean distance (L_2 norm) metric .

Table 6.1. Comparison of Chromatic Distance Measures

Colour 1	Colour 2	Distance Measures						
		Minkowski					Canberra	Cylindrical
		2 2 2 (max= 387)	3 2 2 (max= 1935)	2 3 2 (max= 394)	2 2 3 (max= 394)	1 3 2 (max= 1005)	(max=3)	(max=224)
 R=35, G=19, B=245 H=244, S=57, I=39	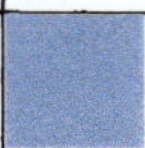 R=7, G=153 B=220 H=198, S=94, I=50	59.97 15.5%	137.23 7.1%	106.89 27.1%	37.38 9.5%	228.43 22.7%	0.469 15.63%	69.04 30.88%
 R=35, G=19 B=245 H=244, S=57, I=39	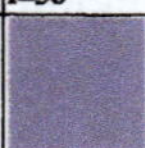 R=45, G=33, B=199 H=244, S=64, I=36	7.676 1.98%	5.737 0.296%	12.59 3.19%	6.341 1.61%	19.205 1.91%	0.09697 3.23%	7.676 3.43%
 R=35, G=19, B=245 H=244, S=57, I=39	 R=14, G=5, B=132 H=244, S=90, I=20	38.190 9.87%	22.696 1.17%	89.759 22.77%	47.881 12.15%	189.93 18.90%	0.5525 18.42%	38.1897 17.08%
 R=35, G=19, B=245 H=244, S=57, I=39	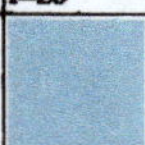 R=19, G=244, B=210 H=172, S=88, I=62	81.285 21.02%	243.448 12.58%	88.492 22.45%	66.405 16.84%	172.71 17.18%	0.61 20.34%	91.443 40.89%

In the case of the achromatic pixels , the same region growing algorithm is used but with all the achromatic pixels in the image considered as level zero seed pixels. There is no seed pixels with a level one or higher. The seed determination algorithm is not used for the achromatic pixels because achromatic pixels constitute a small percentage in most color images. Since intensity is the only justified color attribute that can be used when comparing pixels, the homogeneity criterion used is that if the difference in the intensity values between an unassigned pixel and the seed pixel is less than a threshold value T_{achrom} than the pixel is assigned to the seed pixel's region. That is, if

$$|I_s - I_i| < T_{achrom} \tag{6.42}$$

then pixel i would be assigned to the region of seed pixel s.

6.8.4 Region Merging

The algorithm determines dominant regions from the hue histogram. Dominant regions are classified as regions that have the same color as the peaks in the histogram. Once these dominant regions are determined, each remaining region is compared to them with the same color distance metric used in the region is compared to them with the same color distance metric used in the region growing algorithm (6.38). The region merging algorithm is summarized in the following steps:

1. Determine peaks in the hue histogram of region grown image.
2. Classify regions, that have the same color as these peaks, as dominant regions.
3. Compare each of the non-dominant regions with the dominant regions using the cylindrical distance metric.
4. Assign a non-dominant region to the dominant region if the color distance is less than a threshold T_{merge}.

The color of all the pixels, in regions assigned to a dominant region, are changed to the color of the dominant region.

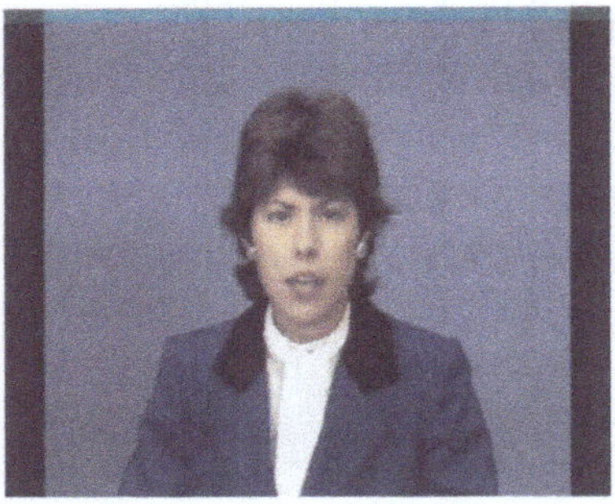

Fig. 6.22. Original 'Claire' image

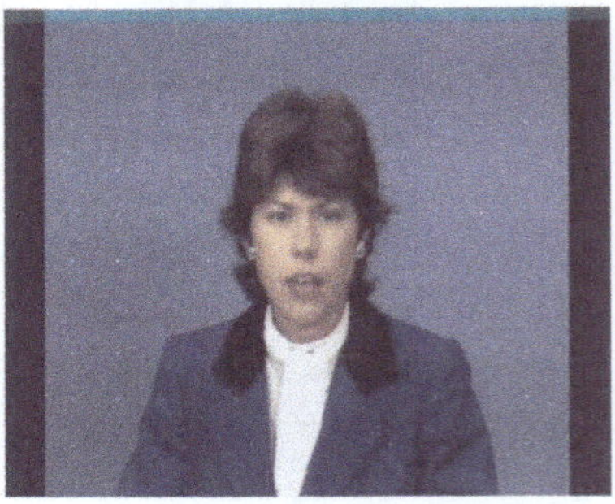

Fig. 6.23. 'Claire' image showing seeds with $VAR = 0.2$

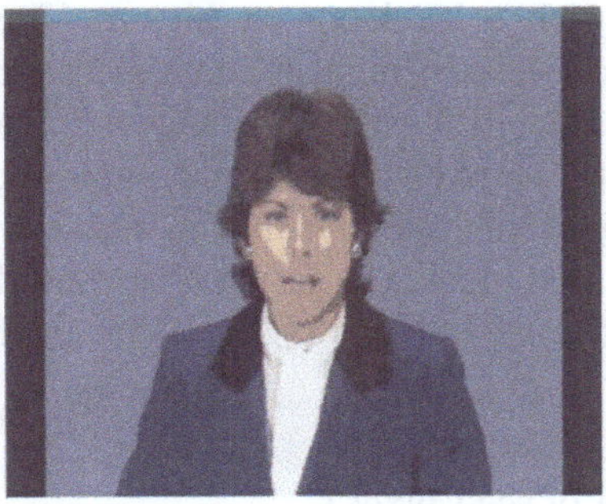

Fig. 6.24. Segmented 'Claire' image (before merging), $T_{chrom} = 0.15$

Fig. 6.25. Segmented 'Claire' image (after merging), $T_{chrom} = 0.15$ and $T_{merge} = 0.2$

Fig. 6.26. Original 'Carphone' image

Fig. 6.27. 'Carphone' image showing seeds with $VAR = 0.2$

Fig. 6.28. Segmented 'Carphone' image (before merging), $T_{chrom} = 0.15$

Fig. 6.29. Segmented 'Carphone' image (after merging), $T_{chrom} = 0.15$ and $T_{merge} = 0.2$

6.8.5 Results

The performance of the proposed color image segmentation scheme was tested with a number of different images. The results on three of these images will be presented here. The original images of 'Claire' , 'Carphone' , and 'Mother_daughter' are displayed in Figs. 6.22, 6.26, and 6.30, respectively. These images are stills from multimedia sequences. More specifically, they are video-phone type images.

The unsupervised seed determination algorithm found seeds in the image that were in the center area of the regions in the image. It was found that increasing the variance threshold T_{VAR} linearly with the level (i.e. $T_{VAR} = VAR * a$) produced the best seed pixels. Fig. 6.23 shows the original 'Claire' image with the level 3 and higher seed pixels found indicated as white pixels. Here VAR was set at 0.2. In particular 1 level 4 and 43 level 3 seed pixels were found. It was found that, for all the images tested, setting VAR to 0.2 gives the best results with no undesirable seeds. Fig. 6.27 shows the original 'Carphone' image with VAR set at 0.2 and the level 2 and higher seed pixels

Fig. 6.30. Original 'Mother-Daughter' image

Fig. 6.31. 'Mother-Daughter' image showing seeds with $VAR = 0.2$

Fig. 6.32. Segmented 'Mother-Daughter' image (before merging), $T_{chrom} = 0.15$

Fig. 6.33. Segmented 'Mother-Daughter' image (after merging), $T_{chrom} = 0.15$ and $T_{merge} = 0.2$

found indicated as white pixels. Here 19 level 2 seed pixels were found. Fig. 6.31 shows the original 'Mother_daughter' image with VAR set at 0.2. Here 1 level 3 (white) and 152 level 2 (black) seed pixels were found.

Figs. 6.24, 6.28, and 6.32 show the three experimental images after the region growing algorithm. It was found that best results were obtained with threshold values of $T_{achrom} = 15$ and $T_{chrom} = 15$ which are, respectively, 15% and 7% of the maximum distance values for the achromatic and the chromatic distance measures. The results show that there are regions in these segmented images that require merging.

Figs. 6.25, 6.29, and 6.33 show the three experimental images after the region merging step. The threshold value (T_{merge}) that gives the best merging results for a varied set of images is 20. This is, approximately, 9% of the maximum chromatic distance value. Most of the regions that were similar in color after the region merging step are now merged.

Table 6.2. Color Image Segmentation Techniques

Techniques		**Summary**
Pixel-based Color-based decision no spatial constraints simple algorithms	Histogram Thresholding	Color regions are determined by thresholding peak(s) in the histogram(s) simple to implement no spatial considerations
	Clustering	Many clustering algorithms K-means & fuzzy K-means pixels in image are assigned to the cluster that is similar in color adjacent clusters frequently overlap in color space, causing incorrect pixel assignment also suffers from no spatial constraints
Edge-based Focus on discontinuity of regions sensitivity to noise	Techniques extended from monochrome techniques	Monochrome techniques applied to each color component independently and then results are combined many first & second derivative operators can be used Sobel, Laplacian, Mexican Hat operators are most popular
	Vector Space Approaches	Views color image as a vector space Vector Gradient, Entropy, Second Derivative operators have been proposed sensitive to noise
Region-based Focus on the continuity of regions consider both color and spatial constraints	Region Growing	Process of growing neighboring pixels or a collection of pixels of similar color properties into larger regions further merging of regions is usually needed
	Split and Merge	Iteratively splitting the image into smaller and smaller regions and merging adjacent regions that satisfy a color homogeneity criterion quadtree data structure is most common used data structure in algorithms
Model-based		Regions modeled as random fields most techniques use the spatial interaction models like MRF or Gibbs Random Field maximum *a posteriori* approach is most common high complexity
Physics-based		Allows the segmentation of color images based on physical models of image formation basic methods are similar to traditional methods above most employ the Dichromatic Reflection Model many assumptions made best results for images taken in controlled environment
Hybrid		Combine the advantages of different techniques most common techniques of color image segmentation today

6.9 Conclusion

Color image segmentation is crucial for multimedia applications. Multimedia databases utilize segmentation for the storage and indexing of images and video. Image segmentation is used for object tracking in the new MPEG-7 video compression standard. And, as shown in the results, image segmentation is used in video conferencing for compression. These are only some of the multimedia applications for image segmentation. It is usually the first task of any image analysis process, and thus, subsequent tasks rely heavily on the quality of segmentation. A number of color image segmentation techniques have been surveyed in this chapter. They are summarized in Table 6.2.

The particular color image segmentation method discussed in the last section of the chapter was shown to be very effective. Classifying pixels as either chromatic or achromatic avoids any color comparison of pixels that are undefined, in terms of color. The seed determination algorithm finds seed pixels that are in the center of regions which is vital when growing regions from these seeds. The cylindrical distance metric gives the best results when color pixels need to be compared. Merging regions that are similar in color is a final means of segmenting the image into even less regions. The segmentation method proposed is interactive [77]. The best threshold values for the segmentation scheme are suggested but these values may be easily changed for different standards. This allows for control of the degree of segmentation.

References

1. Marr, D., (1982): Vision. Freeman, San Francisco, CA.
2. Gonzales, R.C., Wood, R. E., (1992) Digital Image Processing. Addison-Wesley, Boston, Massachusetts.
3. Pal, N., Pal, S.K. (1993): A review on image segmentation techniques. Pattern Recognition, **26**(9), 1277–1294.
4. Skarbek, W., Koschan, A. (1994): Color Image Segmentation: A Survey. Technical University of Berlin, Technical report, **94-32**.
5. Fu, K.S., Mui, J.K. (1981): A survey on image segmentation, Pattern Recognition, **13**, 3–16.
6. Haralick, R.M., Shapiro, L.G. (1985): Survey, image segmentation techniques. Computer Vision Graphics and Image Processing, **29**, 100–132.
7. Pratt, W.K. (1991): Digital Image Processing. Wiley, New York, N.Y.
8. Wyszecki, G., Stiles, W.S. (1982): Color Science. New York, N.Y.
9. Ohlander, R., Price, K., Reddy, D.R. (1978): Picture segmentation using a recursive splitting method. Computer Graphics and Image Processing, **8**, 313–333.
10. Ohta, Y., Kanade, T., Sakai, T. (1980): Color information for region segmentation, Computer Graphics and Image Processing, **13**, 222–241.
11. Holla, K. (1982): Opponent colors as a 2-dimensional feature within a model of the first stages of the human visual system. Proceedings of the 6th Int. Conf. on Pattern Recognition, Munich, Germany, 161–163.
12. von Stein, H.D., Reimers, W. (1983): Segmentation of color pictures with the aid of color information and spatial neighborhoods. Signal Processing II: Theories and Applications, North-Holland, Amsterdam, Netherlands, 271–273.
13. Tominaga S. (1986): Color image segmentation using three perceptual attributes. Proceedings of the Computer Vision and pattern Recognition Conference, **CVPRR'86**, 628–630.
14. Gong, Y. (1998): Intelligent Image Databases: Towards Advanced Image Retrieval. Kluwer Academic Publishers, Boston, Massachusetts.
15. Hartigan, J.A. (1975): Clustering Algorithms. John Wiley and Sons, USA.
16. Tou, J., Gonzalez, R.C. (1974): Pattern Recognition Principles. Addison-Wesley Publishing, Boston, Massachusetts.

17. Tominaga, S. (1990): A color classification method for color images using a uniform color space. Proceedings of the 10th Int. Conf. on Pattern Recognition, **1**, 803–807.
18. Celenk, M. (1988): A recursive clustering technique for color picture segmentation. Proceedings of Int. Conf. on Computer Vision and Pattern Recognition, **CVPR'88**, 437–444.
19. Celenk, M. (1990): A color clustering technique for image segmentation. Computer Vision, Graphics, and Image Processing, **52**, 145–170.
20. McLaren, K. (1976): The development of the CIE (L*,a*,b*) uniform color space. J. Soc. Dyers Colour, 338–341.
21. Gevers, T., Groen, F.C.A. (1990): Segmentation of Color Images. Technical report, Faculty of Mathematics and Computer Science, University of Amsterdam.
22. Weeks, A.R., Hague, G.E. (1997): Color segmentation in the HSI color space using the K-means algorithm. Proceedings of the SPIE, **3026**, 143–154.
23. Heisele, B., Krebel, U., Ritter, W. (1997): Tracking non-rigid objects based on color cluster flow. Proceedings, IEEE Computer Society Conference on Computer Vision and Pattern Recognition, 257–260.
24. Zadeh, L.A. (1965): Fuzzy sets. Information Control, **8**, 338–353.
25. Bezdek, J.C. (1973): Fuzzy Mathematics in Pattern Classification. Ph.D. Thesis, Cornell University, Ithaca, N.Y.
26. Bezdek, J.C. (1981): Pattern Recognition with Fuzzy Objective Function Algorihms. Plenum Press, New York, N.Y.
27. Huntsberger, T.L., Jacobs, C.L., Cannon, R.L. (1985): Iterative fuzzy image segmentation. Pattern Recognition, **18**(2), 131-138.
28. Trivedi, M., Bezdek, J.C. (1986): Low-level segmentation of aerial images with fuzzy clustering. IEEE Transactions on Systems, Man, and Cybernetics, **16**(4), 589-598.
29. Lim, Y.W., Lee, S.U. (1990): On the color image segmentation algorithm based on the thresholding and the fuzzy c-Means techniques. Pattern Recognition, **23**(9), 1235–1252.
30. Goshtasby, A., O'Neill, W. (1994): Curve fitting by a sum of Gaussians. CVGIP: Graphical Models and Image Processing, **56**(4), 281–288.
31. Witkin, A.P. (1984): Scale space filtering: A new approach to multi-scale description. Proceedings of the IEEE Int. Conf. on Acoustics, Speech and Signal Processing, **ICASSP'84**(3), 39A1.1–39A1.4.
32. Koschan, A. (1995): A comparitive study on color edge detection. Proceedings of the 26nd Asian Conference on Computer Vision, **ACCV'95**(III), 574–578.
33. Ikonomakis, N., Plataniotis, K.N., Venetsanopoulos, A.N. (1998): Grey-scale and image segmentation via region growing and region merging. Canadian Journal of Electrical and Computer Engineering, **23**(1), 43-48.
34. Gauch, J., Hsia, C. (1992): A comparison of three color image segmentation algorithms in four color spaces. Visual Communications and Image Processing, **1818**, 1168–1181.
35. Tremeau, A., Borel, N. (1997): A region growing and merging algorithm to color segmentation. Pattern Recognition, **30**(7), 1191–1203.
36. Vlachos, T., Constantinides, A.G. (1992): A graph-theoretic approach to color image segmentation and contour classification. The 4th Int. Conf. on Image Processing and its Applications, **IEE 354**, 298–302.
37. Horowitz, S.L., Pavlidis, T. (1974): Picture segmentation by a directed split-and-merge procedure. Proceedings of the 2nd International Joint Conf. on Pattern Recognition, 424–433.
38. Samet, H. (1984): The quadtree and related hierarchical data structures. Computer Surveys, **16**(2), 187–230.

39. Gevers, T., Kajcovski, V.K. (1994): Image segmentation by directed region subdivision. Proceedings of the 12th IAPR Int. Conf. on Pattern Recognition, **1**, 342–346.
40. Lee, D.L., Schachter, B.J. (1980): Two algorithms for constructing a delaunay triangulation. International Journal of Computer and Information Sciences, **9**(3), 219–242.
41. Abend, K., Harley, T., Kanal, L.N. (1965): Classification of binary random patterns. IEEE Transactions on Information Theory, **IT-11**, 538–544.
42. Besag, J. (1986): On the statistical analysis of dirty pictures. Journal Royal Statistical Society B, **48**, 259–302.
43. Cross, G.R., Jain, A.K. (1983): Markov random field texture models. IEEE Transactions on Pattern Analysis and Machine Intelligence, **PAMI-5**, 25–39.
44. Geman, S., Geman, D. (1984): Stochastic relaxation, Gibbs distributions, and the Bayesian restoration of images. IEEE Transactions on Pattern Analysis and Machine Intelligence, **PAMI-6**, 721–741.
45. Cohen, F.S., Cooper, D.B. (1983): Real time textured image segmentation based on non-causal Markovian random field models. Proceedings of the SPIE, Conference on Intelligent Robots, Cambridge, MA.
46. Cohen, F.S., Cooper, D.B. (1987): Simple, parallel, hierarchical, and relaxation algorithms for segmenting non-causal Markovian random field models. IEEE Transactions on Pattern Analysis and Machine Intelligence, **PAMI-9**(2), 195–219.
47. Derin, H., Elliott, H. (1987): Modeling and segmentation of noisy and textured images using Gibbs random fields. IEEE Transactions on Pattern Analysis and Machine Intelligence, **PAMI-9**(1), 39–55.
48. Lakshmanan, S., Derin, H. (1989): Simultaneous parameter estimation and segmentation of Gibbs random field using simulated annealing. IEEE Transactions on Pattern Analysis and Machine Intelligence, **PAMI-11**(8), 799–813.
49. Panjwani, D.K., Healey, G. (1995): Markov random field models for unsupervised segmentation of textured color images. IEEE Transactions on Pattern Analysis and Machine Intelligence, **PAMI-17**(10), 939–954.
50. Langan, D.A., Modestino, J.W., Zhang, J. (1998): Cluster validation for unsupervised stochastic model-based image segmentation. IEEE Transactions of Image Processing, **7**(2), 180–195.
51. Tekalp, A.M. (1995): Digital Video Processing, Prentice Hall, New Jersey.
52. Liu, J., Yang, Y.-H. (1994): Multiresolution color image segmentation. IEEE Transactions on Pattern Analysis and Machine Intelligence, **PAMI 16**(7), 689–700.
53. Pappas, T.N. (1992): An adaptive clustering algorithm for image segmentation. IEEE Transactions on Signal Processing, **40**(4), 901–914.
54. Chang, M.M., Sezan, M.I., Tekalp A.M. (1994): Adaptive Bayesian segmentation of color images. Journal of Electronic Imaging, **3**(4), 404–414.
55. Baraldi, A., Blonda, P., Parmiggiani, F., Satalino, G. (1998): Contextual clustering for image segmentation. Technical report, **TR-98-009**, International Computer Science Institute, Berkeley, California.
56. Brill, M.H. (1991): Photometric models in multispectral machine vision. in Proceedings, Human Vision, Visual Processing, and Digital Display II, **SPIE 1453**, 369–380.
57. Healey, G.E. (1992): Segmenting images using normalized color. IEEE Transactions on Systems, Man, and Cybernetics, **22**, 64–73.
58. Klinker, G.J., Shafer, S.A., Kanada, T. (1988): Image segmentation and reflection analysis through color. in Proceedings, IUW'88, **II**, 838–853.

59. Klinker, G.J., Shafer, S.A., Kanada, T. (1990): A physical approach to color image understanding. International Journal of Computer Vision, **4**(1), 7–38.
60. Shafer, S.A. (1985): Using color to separate reflection components. Color Research & Applications, **10**(4), 210–218.
61. Tseng, D.-C., Chang, C.H. (1992): Color segmentation using perceptual attributes. Proceedings of the 11th International Conference on Pattern Recognition, **III**, 228–231.
62. Zugaj, D., Lattuati, V. (1998): A new approach of color images segmentation based on fusing region and edge segmentations outputs. Pattern Recognition, **31**(2), 105–113.
63. Moghaddamzadeh, A., Bourbakis, N. (1997): A fuzzy region growing approach for segmentation of color images. Pattern Recognition, **30**(6), 867–881.
64. Ito, N., Kamekura, R., Shimazu, Y., Yokoyama, T. (1996): The combination of edge detection and region extraction in non-parametric color image segmentation. Information Sciences, **92**, 277–294.
65. Saber, E., Tekalp, A.M., Bozdagi, G. (1997): Fusion of color and edge information for improved segmentation and edge linking. Image and Vision Computing, **15**, 769–780.
66. Xerox Color Encoding Standards: (1989). Technical Report, Xerox Systems Institute, Sunnyvale, CA.
67. Beucher, S. and Meyer, F. (1993): The morphological approach to segmentation: The watershed tranformation. Mathematical Morphology in Image Processing, 443-481.
68. Duda, R. O. and Hart, P. E (1973): Pattern Classification and Scene Analysis. Wiley, New York, N.Y.
69. Shafarenko, L., Petrou, M., and Kittler, J. (1998): Histogram-based segmentation in a perceptually uniform color space. IEEE Transactions on Image Processing, **7**(9), 1354-1358.
70. Di Zenzo, S. (1986): A note on the gradient of a multi-image. Computer Vision Graphics, Image Processing, **33**, 116-126.
71. Park, S. H., Yun, I. D., and Lee, S. U. (1998): Color image segmentation based on 3-d clustering: Morphological approach. Pattern Recognition, **31**(8), 1061-1076.
72. Levine, M.D. (1985): Vision in Man and Machine. McGraw-Hill, New York, N.Y.
73. Ikonomakis, N., Plataniotis, K.N., Venetsanopoulos, A.N. (1998): Color image segmentation for multimedia applications. Advances in Intelligent Systems: Concepts, Tools and applications, Tzafestas, S.G. (ed.), 287-298, Kluwer, Dordrecht, Netherlands.
74. Gong, Y., Sakauchi, M. (1995): Detection of regions matching specified chromatic features. Computer Vision and Image Understanding, **61**(2): 263-264.
75. Fisher, N.I. (1993): Statistical Analysis of Circular Data. Cambridge Press, Cambridge, U.K.
76. Ikonomakis, N., Plataniotis, K.N., Venetsanopoulos, A.N. (1999): A region-based color image segmentation scheme. SPIE Visual Communication and Image Processing, **3653**, : 1202-1209.
77. Ikonomakis, N., Plataniotis, K.N., Venetsanopoulos, A.N. (1999): User interaction in region-based color image segmentation. Visual Information Systems. Huijmans, D.P., Smeulders, A.W.M. (eds.), 99-106, Springer, Berlin, Germany.

7. Color Image Compression

7.1 Introduction

Over the past few years the world has witnessed a growing demand for visual based information and communications applications. With the arrival of the 'Information Highway' such applications as tele-conferencing, digital libraries, video-on-demand, cable shopping and multimedia asset management systems are now common place. Hand-to-hand with the introduction of these systems and the simultaneous improvement in the quality of these applications were the improved hardware and techniques for digital signal processing. The improved hardware which offered greater capabilities in terms of computational power, combined with the sophisticated signal processing techniques that allowed for a much greater flexibility in processing and manipulation, gave rise to new information applications, and advances and better quality in existing applications.

As the demand for new applications and higher quality for existing applications continues to rise, the transmission and storage of the visual information becomes a more critical issue [1], [2]. The reason for this is that higher image or video quality requires larger volume of information. However, transmission media have a finite and limited bandwidth. To illustrate the problem, consider a typical (512×512) monochrome (8-bit) image . This image has 2,097,152 bits. By using a 64 Kbit/s communication channel, it would take about 33 seconds to transmit the image. Whereas this might be acceptable for a one time transmission of a single image, it would definitely not be acceptable for tele-conference applications, where some form of continuous motion is required. The large volume of information contained in each image also creates storage difficulties. To store an uncompressed digital version of a 90 minute black and white movie, at 30 frames/sec, with each frame having (512×512×8) bits, would require 3.397386e+11 bits, over 42 GBytes. Obviously, without any form of compression the amount of storage required for a modest size digital library would be staggeringly high. Also, higher image quality, which usually implies use of color and higher image resolution, would be much more demanding in terms of transmission time and storage.

To appreciate the need for compression and coding of visual signals such as color images and video frames, signal characteristics and their storage needs are summarized in Table 7.1.

Table 7.1. Storage requirements

Visual input	Pixels/frame	Bits/pixel	Uncompressed size
VGA image	640×480	8	3.74Mb
XVGA image	1024×768	24	18.87Mb
NTSC frame	480×483	16	3.71Mb
PAL frame	576×576	16	5.31Mb
HDTV frame	1280×720	12	11.05Mb

A $4:2:2$ color sub-sampling scheme employed in NTSC and PAL while a $4:1:1$ color sub-sampling scheme is used in HTDV.

To address the problems of transmission and storage, different image compression algorithms can be employed to: (i) eliminate any information redundancies in the image, and (ii) reduce the amount of information contained in the image. Whereas elimination of information redundancy does not hamper at all the quality of the image, eliminating necessary information does come at the cost of image quality degradation. Images and video signals are amenable to compression due to these factors:

1. **Spatial redundancy**.
2. **Spatial redundancy**. Within a single image or video frame there exists significant correlation among neighboring pixels. Redundancy in an image also includes repeated occurances of base shapes, colors and patterns within the image.
3. **Spectral redundancy**. For visual data, such as color images or multispectral images acquired from multiple sensors, there exists significant correlation amongst samples from the different spectral channels.
4. **Temporal redundancy**. For visual data, such as video streams, there is significant correlation amongst samples in different time instances. The most obvious form is redundancy from repeated objects in consecutive frames of a video stream.
5. **Observable redundancy**. There is considerable information in the visual data that is irrelevant from a perceptual point of view. By taking advantage of the perceptual masking properties of the human visual system and by expressing its insensitivity to various types of distortion as a function of image color, texture and motion, compression schemes can develop a profile of the signal levels that provide just noticeable distortion (JND) in the image and video signals. Thus, it is possible based on this profile to create coding schemes that hide the reduction effects under the JND profile and thereby make the distortion become perceptually invisible.

6. **Meta data redundancy**. Some visual data, such as synthetic images, tend to have high-level features that are redundant across space and time, in other words data that are of a fractal nature.

Depending on the kind of information removed during the compression processes the following forms of compression can be defined:

1. **Lossless compression**. Lossless image compression allows the exact reconstruction of the original image during the decoding (de-compression) process. The problem is that the best lossless image compression schemes are limited to modest compression gains. Lossless compression is mainly of interest in applications where image quality is more important than the compression ratio and visual data must remain unchanged over many consecutive cycles of compression and decompression.
2. **Lossy compression**. Lossy compression algorithms allow only approximate reconstruction of the original image from the compressed data. The lower the quality of the reconstructed image needs to be, the more the original image can be compressed. Examples of lossy compression schemes are the JPEG lossy compression mode used to compress still color images and the MPEG compression standards for video sequences [3–7]. All lossy compression schemes produce artifacts. Although in some applications the degradation may not be perceivable, it may be annoying after several cycles of compression and decompression. Traditionally, image compression techniques were able to achieve compression ratios of about 10:1 without causing a noticeable degradation in the quality of the image. However, any attempt to further reduce the bit rate would invariably result in noticeable distortions in the reconstructed image, usually in the form of block artifacts, color shifts and false contours.
3. **Perceptually lossless compression**. Perceptually lossless image compression deals with lossy compression schemes in which degradation in the image quality is not visible to human observers [8], [9]. Perceptually motivated compression schemes make use of the properties of the human visual system to improve further the compression ratio. In this type of coding, perceptually invisible distortions of the original image are accepted in order to attain very high compression ratios. Since not all signal frequencies in an image have the same importance, an appropriate frequency weighting scheme can be introduced during the encoding process. After the perceptual weighting has been performed, an optimized encoder can be used to minimize an objective distortion measure, such as the mean square error [10].

There are many coding techniques applicable to still, monochrome or color, images and video frames. These techniques can be split into three distinct groups according to the way in which they deal with the source data. In particular, they can be defined as:

1. **Waveform based coding techniques**. These techniques, also called first generation techniques, refer to methods that assume a certain model on the statistics of pixels in the image. The primitives of these techniques are either individual pixels or a block of pixels or a transformed version of their values. These primitives constitute the message to be encoded. There are lossless and lossy waveform based techniques. Lossless techniques include variable length coding techniques, such as arithmetic coding and Lempel Ziv coding, pattern matching and statistical based techniques, such as Fano or Huffman coding. Lossy wave form based techniques include time domain techniques, such as pulse code modulation (PCM) and vector quantization (VQ) [11], where as frequency domain techniques include methodologies based on transforms, such as the Fourier transform, discrete cosine transform (DCT) [12], [13] and the Karhune Loeve (KL) transform as well as techniques based on wavelets and subband analysis/synthesis systems [14], [15], [16].
2. **Second generation coding techniques**. Second generation techniques, model or object based, are techniques attempting to describe an image in terms of visually meaningful primitives, such as distinct color areas, strong edges, contours and texture. Emerging multimedia applications, such a multimedia databases, and video-on-demand will need access to visual data on an object-to-object basis. Visual components, such as color and shape, along with motion for video applications, can be used to support such requirements [17], [18], [19].
   In this group, fractal-based coding techniques can also be included. These techniques are based on the fractal theory, in which an image is reconstructed by means of an affine transformation of its self-similar regions. Fractal based techniques produce outstanding results in terms of compression in images, retaining a high degree of self similarity, e.g. synthetic images [20].

Table 7.2 and 7.3 give a perspective of the available techniques and their classification.

Before reviewing some of the waveform based and second generation techniques in greater detail, the basis on which these techniques are evaluated and compared will be given, along with a few of the important terms that are used throughout the chapter.

7.2 Image Compression Comparison Terminology

Compression methodologies are compared on the basis of the following dimensions of performance [21], [22]:

1. **Image quality.** Image quality refers to the subjective quality of the reconstructed image relative to the original, uncompressed image. This

Table 7.2. A taxonomy of image compression methodologies: First Generation

Waveform based Techniques (1st Generation)				
Lossless		Lossy		
Entropy Coding	Interpixel Redundancy	Spatial Domain	Transform Domain	Hybrid Techniques
Huffman Coding	DPCM	DPCM	Block Transform	BTC/DPCM BTC/VQ
Arithmetic Coding	Runlength Coding	Vector Quantization	Coding BTC	SBC/DPCM SBC/VQ
Lempel-Ziv Coding		Scalar Quantization	(DFT, DCT DST, KL)	Entropy Coded
		Tree Structured Quantization	Subband Coding	Version of Above
		Predictive Vector Quantization	(SBC)	
		Finite-State Vector Quantization Quantization		
		Entropy Coded Version of Above		

Table 7.3. A taxonomy of image compression methodologies: Second Generation

Second Generation Techniques
Object Segmentation Coding
Texture Modeling /Segmentation - Contour Coding - Fractal Coding
Morphological Techniques
Model Based Coding Techniques

subjective assessment refers to an actual image quality rating done by human observers. The rating is done on a five-point scale called the mean opinion score (MOS) with the five points ranging from bad to excellent. Objective distortion measures, such as the mean square error (MSE), the relative mean square error (RMSE), the mean absolute error (MAE) and signal-to-noise ratio (SNR) can quantify the amount of information loss an image has suffered, but in many cases they do not provide an accurate or even correct measure of the actual visual quality degradation.

2. **Coding efficiency.** Two of the more popular measures of an algorithm's efficiency are the compression ratio and the bit rate. Compression ratio is simply the ratio of the number of bits needed to encode the uncompressed image to the number of bits needed to encode the compressed image. An equivalent efficiency measure is the bit rate which gives the average number of bits required to encode one image element (pixel). In the context of image compression, the higher the compression ratio, or the lower the bit rate the more efficient is the algorithm. However, the efficiency measure might be misleading if not considered in unison with the signal quality measure since some image compression algorithms might compress the image by reducing the resolution, both temporal and spatial, or by reducing the number of quantization levels [9]. In other words, to evaluate the efficiency of two algorithms, their respective image quality must be the same. To that end, a measure that incorporates those two dimensions, efficiency and image quality, is the rate distortion function. The rate distortion function describes the minimum bit rate required for a given average distortion.
3. **Complexity.** The complexity of an algorithm refers to the computational effort required to carry out the compression technique. The computational complexity is often given in million instruction per second (MIPS), floating point operation per second (FLOPS), and cost. This performance dimension is important since an algorithm might be preferable to another one which is marginally more efficient but is much more computationally complex.
4. **Communication delay.** A performance dimension of lesser importance, mainly because it is not an important consideration in some applications, is the communication delay. This performance indicator refers to how much delay is allowed before the compressed image is transmitted. In cases where a large delay can be tolerated, e.g. facsimile, more time consuming algorithms can be allowed. On the other hand, in two-way communication applications a long delay is definitely not allowed.

7.3 Image Representation for Compression Applications

When choosing a specific compression method, one should consider the data representation format. Images for compression may be in different formats which are defined by:

- color space used
- the number bits per pixel
- spatial resolution
- temporal resolution for video signals

Initially the image compression techniques were defined in the context of monochrome images. However, most of today's image applications are based on color representation. It was therefore necessary to extend these image compression techniques so that they can accommodate color images.

The extension to color image compression is straight-forward but requires an understanding of the various color models used. Linear RGB is the basic and most widely used color model for color display on monitors. It was mentioned before that in RGB, a color is represented as a composition of the three primary color spectral components of red, green, and blue. A color image can then be represented as three 8-bit planes corresponding to each of the primary colors, for a total of 24 bits/pixel [23]. The value in each of the color planes can then be considered as a gray scale value which would represent the intensity of that particular color at the current pixel. This color representation can then be very easily compressed using the regular spatial domain image compression methods, such as entropy coding. This is simply done by separately compressing each of the three color planes. However, the RGB space is not an efficient representation for compression because there is a significant correlation between the three color components since the image energy is distributed almost equally among them both spatially and spectrally.

A solution is to apply an orthogonal decomposition of the color signals in order to compact the image data into fewer channels. The commonly used YUV, YIQ and YC_bC_r color spaces are examples of color spaces based on this principles. Theoretically, these color coordinate systems can provide nearly as much energy compaction as an optimal decomposition of the Karhunen-Loeve transform. The resulting luminance-chrominance representation exhibits unequal energy distribution favoring the luminance component in which the vast majority of fine detail high frequencies can be found [24].

Since the sensitivity of the human visual system is relatively high for chrominance errors, the chrominance channels need only a fraction of the luminance resolution in order to guarantee sharpness on the perceived image. Therefore, the chrominance components are mostly sub-sampled with respect to the chrominance component when a luminance-chrominance representation, such as the YC_bC_r is used. There are three basic sub-sampling formats for processing color images. In the 4:4:4 format all components have identical vertical and horizontal resolutions. In the 4:2:2 format, also known as CCIR 601 format, the chrominance components have the same vertical resolution as the luminance component, but the horizontal resolution is halved. The most common format is the 4:2:0 used in conjunction with the YC_bC_r color space in the MPEG-1 and MPEG-2 standards. Each MPEG macroblock comprises of four 8×8 luminance blocks and one 8×8 blocks of C_b and C_r color components. A 24 bits/pixel representation is also typical for luminance-chrominance representation of digital video frames. However, 10-bit representation of the components is used in some high-fidelity applications.

7.4 Lossless Waveform-based Image Compression Techniques

Waveform-based image compression techniques can reduce the bit rate by efficiently coding the image. The coding is done without considering the global importance of the pixel, segment, or block being coded. Conventional waveform-based techniques can be identified either as lossless techniques or lossy techniques. Both these classes will be described in detail.

There are two main ways in which the bit rate can be reduced without losing any information. The first method is to simply use efficient codes to code the image. The second method is to try and reduce some of the redundant information that exist in the image.

7.4.1 Entropy Coding

In entropy coding bit rate reduction is based solely on codeword assignment. Entropy is the amount of information based on the probabilistic occurrence of picture elements. Mathematically, entropy is defined as:

$$H(X) = -\sum P(x_i) \log P(x_i) \tag{7.1}$$

where $P(x_i)$ is the probability that the monochrome value (x_i) will occur, and $H(X)$ is the entropy of the source measured in bits [25]. These probabilities can be found from the image's histogram. In this sense, the entropy describes what is the average information or uncertainty of every pixel. Since it is very unlikely that each of the possible gray-levels will occur with equal probability, variable length codewords can be assigned to describe specific pixel values with the more probable pixel values being assigned shorter codewords, thus achieving shorter average codeword length. This coding (compression) principle is employed by the following coding methods:

1. **Huffman coding.** This is one of the most straightforward and practical encoding methods. Huffman coding assigns fixed codewords to the source

words (in this case the source words being the pixel values). The least probable source words are assigned the longest codewords whereas the most probable are assigned the shortest codewords. This method requires knowledge of the image's histogram. With this codeword assignment rule, Huffman coding approaches the source's entropy. The main advantage of this method is the ease of implementation. A table is simply used to assign source words their corresponding codewords. The main disadvantages are that the size of the table is equal to the number of source words, and the table with all the codeword assignments also has to be made known to the receiver.

2. **Arithmetic coding.** Arithmetic coding can approach the entropy of the image more closely than can be done with Huffman coding. Unlike Huffman coding, there is no one-to-one correspondence between the source words and the codewords [26]. In arithmetic coding, the codeword defines an interval between 0 and 1. The specific interval is based on the probability of occurrence of the source word. The main idea of arithmetic coding is that blocks of source symbols can be coded together by simply representing them with smaller and more refined intervals (as the block of source symbol increases, more bits would be required to represent the corresponding interval) [26]. Compared to Huffman coding, the main advantage of this method is that less bits are required to encode the image since it is more economical to encode blocks of source symbols than individual source symbols. Also, no codeword table is required in this method, and thus arithmetic coding does not have the problem of memory overhead. However, the computational complexity required in arithmetic coding is considerably higher than in Huffman coding.
3. **Lempel-Ziv coding.** Lempel-Ziv coding is a universal coding scheme. In other words a coding scheme which approaches entropy without having prior knowledge of the probability of occurrence of the source symbols. Unlike the two entropy methods mentioned above, the Lempel-Ziv coding method assigns blocks of source symbols of varying length to fixed length codewords. In this coding method the source input is parsed into strings that have not been encountered thus far. For example, if the strings '0', '1', and '10' are the only strings that have been encountered so far, then the strings '11', '100', '101' are examples of strings that are yet to be encountered and recorded. When a new string is encountered, it is recorded by indexing its prefix (which will correspond to a string that has already appeared) and its last bit. The main advantage of this coding method is that absolutely no prior knowledge of the source symbol probabilities is needed. The main disadvantage is that since all codewords are of fixed length, short input source sequences, such as low resolution images might be encoded into longer output sequences. However, this method does approach entropy for long input sequences.

It is important to note that that entropy coding can always be used to supplement other more sophisticated and efficient algorithms, by assigning variable codeword length to the output of those algorithms. It should also be emphasized that entropy coding utilizes only the probability of occurrence of the different pixel value but not the correlation between the values of neighboring pixels. Entropy coding can therefore reduce the bit rate by usually no more than 20-30% resulting in a compression ratio of up to 1.4 : 1.

7.4.2 Lossless Compression Using Spatial Redundancy

More significant bit rate reduction can be realized if the interpixel redundancy that exists in the image is reduced. Since images are generally characterized by large regions of constant or near constant pixel values, there is considerable spatial redundancy that can be removed. The following is a description of some of the common methods that can be used to remove this redundancy without losing any information.

1. **Predictive coding**. One way of reducing the spatial redundancy is to use neighboring pixel values as an estimate to the current pixel [25], [28]. Therefore, instead of encoding the actual value of the current pixel, the difference between the predicted value, predicted from pixel that were already traversed, and the actual pixel value is encoded. The coding method is called differential pulse code modulation (DPCM). Since that difference would generally be small, the dynamic range of the error will be much smaller than the dynamic range of the pixel values, and therefore and entropy coding method can be used very effectively to encode the error. The overall coding procedure can be summarized as follows:
   - Find a linear estimate of the current pixel from its neighbors according to:
   $$\hat{f}(m,n) = \sum_i \sum_j a(i,j) f(m-i, n-j) \tag{7.2}$$
   In many cases the estimate is rounded off to the closest integer so that there will not be a need to deal with decimals. In addition, the only pixels allowed to be used in the estimation are those that occur prior to the current one since these are the pixels that will be available during the image reconstruction.
   - Find the error between the actual value of the current pixel and the corresponding estimate according to:
   $$e(m,n) = f(m,n) - \hat{f}(m,n) \tag{7.3}$$
   Encode the error value using one of the several entropy coding techniques described before.
   - At the decoder end, an estimate of the current pixel is again derived using the same prediction model, and the decoded error that was transmitted is added to the estimate to obtain the original value of the pixel according to:

$$f(m,n) = \hat{f}(m,n) + e(m,n) \tag{7.4}$$

This compression scheme can achieve much better compression ratios than those obtained by only using entropy coding schemes. The compression ratios tend to vary from 2:1 to 3:1 [27]. The variation in compression ratio is due to several factors. One of the main factors is the particular parameters that are chosen to estimate the pixels. Indeed, better prediction parameters will result in closer estimates and by extension will reduce the bit rate. Moreover, if adaptive linear prediction parameters are chosen by splitting the image into smaller blocks and computing the prediction parameters for each block, the compression ratio can be further improved [25]. Another way of improving the compression ratio is by scanning the image using a different pattern, such as Peano scan or Worm path patterns [28] rather than using the regular raster scan pattern from left to right and top to bottom. By traversing the image in a different order, estimates of the current pixel can also be derived from pixels which are below it and thus further reduce the interpixel redundancy.

2. **Runlength coding**. This coding algorithm is intended mainly for the compact compression of bi-level images and is widely used for fax transmissions. This scheme centers on the fact that there are only two types of pixels, namely black and white. Also, since high correlation between neighboring pixels exists it would be enough to simply indicate where and how long a black or white run of pixels is in order to perfectly reconstruct the image from that information. The runlength coding method most often used is based on the relative address coding (RAC) approach [26]. This specific method codes the runs of black or white pixels on the current line relative to the black and white runs of the previous line. This way both the correlation between vertical neighbors and horizontal neighbors is exploited to reduce the interpixel redundancy. The coding algorithm is as follows:
   - Two coding modes are defined. The first one is the horizontal mode which codes the black and white runs without referring to the previous line, and the vertical mode where the previous line is taken into account in order to take advantage of the vertical correlation of the image [29].
   - Horizontal mode uses a Huffman coding method to assign the various black and white runlength variable length codewords (based on the probability of occurrence of a specific runlength in a typical image). In vertical mode the information coded just indicates the beginning and ending position of the current runlength relative to the corresponding runlength in the previous line.
   - The first line is always coded using horizontal mode. Furthermore, one in every few lines also has to be coded using horizontal mode to reduce the susceptibility of this scheme to error. All other lines are coded using the vertical mode [29].

Compression ratios of 9:1 to 11:1 are achieved using this technique on bi-level images [29]. However, gray scale and color images are usually ill-suited for this type of compression method. This is because to code a monochrome image using runlength coding requires bit-plane decomposition on the image, namely breaking down the m-bit gray scale image to m 1-bit planes [26]. While it is found that high correlation exist between the pixels of the most significant bit-planes, there is significantly less correlation in the less significant bit-planes and thus the overall compression ratios achieved are not as high as those achieved using predictive coding [30].

7.5 Lossy Waveform-based Image Compression Techniques

Lossy compression techniques allows for some form of information loss and possibly some degradation in the quality of the image. As was mentioned above, the best that can be achieved in terms of compression when perfect reconstruction is required is about a 2:1 to a 3:1 compression ratio. However, when the perfect reconstruction constraint is dropped, much higher compression ratios can be achieved. The tradeoff is of course in the quality of the image and the complexity of the algorithm. Lossy domain can be done by using either spatial domain, or transform domain methods. The following section will consider both.

7.5.1 Spatial Domain Methodologies

Lossy spatial domain coding methods, much like their lossless counterparts, exploit the spatial redundancy in an image. However, in lossy coding, the accuracy of representing the residual information, that is the information that remains once the basic redundancy is removed, is compromised in order to obtain higher compression ratios. The compressed image cannot then be perfectly reconstructed due to this inaccurate lossy representation. Some of the common spatial domain coding methods are described below.

1. **Predictive coding.** Lossy predictive coding essentially follows the same steps as the lossless predictive coding with the exception that a quantizer is used to quantize the error between the actual and predicted values of the current pixel [26]. When a quantizer is used, there are only several discrete values that the encoded error value can take and thus there is an improvement in the compression ratio. However, use of a quantizer results in quantization error, and the image cannot be perfectly reconstructed since the actual error values are no longer available. The performance of this coding method in terms of coding efficiency and reconstructed image quality depends on the:

- **Prediction model**. The proper choice of prediction parameters, either adaptive or global, will minimize the prediction error and improve the compression ratio. Also, the number of previous pixels that are used to predict the value of the current pixel will also affect the effectiveness of the prediction. The scanning pattern, raster scan or Peano scan, also affects the performance of the this coding method.
- **Quantizer**. Choice of the number of quantizer levels and the actual levels used. Given a specified number of quantizer levels, the problem is reduced to finding the decision levels, and reconstruction levels, which are the unique values into which the decision level intervals are mapped in a many-to-one fashion, that will minimize the given error criterion, objective or subjective. An example of a quantizer that minimizes the mean-square error is the Lloyd-Max quantizer.

2. **Vector quantization**. This compression technique operates on blocks rather than on individual pixels. It can decompress visual information in real time using software, without the use of special hardware and does so with reasonable quality. The main idea of vector quantization (VQ) is that a block of k image pixels, which henceforth will be referred to as a block of dimension k, can be represented by a k dimensional template chosen from a table of pre-defined templates [11]. The template to represent a particular k-dimensional block is chosen on the basis of minimizing some error criterion, such as the template closest to the block in some sense. A code representing the chosen template is then transmitted. The encoder and the decoder use the same codebook. To optimize performance, a training method involving the use of test sequences is utilized to generate the codebook in an automatic manner. At the receiver end, the decoder can use the index to fetch the corresponding codeword and use it as the decompressed output. The decompression is not as computationally intensive as that employed in transform based schemes, such as the JPEG [31]. The coding efficiency, typically up to a 10:1 compression ratio, and image quality will depend on the following:
   - **Template table size**. Large tables and large number of templates result in smaller quantization errors. This will translate to a higher and better reconstructed image quality. However, large template tables require longer codes to represent the selected template and so the bit rate increases.
   - **Choice of templates**. The main problem with the VQ method is that the specific templates chosen to represent the blocks are usually image dependent. Hence, it is hard to construct a table that will yield a consistent image quality performance which is independent of the image. Also, to improve the subjective quality of the image it is sometimes necessary to construct context dependent templates. For example specific templates are needed for situations in which the k-dimensional block has an edge, and different templates should be considered for situations

where the block is a shade [11]. This inevitably increases the size of the template table and with it the computational complexity and bit rate.

7.5.2 Transform Domain Methodologies

Transform domain coding methods have become by far the most popular and widely used conventional compression techniques. In this type of coding the image is transformed into an equivalent image representation. Common linear transformations that are used in transform coding are the Karhunen-Loeve (KL), discrete Fourier transform (DFT), discrete cosine transform (DCT), and others. The main advantages in this kind of representation is that the transformed coefficients are fairly de-correlated and most of the energy, therefore most of the information, of the image is concentrated in only a small number of these coefficients. Hence, by proper selection of these few important coefficients, the image can be greatly compressed. There are two transform domain coding techniques that warrant special attention. These two techniques are the discrete cosine transform (DCT) coding and subband coding [32].

The DCT transform and the JPEG compression standard. Of the many linear transforms known, the DCT has become the most widely used. The two dimensional DCT pair (forward and inverse transform), used for image compression, can be expressed as follows [34], [31], [33]:

$$C(u,v) = \frac{2}{N}\sum_{x=0}^{N-1}\sum_{y=0}^{N-1} f(x,y)\cos\left[\frac{(2x+1)u\pi}{2N}\right]\cos\left[\frac{(2y+1)v\pi}{2N}\right] \tag{7.5}$$

for u, $v = 0, 1, ..., N-1$ (for $u, v = 0$, the scaling factor is $\frac{1}{N}$)

$$f(x,y) = \frac{2}{N}\sum_{u=0}^{N-1}\sum_{v=0}^{N-1} C(u,v)\cos\left[\frac{(2x+1)u\pi}{2N}\right]\cos\left[\frac{(2y+1)v\pi}{2N}\right] \tag{7.6}$$

for $x, y = 0, 1, ..., N-1$ (for $x, y = 0$, the scaling factor is $\frac{1}{N}$)

In principle, DCT introduces no loss to the original image samples. It simply transforms the image pixels to a domain in which they can be more efficiently encoded. In other words, if there are no additional steps, such as quantization of the coefficients, the original image block can be recovered exactly. However, as it can be seen from (7.5) and (7.6) the calculations contain transcendental functions. Therefore, no finite time implementation can compute them with perfect accuracy. Because of the finite precision used for the DCT inputs and outputs, coefficients calculated by different algorithms or by discrete implementations of the same algorithm will result in slightly different output for identical input. Nevertheless, DCT offers a good and practical compromise between information packing abilities, that is packing a lot of information into a small number of coefficients, computational complexity, and minimization of block artifact image distortion [26]. These attributes are what prompted the International Standards Organization (ISO) and the Joint Photographic Expert Group (JPEG) to base their international standard for still image compression on the DCT.

The JPEG standard is used for compressing and decompressing continuous tone monochrome as well as color images. Applications range from compressing images for audio-graphical presentations to desktop publishing, to multimedia database browsing and tele-medicine. JPEG is of reasonably low computational complexity, is capable of producing compressed images of high quality and can provide both lossless and lossy compression of arbitrary sized images. JPEG converts a block of an image in the time domain into the frequency domain using the DCT transform. Since the human vision system is not sensitive to high spatial frequencies, coarser quantization levels can be used to generate a rough representation of the high spatial frequency portion of the image. Because the coarser representation requires a fewer number of bits, the process reduces the amount of information needed to be stored or communicated.

The JPEG standard does not specify any specific color model to be used for the color image representation. However, in most cases JPEG handles colors as independent components so that each component can be processed as a monochrome image. The necessary color space transforms were performed before and after the JPEG algorithm. As there are many ways to represent color images, the standard does not specify any color space for the representation of the color images. Currently the JPEG standard is set up for use with any three-variate color space. Common color representations used in conjunction with the standard include color models, such as the linear RGB, the YIQ, the YUV and the YC_bC_r color spaces. Experimentation with different color spaces indicate that tristimulus color models are not very efficient for use as a color compression space. For example, the major weakness of the linear RGB color space, from a compression point of view, is the spectral redundancy in the three channels. Simulation studies had revealed that the color information is encoded much less efficiently for the RGB color space than it is for other color spaces. Similarly, studies show that perceptually uniform color spaces, such as the CIE $L^*a^*b^*$ space, are good color compression spaces. Color spaces derived linearly form RGB, such as the YIQ, YUV and YC_bC_r also provide excellent results. On the contrary, perceptually motivated spaces, such as the HVS, do not constitute an efficient color space for compression purposes. The poor performance should be attributed mainly in the poor quantization of the hue values using default quantization tables [24]. In summary, it can be said that the JPEG algorithm is a color space dependent procedure and that both numerical measures and psychological techniques indicate that uncorrelated color spaces, such as the YC_bC_r should be used to maximize the coding again.

The major objective of the JPEG committee was to establish a basic compression technique for use throughout industry. For that reason the JPEG standard was constructed to be compatible with all the various types of hardware and software that would be used for image compression. To accomplish this task a baseline JPEG algorithm was developed. Changes could be made to the baseline algorithm according to individual users' preference but only the baseline algorithm would be universally implemented and utilized. Compression ratios that range from 5:1 to 32:1 can be obtained using this method, depending on the desired quality of the reconstructed image and the specific characteristics of the image.

The JPEG provides four encoding processes for applications with communications or storage constraints [3]. Namely,

1. **Sequential mode.** In the JPEG sequential mode or baseline system the color image is encoded in a raster scan manner from left to right and top to bottom. It uses a single pass through the data to encode the image and employs 8-bit representation per channel for each input.
2. **Lossless mode.** An exact replica of the original color image can be obtained using the JPEG lossless mode. This mode is intended for applications requiring lossless compression, such as medical systems where scans are stored, indexed, accessed and transmitted from site to site on demand, and multimedia systems processing photographs for accident claims, banking forms or insurance claims. In this mode the image pixels are handled separately. Each pixel is predicted based on three adjacent pixels using one of eight possible predictor models. An entropy encoder is then used to losslessly encode the predicted pixels.
3. **Progressive mode.** The color image is encoded in multiple scans and each scan improves the quality of the reconstructed image be encoding additional information. Progressive encoding depends on being able to store the quantized DCT coefficients for an entire image. There are two forms of progressive encoding for JPEG. The spectral selection and the successive approximation methodologies. In the first approach the image is encoded from a low frequency representation to a high frequency sharp image. In JPEG spectral selection progressive mode the image is transformed to the frequency domain using the DCT transform. The initial transmission sends low frequency DCT coefficients followed by the higher frequency coefficients until all the DCT coefficients have been transmitted. Reconstructed images from the early scans are blurred since each image lacks the high frequency components until the end layers are transmitted. In the JPEG successive approximation mode all the DCT coefficients for each image block are sent in each scan. However, only the most significant bits of each coefficient are sent in the first scan, followed by the next most significant bits until all the bits are sent. The resulting reconstructed images are of reasonably good quality, even for the very early scans, since the high-frequency components of the image

are preserved in all scans. The progressive mode is ideal for transmitting images over bandwidth limited communication channels since end-users can view the coarse version of the image first and then decide if a finer version is necessary. Progressive mode is also convenient for browsing applications in electronic commerce or real estate applications where a low resolution image is more than adequate if the property is of no interest to the customer.

4. **Hierarchical mode.** The color image is encoded at multiple resolutions. In a JPEG hierarchical mode the low resolution image is used as the basis for encoding a higher resolution of the same image by encoding the difference between the interpolated low resolution and higher resolution versions. Lower resolution versions can be accessed without first having to reconstruct the full resolution image. The different resolution modes can be achieved by filtering and down sampling the image, usually in multiples of two in each dimension. The resulting decoded image is up sampled and from the next level, which is then coded and transmitted as the next layer. The process is repeated until all layers have been coded and transmitted. The hierarchical mode can be used to optimize equipments with different resolutions and display capabilities.

JPEG utilizes a methodology based on DCT for compression. It is a symmetrical process with the same complexity for coding and decoding. The baseline JPEG algorithm is composed of three compression steps and three decompression steps. The compression procedure as specified by the JPEG standard is as follows [34]:

- Each color image pixel is transformed to three color values corresponding to luminance and two chrominance signals, e.g. YC_bC_r. Each transformed chrominance channel is down sampled by a predetermined factor.
- The transform is performed on a sub-block of each channel image rather than on the entire image. The block size chosen by the JPEG standard is 8×8 pixels resulting in 64 coefficients after the transform is applied. The blocks are typically inputed block-by-block from left-to-right and then block row by block row top-to-bottom.
- The resultant 64 coefficients are quantized according to a predefined table. Different quantization tables are used for each color component of an image. Tables 7.4 and 7.5 are typical quantization tables for the luminance and chrominance components used in the JPEG standard.
  The quantization is an irreversible lossy compression operation in the DCT domain. The extent of this quantization is what determines the eventual compression ratio. This quantization controls the bit accuracy of the respective coefficients and therefore determines the degree of image degradation, both objective and subjective. Because much of the block's energy is contained in the direct current, zero frequency, (DC) coefficient, this coefficient receives the highest quantization precision. Other coefficients that hold little of the block's energy can be discarded altogether.

Table 7.4. Quantization table for the luminance component

16	11	10	16	24	40	51	61
11	12	14	19	26	58	60	55
14	13	16	24	40	57	69	56
14	17	22	29	51	87	80	62
18	22	37	56	68	109	103	77
24	35	55	64	81	104	113	92
49	64	78	87	103	121	120	101
72	92	95	98	112	100	103	99

Table 7.5. Quantization table for the chrominance components

17	18	24	47	99	99	99	99
18	21	26	66	99	99	99	99
24	26	56	99	99	99	99	99
47	66	99	99	99	99	99	99
99	99	99	99	99	99	99	99
99	99	99	99	99	99	99	99
99	99	99	99	99	99	99	99
99	99	99	99	99	99	99	99

- After quantization only the low frequency portion of the block contains non-zero coefficients. In order to reduce the number of bits required for storage and communication, as many zeros as possible are placed together so that rather than dealing with each individual zero, representation is in terms of the number of zeros. This representation is accomplished through the zig-zag scan shown in Fig. 7.1.
  The ordering converts the matrix of transform coefficients into a sequence of coefficients along the line of increasing spatial frequency magnitude. The scan pertains only to the 63 AC coefficients. In other words it omits the DC coefficient in the upper left corner of the diagram. The DC coefficient represents the average sample value in the block and is predicted from the previously encoded block to save bits. Only the difference from the previous DC coefficient is encoded, a value much smaller than the absolute value of the coefficient. The quantized coefficients are encoded using an entropy coding method, typically Huffman coding, to achieve further compression [34]. JPEG provides the Huffman code tables used with DC and AC coefficients for both luminance and chrominance. For hierarchical or lossless coding, arithmetic coding tables can be used instead of Huffman coding tables. Once encoded, the coefficients are transmitted to the receiver where they are decoded, and an inverse transformation is performed on them to obtain the reconstructed image.

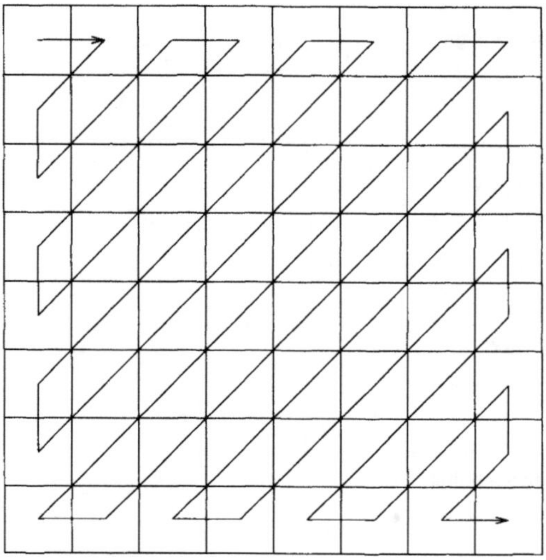

Fig. 7.1. The zig-zag scan

These steps should be repeated until the entire image is in a compressed form. At this point the image can be stored or transmitted as needed. The overall scheme is depicted in Fig. 7.2

The steps in the decompression part of the standard are: (i) decoding the bit stream, (ii) de-quantization, (iii) transforming from a frequency domain back to a spatial image representation, (iv) up-sampling each chrominance channel, and (v) inverse transformation of each color pixel to recover the reconstructed color image. De-quantization is performed by multiplying the coefficients by the respective quantization step. The basic unit is an (8×8) block. The values of the pixels in the individual image blocks are reconstructed via the inverse discrete cosine transformation (IDCT) of (7.6). When the last three steps have been repeated for all the data, an image will be reconstructed.

To illustrate the preceding discussion, Fig. 7.3-7.8 show the original RGB color image 'Peppers' and results coded with JPEG at different quality levels. The distortions introduced by the coder at lower quality level are obvious.

Recently, a new standard, the so called JPEG 2000, was introduced as an attempt to focus existing research efforts in the area of still color image compression. The new standard is intended to provide low bit rate operation with subjective image quality superior to existing standards, without sacrificing performance at higher bit rates. The scope of JPEG 2000 includes not only potential new compression algorithms but also flexible compression architectures and formats. Although it will be completed by the year 2000, it will offer state-of-the art compression for many years beyond. It will also serve image compression needs that are currently not served and it will provide

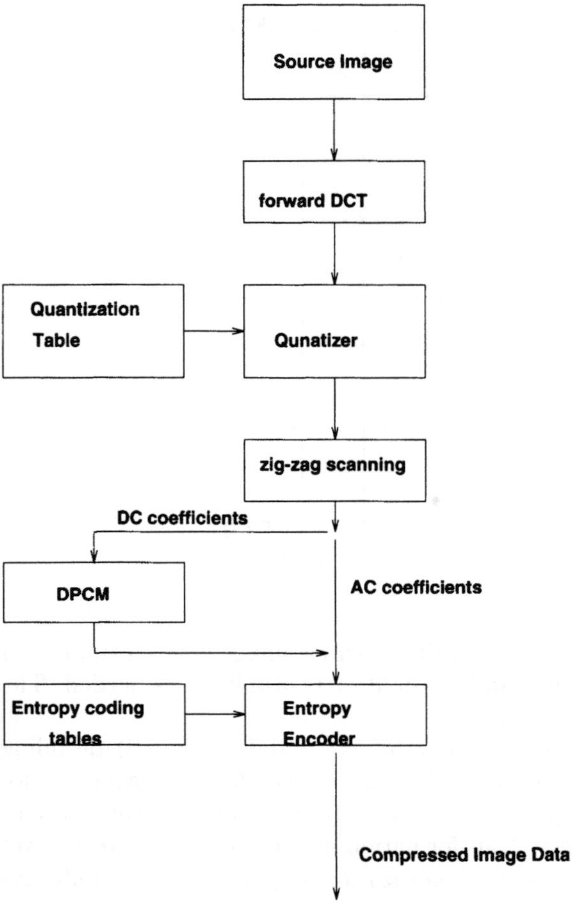

Fig. 7.2. DCT based coding

access to markets that currently do not consider compression as useful for their applications. It is anticipated that the new standard will address open issues, such as [4]:

- **Variable image formats.** The current JPEG standard does not allow large image sizes. However, with the lowering cost of display technologies visual information will be widely available in the HDTV format and thus the compression standards should support such representation.
- **Content-based description.** Visual information is difficult to handle both in terms of its size and the scarcity of tools available for navigation and retrieval. A key problem is the effective representation of this data in an environment in which users from different backgrounds can retrieve and handle information without specialized training. A content-based approach based on visual indices, such as color, shape and texture seem to be a natural choice. Such an approach might be available as part of the evolving JPEG-2000 standard.

Fig. 7.3. Original color image 'Peppers'

Fig. 7.4. Image coded at a compression ratio 5 : 1

Fig. 7.5. Image coded at a compression ratio 6 : 1

Fig. 7.6. Image coded at a compression ratio 6.3 : 1

Fig. 7.7. Image coded at a compression ratio 6.35 : 1

Fig. 7.8. Image coded at a compression ratio 6.75 : 1

- **Low bit rate compression.** The performance of the current JPEG standard is unacceptable in very low bit rates mainly due to the distortions introduced by the transformation module. It is anticipated that research will be undertaken in order to guarantee that the new standard will provide excellent compression performance in very low bit rate applications.
- **Progressive transmission by pixel accuracy and resolution.** Progressive transmission that allows images to be reconstructed with increasing pixel accuracy or spatial resolution as more bits are received is essential in many emerging applications, such as the World Wide Web, image archiving and high resolution color printers. This new feature allows the reconstruction of images with different resolutions and pixel accuracy, as needed and desired, for different target and devices.
- **Open architecture.** JPEG 2000 follows an open architecture design in order to optimize the system for different image types and applications. To this end, research is focused in the development of new highly flexible coding schemes and the development of a structure which should allow the dissemination and integration of those new compression tools. With this capability, the end-user can select tools appropriate to the application and provide for future growth. This feature allows for a decoder that is only required to implement the core tool set plus a parser that understands and executes downloadable software in the bit stream. If needed, unknown tools are requested by the decoder and sent from the source.
- **Robustness to bit errors.** JPEG 2000 is designed to provide robustness against bit errors. One application where this is important is wireless communication channels. Some portions of the bit stream may be more important than others in determining decoded image quality. Proper design of the bit stream can prevent catastrophic decoding failures. Usage of confinement, or concealment, restart capabilities, or source-channel coding schemes can help minimize the effects of bit error.
- **Protective image security**. Protection of the property rights of a digital image is of paramount importance in emerging multimedia applications, such as web-based networks and electronic commerce. The new standard should protect digital images by utilizing one or more of four methods, namely: (i) watermarking, (ii) labeling, (iii) stamping, and (iv) encryption. All of these methods should be applied to the whole image file or limited to part of it to avoid unauthorized use of the image.
- **Backwards compatibility** It is desirable for JPEG 2000 to provide for backwards compatibility with the current JPEG standard.
- **Interface with MPEG-4.** It is anticipated that the JPEG 2000 compression suite will be provided with an appropriate interface allowing the interchange and the integration of the still image coding tools into the framework of content-based video standards, such as MPEG-4 and MPEG-7.

In summary, the proposed compression standard for still color images includes many modern features in order to provide low bit rate operation with subjective image quality performance superior to existing standards. By taking advantage of new technologies the standard is intended to advance standardized image coding systems to serve applications into the next millennium [4].

Subband coding techniques. Subband coding of images has been the subject of intensive research in the last few years [14], [35], [36]. This coding scheme divides the frequency representation of the image into several bands. Selection of bands is done by using a bank of bandpass filters. This scheme is similar to the DCT coding in that it divides the image's spectrum into frequency bands and then codes and transmits the bands according to the portion of the image's energy that they contain. However, implementation of subband coding is done via actual passband filters while, DCT is done computationally using a discrete linear transform. This method of implementation could affect the performance of subband coding in terms of complexity, matching to the human perceptual system and robustness to transmission error [21].

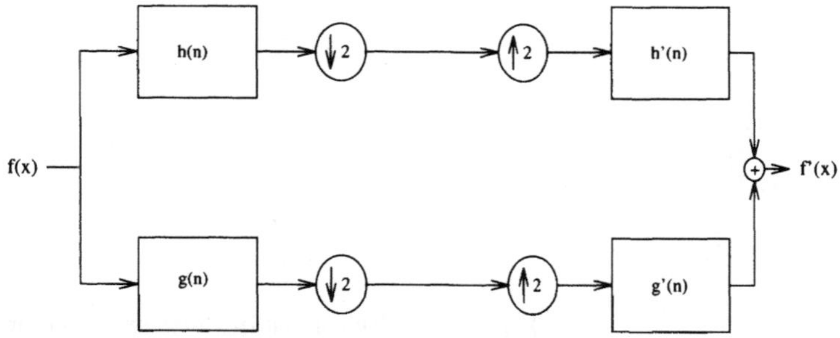

Fig. 7.9. Subband coding scheme

Specifically, most emphasis on subband coding techniques is given to the wavelet decomposition, a subset of subband decomposition, in which the transformed representation provides a multiresolution data structure [15], [37].

The first step in the wavelet scheme is to perform a wavelet transformation of the image. One of the more practical ways of implementing the wavelet transform is to carry out a multiresolution analysis (MRA) decomposition on the image. MRA performs the decomposition by an iterative process of low-pass, and high-pass filtering, followed by sub-sampling the resultant output signals. This type of iterative process yields a pyramid-like structure of signal components, which includes a single low-resolution component, and a series of added detail components, which can be used to perfectly reconstruct the original image.

The scheme is shown in Fig. 7.9. The actual decomposition algorithm is based on the following classical methodology:

1. Starting with the actual image, row low-pass filtering is performed, using the low-pass filter $g(n)$, by means of convolution operations.
2. The above is followed by performing column low-pass filtering on the low-passed rows to produce the *ll* subimage.
3. Column high-pass filtering, using the high-pass filter $h(n)$, is now performed on the low-passed rows to produce the *lh* subimage.
4. Row high-pass filtering, using $h(n)$, is performed on the input image.
5. Column low-pass filtering is performed on the high-passed rows to produce the *hl* subimage.
6. Column high-pass filtering is performed on the high-passed rows to produce the *hh* subimage.
7. The entire procedure is now repeated $(l - 1)$ more times, where l is the specified number of desired decomposition levels on the resultant *ll* subimage. In other words, the *ll* subimage now serves as the input image for the next decomposition level.

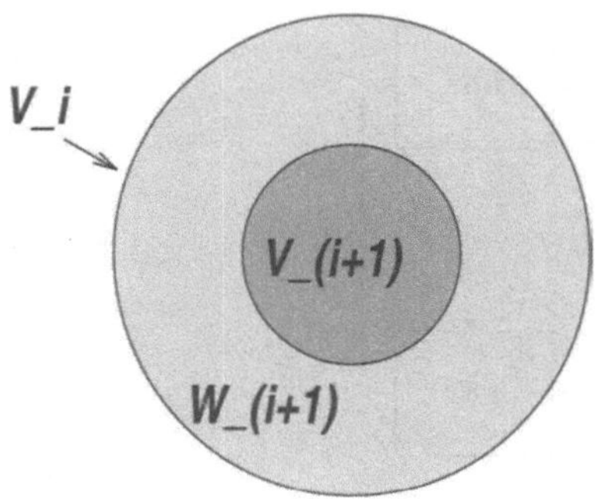

Fig. 7.10. Relationship between different scale subspaces

The MRA decomposition is implemented as a linear convolution procedure using a particular wavelet filter. It is depicted in Fig. 7.11. Since it is not possible to know a priori which filter would be the best basis in terms of information compactness for the image, the wavelet scheme must try to find the best filter by essentially trying out each of the available filters, and selecting the filter that gives the smallest number of non-zero coefficients (by extension, that filter will very often also result in the highest compression ratio). The chosen filter is consequently used in the MRA decomposition procedure.

In spite of the simplicity and straightforwardness of the MRA decomposition algorithm, there are two critical choices that have to made with respect to the algorithm, that greatly affect compression performance for a given image. These two choices, or factors, are the choice of wavelet filter, and the

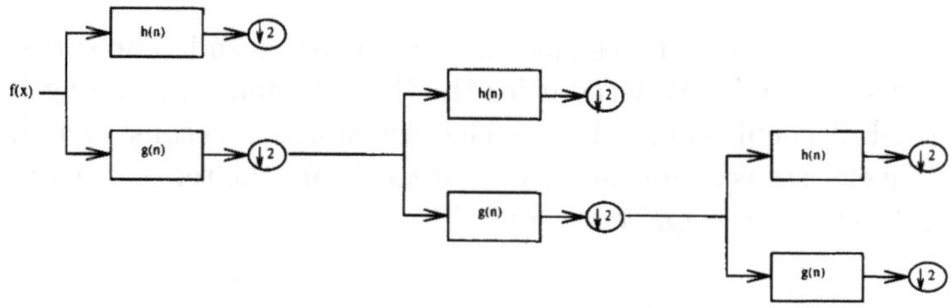

Fig. 7.11. Multiresolution analysis decomposition

number of MRA decomposition levels. The most crucial aspect of carrying out the wavelet transform is the choice of wavelet filter. Unlike the JPEG, and other Block Transform schemes where the transformation is performed onto one particular basis, the DCT base in JPEG, in the wavelet transform there is not a clearly defined base onto which every image is transformed. Rather, every wavelet filter represents a different base. A method for calculating the optimal base, and with it the filter coefficients, is to select the 'best' (rather than optimal) wavelet filter from a reservoir of available filters. Another very crucial consideration in the implementation of the MRA Decomposition procedure is the number of decomposition levels that are to be used. In this coding scheme the *ll* subimage is coded using a lossless scheme that does not compress the subimage to a high degree. That means that a large *ll* component will adversely affect the achieved compression ratio. On the other hand, a small *ll* subimage will adversely affect the resultant image quality.

Once the wavelet transform representation is obtained, a quantizer is used to quantize the coefficients. The quantization levels can be fixed or they can be determined adaptively according to the perceptual importance of the coefficients, and according to the complexity of the given image (images with a higher complexity normally have to be quantized more coarsely to achieve reasonable bit rates). The use of human visual system properties to quantize the wavelet coefficients enables the scheme to coarsely quantize coefficients which are visually unimportant. In many of the cases, those visually unimportant coefficients are simply set to zero. By contrast, wavelet coefficients which are deemed to be visually important are quantized more finely. The quantization and the overall reduced number of wavelet coefficients ultimately give better compression ratios, at very high image quality levels. The actual coding stage follows the quantization step. The coding stage consists of differential pulse code modulation (DPCM) to code the (*ll*) subimage, and a zero run-length coder to code the added detail wavelet coefficients. DPCM, which is a lossless coding technique, is used in order to preserve the (*ll*) subimage perfectly. The run-length scheme, on the other hand, is ideally suited for coding data in which many of the coefficients are zero-valued. The DPCM / zero run-length coding combination achieves bit rates that are slightly better than the bit rates achieved by JPEG.

A completely reverse process takes place at the receiver's end. The coded data stream is decoded, a DPCM decoder for the (ll) subimage, a run-length decoder for the detail coefficients. The wavelet transform is reconstructed, and an inverse Wavelet transform is applied on it to obtain the reconstructed image. The overall scheme is depicted in Fig. 7.12.

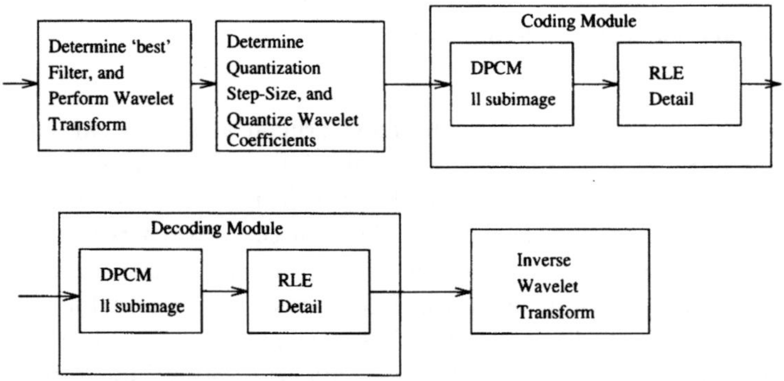

Fig. 7.12. The wavelet-based scheme

7.6 Second Generation Image Compression Techniques

The main characteristic of first generation techniques is that most of the emphasis is placed on deciding how to code the image. In contrast, in second generation or model based techniques the emphasis is placed on deciding what should be coded, with the choice of coding the information becoming a secondary issue [17]. Hence, the methodology of second generation techniques can be broken down into two parts (as seen in Fig. 7.13), where the first part selects the information from the image (Message Select module), and the second part codes the selected messages (Message Coder module). It was mentioned before that the human visual system (HVS) perceives visual information in a very selective manner. That is, the HVS picks up specific features from the overall image that it perceives. Therefore, second generation techniques can be very useful for perceptually lossless coding since they can select features that are more relevant to the HVS and then code those features.

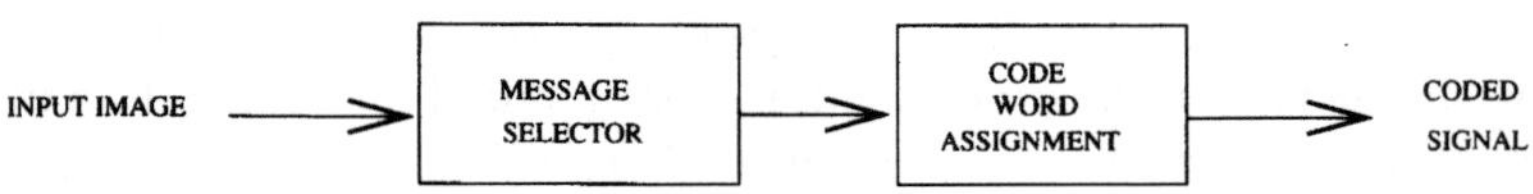

Fig. 7.13. Second generation coding schemes

In general, second generation techniques will pre-process an image in an attempt to extract visual primitives such as contours, and the textural contents surrounding the contours. Since contours and textures can be coded very efficiently, compression ratios in excess of 50:1 can be obtained. Below is a short rundown of some common second generation techniques.

1. **Pyramidal coding**. An image is successively passed through a low-pass filter a number of times. At each iteration, the error between the resultant low-pass image and the initial image is found. A low-pass filtering operation is now performed on the resultant image from the previous step and again the output image is used to find error between the output and input images of that stage. This recursive relationship can be expressed as:

$$e_p(m,n) = x_{p-1}(m,n) - x_p(m,n), \text{ for } \quad p = 1, 2, .., P \tag{7.7}$$

The error values at each iteration constitute high frequency information. Since the human vision system prefers high frequency information, but at the same time it does not have a high contrast sensitivity to it, a small number of bits/pixel are required to code the error information. Also coded is the low frequency information $x_P(m,n)$ which does not require a large number of bits to code. This technique achieves a modest compression ratio of 10:1, but with perceptually lossless image quality.
2. **Visual pattern image coding**. VPIC compression technique is similar to VQ in that the technique attempts to match a block to a pattern from a pre-defined set of patterns, and then transmit the index corresponding to that pattern. The main difference between the two is in the principle used to match the pattern. In VQ, an arbitrary block is matched to the pattern closest to it in some error sense, usually MSE. In VPIC, a block is broken down to its low-pass component which is just the intensity average of the block, and into its spatial variation, or edge information (the high frequency component). Therefore, the mapping criterion adheres to the behavior of the HVS and not to some absolute mathematical error measure. The edge information is then matched to edge patterns from a pre-defined table. The information transmitted is then the average intensity of the block along with an index which corresponds to the pattern selected. This technique is characterized by very low complexity, high compression ratio (11:1-20:1), and excellent image quality [38].
3. **Region growing based coding**. In this technique the image is first segmented into a collection of closed contours that are perceptible to human observers. Once the image is completely partitioned, the closed contours and the visual attributes, such as color and texture inside the closed segments are coded separately and transmitted. Efficient coding of the contours and the visual contents can translate into impressive compression ratios (in excess of 70:1). The image quality then becomes a function of how coarse the segmentation process and coding processes

are. The different algorithms and methodologies discussed in Chap. 6 can be used to guide the coding process.

4. **Fractal coding.** Fractal coding also operates on blocks rather than on the image as a whole. The main idea in fractal coding is to extract the basic geometrical properties of a block. This extraction is done by means of applying contractive affine transformation on the image blocks. Fractal image compression is based on the observation that all real-world images are rich in affine redundancy. That is, under suitable affine transformations, larger blocks of the image look like smaller blocks in the same image. These affine maps give a compact representation of the original image and are used to regenerate that image, usually with some amount of loss. Therefore, a fractal compressed image is represented in terms of the self similarity of essential features and not in terms of pixel resolution. This is a unique property of the fractal transform and therefore an image can be represented in any resolution without encountering the artifacts that are prevalent when using transform based techniques, such as JPEG.

   Most fractal image coding techniques are based on iterated function systems (IFS) [39]. An IFS is a set of transformations, each of which represents the relationship between a part of the image and the entire image. The objective of the coding scheme is to partition the image into several subimages and find transformations that can map the entire image into these subimages. When these transformations are found, they represent the entire image. In this way, images with global self-similarity can be coded efficiently. However, it is difficult to find such transformations in real life images since natural images are rarely globally self-similar. To this end, a coding scheme based on the so-called partitioned IFS technique was proposed in [40]. In the partitioned IFS approach the objective is to find transformations that map a part of the image into another part of the image. Such transformations can easily be found in natural images. However, the compression ratio of the partitioned IFS is not as high as that of the direct IFS coding scheme.

   Fractal compression techniques that can be implemented in software are resolution independent and can achieve high compression efficiency [41]. However, unlike the DCT based compression algorithms which are symmetric, with decompression being the reverse of compression in terms of computational complexity, fractal compression is computationally intensive, while decompression is simple and so fast that it can be performed using software alone. This is because encoding involves many transformations and comparisons to search for a set of fractals while the decoder simply generates images according to the fractal transformation received. These features make fractal coding well suited to CD-ROM mass storage systems and HDTV broadcasting systems. In summary, the main advantages of fractal based coding schemes are the large compression efficiency (up to 40:1), with usually a relatively good image quality, and resolution independence. The main disadvantage of the scheme is its complexity in terms of the computational effort [20].

7.7 Perceptually Motivated Compression Techniques

As was described in the previous sections, efficiency, in terms of the achieved bit rate or compression ratio, and image quality are dependent on each other. Lower bit rates can be achieved at the expense of a higher distortion. The main problem in devising efficient image compression techniques is that it is not clear what distortion measures should be used. Traditional objective distortion measures, such as the MSE, do not appear to be very useful in establishing an accurate relationship between efficiency and image quality. This is because objective distortion measures do not correlate well with the distortion perceived by the human visual system (HVS). That is, low MSE distortion might result in degraded images which human observers will not find pleasing, and vice versa. Therefore, in order to improve the performance of image compression techniques it is first necessary to get a better understanding of the human visual system. This section will describe some of the important features of the human visual system that have a direct impact on how images are perceived. Once the human visual system is better understood, its features and behavior can be more successfully incorporated into various compression methods.

7.7.1 Modeling the Human Visual System

Perhaps the most difficult part in designing an effective compression method is coming up with a good robust model for the HVS. The difficulty arises because of the complexity and multi-facet behavior of of the HVS and human perception. Although a model that accounts for all aspects of the HVS is not available, a simplified model that attempts to approximate and explain the behavior of the human visual system exists. The general HVS model is presented in Fig. 7.14

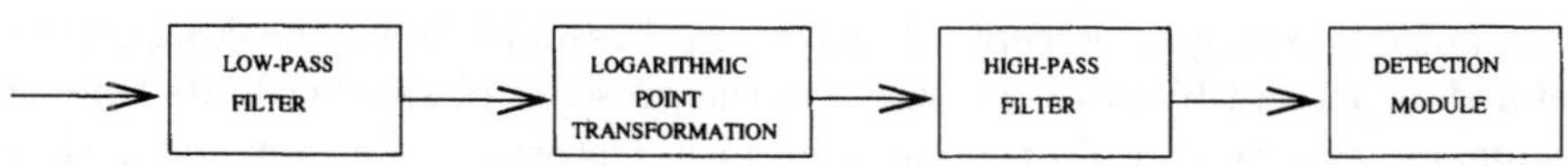

Fig. 7.14. The human visual system

This simplified model of the HVS consists of four components: a low-pass filter, a logarithmic point transformation to account for some of the nonlinearities of the HVS, a high-pass filter and a detection module. The low-pass filtering is the first operation that the HVS performs. This operation corresponds to filtering done by the optical system before the visual information is converted to neural signals [12]. The logarithmic point transformation module models the system's ability to operate over a large intensity range. The high-pass filter block relates to the 'lateral inhibition' phenomenon and comes about from the interconnections of the various receptor regions (in lateral inhibition the excitation of a light sensor inhibits the excitation of a neighbor sensor) [12]. These three blocks model elements of the HVS that are more physical in nature. More specifically, both the low-pass and high-pass filtering operations arise because of the actual physiological structure of the eye, while the need to model the logarithmic non-linearity relates to the physiological ability of the eye to adapt to a huge light intensity range. These operations are relatively straightforward and are therefore easy to represent by this model. The detection module, on the other hand, is considerably harder to model since its functions are more psychophysical in nature.

Even though it is extremely hard to accurately and completely model the detection block, an attempt should be made to include as many human perceptual features as possible in such a model. Examples of some of those features are feedback from higher to lower levels in perception, interaction between audio and visual channels, descriptions of non-linear behavior and peripheral, and other high-level effects [42]. At this point, it is of course not possible to include all of the above features. However, some human perceptual phenomena on which more is known can be incorporated into the detection model and later be used in image coding. Specifically, there are four dimensions of operations that are relevant to perceptual image coding. These are: (i) intensity, (ii) color, (iii) variation in spatial detail, and (iv) variation in temporal detail. Since the focus of this section is on compression of still images, the first three properties are of more importance.

A good starting point for devising a model for the detection block is recognizing that the perceptual process is actually made of two distinct steps. In the first step, the HVS performs a spatial band-pass filtering operation [42]. This operation does, in fact, accurately model and explain the spatial frequency response curve of the eye. The curve shows, that the eye has varying sensitivity response to different spatial frequencies, and thus the human visual system itself splits an image into several bands before processing it, rather than processing the image as a whole. The second step is what is referred to as noise-masking or perceptual distortion threshold. Noise-masking can be defined as perceptibility of one signal in the presence of another in its time or frequency vicinity [12]. As the name implies, distortion of an image which is below some perceptual threshold can not be detected by the detection block of the human eye. This perceptual threshold, or more precisely, the point at which a distortion will become noticeable, is the so called 'just noticeable distortion' (JND). Following the perceptual distortion processing, the image can be encoded in a manner that considers only information that exceeds the JND threshold. This step is referred to as perceptual entropy. Perceptual entropy coding used alone will produce perceptually lossless image quality. A more general but flexible extension of the JND is the minimally noticeable distortion (MND). Again, as the name suggest, coding an image using an MND threshold will result in a noticeable distortion, but will reduce the bit rate [42].

Next, a few well known perceptual distortion threshold phenomena will be described. These phenomena relate to intensity and variation in spatial detail, which are two of the features that can be incorporated into the image detection and encoding step. Specifically:

1. **Intensity.** The human eye can only distinguish a small set of intensities out of a range at any given point. Moreover, the ability to detect a particular intensity level depends almost exclusively on the background intensity. Even within that small range the eye cannot detect every possible intensity. In fact, it turns out that a small variation in intensity between the target area and the surrounding area of the image cannot be noticed. In effect, the surrounding area masks small variations in intensities of the target area. More specifically, if the surrounding area has the same intensity as the background (i.e. $L = L_B$ where L denotes the intensity of the surrounding area and L_B denotes the background intensity) then the just noticeable distortion in intensity variation, ΔL, is about 2% of the surrounding area intensity [12]. Mathematically, this relation can be expressed as:
$$\frac{\Delta L}{L} \approx 2\% \tag{7.8}$$
The above ratio is known as the 'Weber Ratio'. This ratio and the JND contrast threshold increases if L is not equal to L_B or if L is particularly high or low. The implications of this for perceptual image coding are that small variations in intensity of a target area relative to its neighbors do not have much importance since the human visual system will not be able to detect these small variations. This property can lend itself nicely for reducing perceptual entropy and the bit rate.
2. **Color.** The human visual system is less sensitive to chrominance than to luminance. When color images are represented as luminance and chrominance components, for example YC_bC_r, the chrominance C_b, C_r can be coded coarsely and fewer bits used. That is to say, the chrominance components can be sub-sampled at a higher ratio and quantized more coarsely. Despite its simplicity the method is quite efficient and it is widely used as preprocessing step, prior to applying spatial and temporal compression methods in coding standards, such as JPEG and MPEG.
3. **Variation in spatial detail**. Two other masking properties that can be useful for perceptual image coding relate to the ability of the eye to detect variation in spatial detail. These two properties are the simultaneous contrast and Mach bands effects and both occur as a result of the lateral

inhibition phenomenon. In the simultaneous contrast phenomenon, the perceived brightness of a target area changes as the luminance (or intensity) of the surrounding area changes. The target area appears to become darker as the surrounding area becomes brighter, and vice versa, the target area appears to become brighter as the surrounding area becomes darker [26]. Also, if the illumination on both the target and surrounding area is increased, then the target area will appear to have become brighter if the contrast (ΔL) is low, but will appear to have become darker if the contrast is high. The Mach bands effect refers to the eye's tendency to accentuate the actual contrast sharpness at boundaries or edges. That is, regions with a high constant luminance will cause a neighboring region of lower luminance to appear to have even lower luminance. Another example of this effect is that when two regions of high contrast are separated by a transition region in which the luminance changes gradually and smoothly, the transition luminance levels will hardly be noticeable, while the two main regions will still appear to have a high contrast. This effect illustrates the fact that the human eye prefers edge information and that transition regions between regions of high contrast are not detected. In the context of masking, it can be said that luminance values at transition regions are masked by the main boundary regions. Consequently, in lossy compression, edge information should be preserved since the human visual system is very sensitive to its presence, while transition regions do not have to be faithfully preserved and transmitted.

These well known phenomena can be used to remove visual information that cannot be detected by the human visual system. The following is a summary of more specific ways of how to use these and other properties to efficiently encode still images, as well as image sequences.

1. **Contrast sensitivity**. This is one of the most obvious places where considerable bit rate reduction can be obtained. Human observers react more to high frequency information and sharp spatial variation (like edges). However, they cannot detect those spatial variations if the contrast, change in spatial variation, falls below a certain threshold. Also, it has been shown experimentally that the contrast sensitivity is a function of spatial frequency [12], [9]. Specifically, the highest contrast sensitivity is for spatial frequencies at about 5-7 cycles/degree, with the sensitivity dropping off rapidly for higher spatial frequencies [9]. A good way to take advantage of this property would be to concentrate mainly on high frequency information and code it coarsely because of the low sensitivity to the exact value of high frequency information.
2. **Dynamic contrast sensitivity**. This property is an extension of the contrast sensitivity function to image sequences. Low resolution luminance have their highest sensitivity at low temporal frequencies, with the sensitivity rapidly falling off at about 20 Hz. This implies that less

precision is required for encoding information for high temporal frequencies than is required at low temporal frequencies [9].

3. **Luminance masking**. Another place where the bit rate can be reduced is by using the luminance masking. Since the eye cannot detect an intensity change ΔL which is below the Weber ratio, areas in the image that have small intensity variations relative to the surrounding areas do not have to be faithfully or accurately transmitted. This property can be useful in coding low frequency information, where only a small number of bits would be needed to code the low frequency contents of a large image area.

Lastly, to conclude this section, some image compression implementations in which the human visual system is incorporated will be described.

7.7.2 Perceptually Motivated DCT Image Coding

The approach presented in [43] is essentially based on determining the appropriate quantization values for the quantization matrices so that they match well with the contrast sensitivity function (CSF). Normalizing the DCT coefficients will automatically eliminate small contrast variations, and will yield low bit rates.

As was described earlier, the overall compression is determined by the extent of quantizing of each of the DCT coefficients as defined in a quantization table. After transforming an $(n \times n)$ block of pixels to its DCT form, the DCT coefficients are normalized using the normalization matrix, according to the relation:

$$\hat{T}(u,v) = \text{round}\left[\frac{T(u,v)}{Z(u,v)}\right] \tag{7.9}$$

where $T(u,v)$ is the DCT coefficient, $Z(u,v)$ is the corresponding normalizing value, and $\hat{T}(u,v)$ is the normalized DCT coefficient.

Since different DCT coefficients have higher contrast sensitivity than other coefficients, greater precision (in the form of more bits/coefficient) is required, and their corresponding normalization value will be lower than that of the other less important coefficients. For example, the suggested JPEG normalization table normalizes the low frequency coefficients, such as the DC value by relatively small values.

It now seems like a straightforward task to recompute the quantization values in accordance with the CSF. According to the rule, low quantization values, or more precision, should be assigned to those spatial frequencies to which the eye is more sensitive. However, the task is a bit more involved than that. To begin with, the CSF is based on the visibility of the human visual system to a full field sinusoid. The DCT, on the other hand, is not completely physically compatible with the Fourier transform. In other words, in order to use the CSF with the DCT, a correction factor must be applied to the DCT coefficients [43], [44]. Another problem that complicates the DCT coding method based on CSF is that of sub-threshold summation. Namely, there are some situations in which some of the DCT frequencies might be below the contrast threshold as ascribed by the CSF, but the summation of these frequencies is very much visible. Other factors that have to be taken into account are the visibility of the DCT basis function due to the oblique effect, the effects of contrast masking, orientation masking, and the effects of mean luminance, and the size of the pixel in the particular monitor being used [43].

By considering several of these effects, quantization tables that are compatible with the human visual system were introduced. Tables 7.6 and 7.7 show the basic normalization tables, for the luminance component, suggested by JPEG next to the normalization table that incorporates the attributes of the human visual system.

Table 7.6. The JPEG suggested quantization table

16	11	10	16	24	40	51	61
12	12	14	19	26	58	60	55
14	13	16	24	40	57	69	56
14	17	22	29	51	87	80	62
18	22	37	56	68	109	103	77
24	35	55	64	81	104	113	92
49	64	78	87	103	121	120	101
72	92	95	98	112	100	103	99

Table 7.7. Quantization matrix based on the contrast sensitivity function for 1.0 min/pixel

10	12	14	19	26	38	57	86
12	18	21	28	35	41	54	76
14	21	25	32	44	63	92	136
19	28	32	41	54	75	107	157
26	35	44	54	70	95	132	190
38	41	63	75	95	125	170	239
57	54	92	107	132	170	227	312
86	76	136	157	190	239	312	419

With the above quantization table, the bit rate can be reduced from 8 bits/pixel to less than 0.5 bit/pixel, while maintaining very high, perceptually lossless, image quality.

An important characteristic of the perceptually motivated coder is that all the perceptual overhead are done in the encoder only. The decoding performance of the perceptually motivated JPEG is the same as that for a baseline JPEG. Therefore, such an approach is ideal for decoding-heavy applications.

A specific example is in multimedia communications. With the continuous advancement of computer processing power and display device technology, and the rapidly increasing popularity of the Internet, visual information is now very much within reach for end-users. One characteristic of visual information over the Internet is that it is, in most cases, accessed by decoding-heavy applications or systems. For instance, front pages of information providers are accessed by millions of hits every day, but the images and video streams are only created once. The same is true for thousands of JPEG files and MPEG video clips on the Internet. A perceptually motivated scheme designed for reducing storage costs for image and video information encoded using transform based techniques, such as JPEG, or MPEG, offers an attractive option.

7.7.3 Perceptually Motivated Wavelet-based Coding

Application of the MRA decomposition stage on the image produces several wavelet coefficient subimages and a single *ll* subimage which is a scaled-up version of the original image. Although many of the wavelet coefficients are zero-valued, the vast majority of them have a non-zero value. Hence, it becomes very inefficient to try and compress the wavelet coefficients using the Zero Run-Length coding technique, which is based on the premise that most of the coefficients are zero-valued. Fortunately, many of the non-zero coefficients do not contribute much to the overall perceptual quality of the image, and consequently can be greatly quantized, or discarded altogether. To achieve that, the wavelet coefficient subimages are processed with a Processing Module that uses properties of the human visual system (HVS) to determine the extent of the quantization to be applied on a given wavelet coefficient. Coefficients which are visually insignificant would ordinarily be quantized more coarsely (possibly being set to 0), while the visually significant coefficients would be more finely quantized.

As it was explained before, there are several common HVS properties that can be incorporated into the processing module.

1. The HVS exhibits relatively low sensitivity to high resolution bands, and has a heightened sensitivity to lower resolution bands.
2. Certain spatial features in an image are more important to the HVS than other. More specifically, features such as edges and texture are visually more significant than background features that have a near constant value.
3. There are several masking properties that mask small perturbations in the image.

A number of HVS based schemes to process wavelet coefficients have been developed over the past few years. Most notably, an elegant method that combines the band sensitivity, luminance masking, texture masking, and edge height properties into a single formula that yields the quantization step-size for a particular wavelet coefficient was developed in [37]. The formula is given as:

$$q_{step}(r, s, x, y) =$$

$$q_0 * frequency(r, s) * luminance(r, x, y) * texture(r, x, y)^{0.034} \quad (7.10)$$

In the above equation, q_0 is a normalization constant that can be a decomposition level, s represents the particular subimage within a decomposition level. For example hl, lh, hh, and x and y are the spatial coordinates within every subimage. The *frequency*, *luminance* and *texture* components are calculated as follows:

$$frequency(r, s) = \left\{ \begin{array}{ll} \sqrt{2}, & \text{if s = HH} \\ 1, & \text{otherwise} \end{array} \right\} * \left\{ \begin{array}{l} 1.00, \text{ if r = 0} \\ 0.32, \text{ if r = 1} \\ 0.16, \text{ if r = 2} \end{array} \right\} \quad (7.11)$$

$$luminance(r, x, y) =$$

$$3 + \frac{1}{256} \sum_{i=0}^{1} \sum_{j=0}^{1} I^{2,ll}(i + 1 + x/2^{2-r}, j + 1 + y/2^{2-r}) \quad (7.12)$$

$$texture(r, x, y) = \sum_{k=1}^{2-r} 16^{-k} \sum_{s}^{hh,lh,hl} \sum_{i=0}^{1} \sum_{j=0}^{1} (I^{k+r,s}(i + x/2^k, j + y/2^k))^2$$

$$+16^{2-r} var(I^{2,ll}(\{1, 2\} + x/2^{2-r}, \{1, 2\} + y/2^{2-r})) \quad (7.13)$$

In (7.12) the notation $(I^{2,ll}(x, y))$ denotes the coefficient values of the ll subimage at the third MRA decomposition level.

These equations ((7.11)-(7.13)) are essentially heuristic formulas in the sense that they do not necessarily give optimal results, but rather give good results for most images (with respect to the image quality and the bit rate). Better image quality at the expense of the bit rate can always be obtained by altering the parameter values in the above equations. As it was pointed out in [46], the main problem with the method in [37] was the relatively high computational effort involved in computing the quantization step size values.

The method in [46] requires computation of the texture component, which as (7.13) shows, is based on all spatially related coefficient values in the lower level subimages. An alternative way of computing the quantization step size, in which the texture component is not used, was proposed in [46]. Rather, computation of the quantization step size is based only on the luminance level and edge height associated with a particular wavelet coefficient. Both the luminance level and edge height values are computed from the ll subimage, which greatly reduces the computational effort [46]. The quantization step size is then calculated as:

$$q_{step}(s, r, x, y)$$

$$= q_0 * frequency(s,r) * \min\{BS(s,x,y), ES(s,x,y)\} \tag{7.14}$$

In the above equation $BS(s,x,y)$ is the background sensitivity function which is solely based on the luminance values derived from the ll subimage. Similarly, $ES(s,x,y)$ is the sensitivity function which is solely based on the edge height values that are also derived from the ll subimage. Since computational efficiency is a paramount consideration in [46], the quantization procedure used was similar to the methodology proposed. Several modifications that enabled the overall performance of the wavelet scheme to exceed that of JPEG are also suggested in [46].

Like these methods, the implemented processing module also computes a quantization level for each wavelet coefficient based on the local luminance level, and the edge height in the vicinity of the particular wavelet coefficient. There are, however, two defining differences between the implemented method and method introduced in [46]. The first difference is that the implemented scheme does take into account the fact that sharp edges are visually more significant than other spatial features. This HVS property is incorporated into the scheme by quantizing visually insignificant features more coarsely than visually significant features. The coarseness of the quantization is controlled through the normalization factor q_0. In other words, the scheme uses two normalization factors; one for edges, and one for non-edge features (normally referred to as background information). The second difference is that the two normalization factors, q_0^{edge}, and q_0^{back}, are made into adaptive parameters that are dependent on the complexity of the particular image. More specifically, it is found that for high complexity images, images with a lot of edge information, the normalization factors have to be increased in order to achieve compression ratios that are comparable to that of JPEG.

The implemented processing module works as follows:

1. All the background luminance values are computed using the ll subimage pixel values. The luminance values are computed according to [37]:

$$luminance(x_{ll}, y_{ll}) = 3 + \frac{1}{256}\sum_{i=0}^{1}\sum_{j=0}^{1} I^{ll}(x_{ll}+i, y_{ll}+j) \tag{7.15}$$

   In the implemented scheme, the Processing Module stores the luminance values in memory, and then retrieves these values as they are needed.

2. All the edge height values are computed using the ll subimage pixel values. The edge height values are computed according to [46]:

$$EH(x_{ll}, y_{ll}) = 0.37|D_{vert}| + 0.37|D_{hori}| + 0.26|D_{diag}| \tag{7.16}$$

   where

$$D_{vert} = I^{ll}(x_{ll}, y_{ll}) - I^{ll}(x_{ll}, y_{ll}+1)$$
$$D_{hori} = I^{ll}(x_{ll}, y_{ll}) - I^{ll}(x_{ll}+1, y_{ll})$$
$$D_{diag} = I^{ll}(x_{ll}, y_{ll}) - I^{ll}(x_{ll}+1, y_{ll}+1)$$

In the suggested scheme, the processing module stores the edge height values in memory, and then retrieves these values as they are needed.

3. The next step is to determine the quantization parameter values that correspond to the particular image being compressed. Besides the quantization parameters q_0^{edge} and q_0^{back} which control the quantization values for the edges and background features respectively, an additional parameter q_{thresh} is needed. Features with edge height values above this threshold value would be considered as edges. As was mentioned above, the quantization parameter values are adjusted to reflect the complexity of the particular image. Images with high complexity require parameters with large values in order to be compressed efficiently. A good measure of an image complexity is provided by the number of wavelet coefficients retained during the filter selection stage. Complex images invariably produce more retained coefficients than simpler images.
In determining what quantization parameter values to use for each image, the only guiding criterion is to find the parameters which would give results that are better than what is achieved with JPEG. Hence, by a process of trial and error, the parameter values are continuously adjusted until the best results (PSNR and the corresponding compression ratio) for a particular image are obtained. A particular result is considered to be good if both the PSNR and compression ratio exceeded the JPEG values. For images where it is not possible to exceed the performance of JPEG, the best compromise of PSNR and compression ratio is used. Following this method of trial and error, the quantization parameter values for several trial images are determined.
Using these manually determined parameter values, a linear function is derived for each parameter using a simple linear regression procedure. In each linear function, each quantization parameter is expressed as a function of the number of retained coefficients. The three derived linear functions are given as:

$$q_{thresh} = -1.1308 + 0.0013 * \#\text{ retained coefficients} \tag{7.17}$$

$$q_0^{edge} = 0.8170 + 0.00001079 * \#\text{ retained coefficients} \tag{7.18}$$

$$q_0^{back} = 1.0223 + 0.00002263 * \#\text{ retained coefficients} \tag{7.19}$$

4. The last part of the processing stage is to process each of the wavelet coefficients in the various detail subimage. The processing procedure is simple and takes place as follows:
   a) For a particular wavelet coefficient, use the spatial coordinates of that coefficient to find the corresponding ll subimage spatial coordinates. Use the ll spatial coordinates (i.e., x_{ll} and y_{ll}) to fetch the corresponding edge height value stored in memory.
   b) If the edge height value exceeds the q_{thresh} parameter value, the coefficient is an edge coefficient. In that case, use the q_0^{edge} quantiza-

tion parameter to calculate the quantization step size for the current wavelet coefficient using the formula:

$$q_{step} = \text{floor}\left(q_0^{edge} * frequency(r,s) * luminance(x_{ll}, y_{ll}) + 0.5\right) \quad (7.20)$$

where $luminance(x_{ll}, y_{ll})$ is the luminance value calculated in the first step.

c) If the edge height value is lower than the q_{thresh} parameter value, the coefficient is a background coefficient. In that case, use the q_0^{back} quantization parameter to calculate the quantization step size for the current wavelet coefficient using the formula:

$$q_{step} = \text{floor}\left(q_0^{back} * frequency(r,s) * luminance(x_{ll}, y_{ll}) + 0.5\right) \quad (7.21)$$

d) Quantize the wavelet coefficient using q_{step}.

The operation of the perceptual processing module is depicted in Fig. 7.15.

7.7.4 Perceptually Motivated Region-based Coding

Region growing based coding is a second generation image compression technique that operates in the spatial domain [17], [18]. As such, the emphasis of this method is placed on initial selection of image information. The information selection part is followed by an efficient coding procedure. The technique is based on segmenting an image into contour regions in which the contrast variation is small. The texture in each segment is also coded in accordance to some error criterion. This segmentation is consistent with the human vision system behavior which prefers edge (or contour) information and cannot distinguish well between small contrast variations. The segmentation procedure is carried out by first selecting a segmentation parameter which could be a specific color, color variation, or any other appropriate measure of discrimination, such as texture. Because noise can adversely affect a segmentation process based on color, it is desirable that the noise be first removed by means of vector filtering. In this particular case the segmentation parameter is chosen to be color variation. Hence, the rule for segmenting the image would be that neighboring pixels that are within a certain color range will be grouped into the same segment. Depending on the compression ratio desired, it might become necessary to reduce the number of contour segments that are obtained at the end of this step. This can be achieved by joining neighboring segments to one another, or using higher threshold value for the segmentation parameter. The different procedures discussed in Chap. 6 can be utilized for this task.

Once the color image is partitioned, the contours and texture of each segment have to be coded and transmitted. The contours themselves have to be carefully and accurately coded since the human visual system is particularly sensitive to edge information. In such an approach, contours are coded

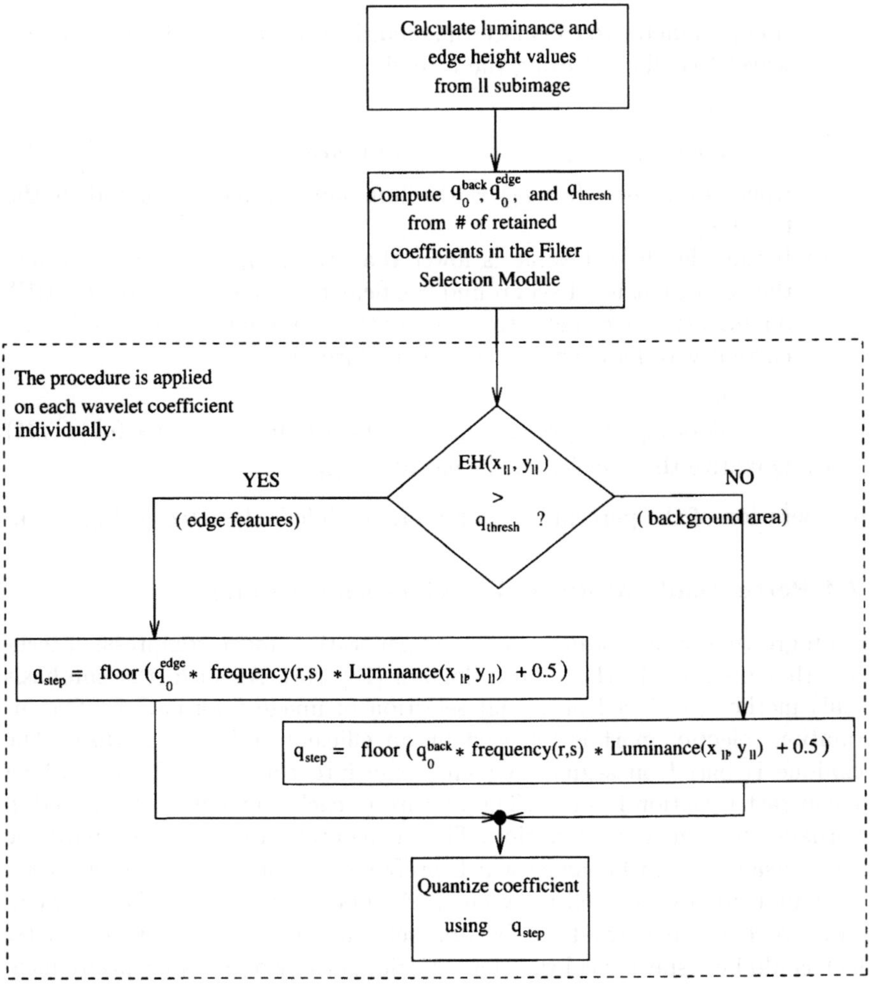

Fig. 7.15. Overall operation of the processing module

by using line and circle segments wherever possible. It should also be noted that adjacent segments will share contours, and therefore further coding reduction can be realized by coding these contours only once. Although the human visual system is less sensitive to textural variations than it is to the existence of contours, care should be taken that the textural contents of each segment are not overly distorted. The contrast variation within every segment is kept below the segmentation parameter. Therefore, it is usually enough to approximate the texture by using a 2-D polynomial. It is then enough to simply transmit the polynomial's coefficients in order to reconstruct the shape of the texture inside every contour segment.

The technique gives varying degrees of compression ratios and image quality. Good image quality can be obtained at the expense of a larger bit rate by simply allowing for closed contour segments and higher order polynomials to approximate the textural contents within each segment. As an example, compression ratios of the order of 50:1 with relatively good image quality have been obtained using the proposed methodology.

7.8 Color Video Compression

Compressing video signals means that the algorithm should have the ability to exploit temporal masking as well as spectral masking. In video coders, such as the industry standard MPEG, the components of digital color video signals are compressed separately with shared control and motion estimation mechanisms.

Existing compression techniques for still images can serve as the basis for the development of color video coding techniques. However, digital video signals have an associated frame rate from 15 to 60 frames per second which provides the illusion of motion in the displayed signal. A moving object in a video sequence tends to mask the background that emerges when the object moves, making it easier to compress the part of the uncovered image. In addition, since most video objects move in predictable patterns the motion trajectory can be predicted and used to enhance the compression gain. Motion estimation is computationally expensive and only luminance pixels are regarded in the calculations. For a block of (16×16) luminance pixels from the current frame the most similar block in the previous frame is searched for. Differences in the coordinates of these blocks define the elements of the so called motion vector. The current frame is predicted from the previous frame with its blocks of data in all the color components shifted according to the motion vectors which have to be transmitted to the decoder as side information.

Although still color images are sized primarily to fit workstations equipped with (640×480) VGA or (1024×768) XVGA color monitors, video signals can be of many sizes. For example, the input video data for very low bit rate applications is composed of small sized color images in the quarter common intermediate format (QCIF) with (144×176) pixels in luminance and a quarter of this resolution in the chrominance components. the frame rate for this application is approximately 5 to 15 frames per second. Medium bit rate video applications deal with images of average size, approximately (288×352) pixels in luminance and a quarter of this resolution in chrominances at a frame rate of 25 or 30 frames per second. Alternatively the ITU-R 601 standard with interlaced (576×720) pixels in luminance and half-horizontal resolution in chrominances is also used.

A number of standards are available today. Depending on the intended application they can be defined as:

1. **Standards for video conferencing applications.** This family includes the ITU standard H.261 for ISDN video conferencing, the H.263 standard for POTS video conferencing and the H.262 standard for ATM based, broad band video conferencing. H.261 is a video codec capable of operation at affordable telecom bit rates. It is a motion-compensated, transform-based coding scheme, that utilizes (16×16) macroblock motion compensation, (8×8) block DCT, scalar quantization and two-dimensional run level, variable length, entropy coding. H.263 is designed to handle very low bit rate video with a target bit rate range of 10-30 Kbits per second. The key technical features of H.263 are variable block size motion compensation, overlapped block motion compensation, picture extrapolating motion vectors, median-based motion vector prediction and more efficient header information signaling.
2. **Standards for multimedia applications.** This family includes the ISO MPEG-1 standard intended for storing movies on CD read-only memory with 1.2 Mb/s allocated to video coding and 256 Kb/s allocated to audio coding. The MPEG-2 standard was developed for storing broadcast video on DVD with 2 to 15 Mb/s allocated to video and audio coding. In the most recent member of the family, the emphasis has shifted from pixel coding to object-based coding at rates of 8Kb/s or lower and 1 Mb/s or higher. The MPEG-4 visual standard will include most technical features of the priori video and still image coding schemes and will also include a number of new features, such as wavelet-based coding of still images, segmented shape coding of objects and hybrids of synthetic and natural video coding.

Most standards use versions of a motion compensated DCT-based block hybrid coder. The main idea is to combine transform coding, primarily in the form of the DCT 8×8 pixel blocks with predictive coding in the form of DPCM in order to reduce storage and computation of the compressed image. Since motion compensation is difficult to perform in the transform domain the first step in the video coder is to create a motion compensated prediction error using macroblocks of (16×16) pixels. The resulting error signal is transformed using a DCT, quantized by an adaptive quantizer, entropy encoded using a variable length coder, and buffered for transmission over a fixed rate channel.

The MPEG family of standards is based on the above principle. The MPEG-1 system performs spatial coding using a DCT of (8×8) pixel blocks, quantizes the DCT coefficients using fixed or perceptually motivated tables, stores the DCT coefficients using the zig-zag scan and process the coefficients using variable run-length coding. Temporal coding is achieved by using uni- and bi-directional motion compensated prediction with three types of frames. Namely,

1. **Intraframe (I).** The I-frames from a video sequence are compressed independently from all previous or future frames using using a procedure

similar to JPEG. The resulting coefficients are passed through the inversed DCT transform in order to generate the reference frame, which is then stored in memory. This I frame is used for motion estimation for generating the P- and B- frames.

2. **Predictive (P)**. The P-frames are coded based on the previous I-frames or P-frames. The motion compensated for forward predicted P-frame is generated using the motion vectors and the referenced frame. The DCT coefficients from the difference between the input P-frame and the predicted frame are quantized and coded using variable length and Huffman coding. The P-frame is generated by performing the inverse quantization, taking the inverse DCT of the difference between the predicted frame and the input frame and finally adding this difference to the forward predicted frame.
3. **Bi-directionally frames (B)**. The B-frames are coded based on the next and/or the previous frames. The motion estimation module is used to bi-directionally estimate the motion vectors based on the nearest referenced I and P frames. The motion-compensated frame is generated using the pair of nearest referenced frames and the bi-directionally estimated motion vectors.

The video coder generates a bit stream with variable bit rate. In order to match this bit rate to the channel capacity, the coder parameters are controlled according to the output buffer occupancy. Bit rate control is performed by adjusting parameters, such as the quantization step used in the DCT component and the distance between intra frame and predictive frames.

The compression procedure as specified by the MPEG standard is as follows:

1. Preprocessing the input frames. Namely, color space conversion and spatial resolution adjustment. Frame types are decided for each input frame. If bi-directional frames are used in the video sequence, the frames are reordered.
2. Each frame is divided into macroblocks of (16×16) pixels. Macroblocks in I-frames are intra coded. Macroblocks in P-frames are either intra coded or forward predictive coded based on previous I-frames or P-frames, depending on coding efficiency. Macroblocks in B-frames are intra coded, forward predictive coded, backward predictive coded, or bi-directionally predictive coded. For predictive coded macroblocks motion vectors are found and predictive errors are calculated.
3. The intra coded macroblocks and the predictive errors of the predictive coded macroblocks are divided into six (4 luminance and 2 chrominance) blocks of (8×8) pixels each. Two-dimensional DCT is applied to each block to obtain transform coefficients, which are quantized and zig-zag scanned.

4. The quantized transform coefficients and overhead information, such as frame type, macroblock address and motion vectors are variable length coded using predefined tables.

The operation of the coding module is depicted in Fig. 7.16. The decoder is depicted in Fig. 7.17.

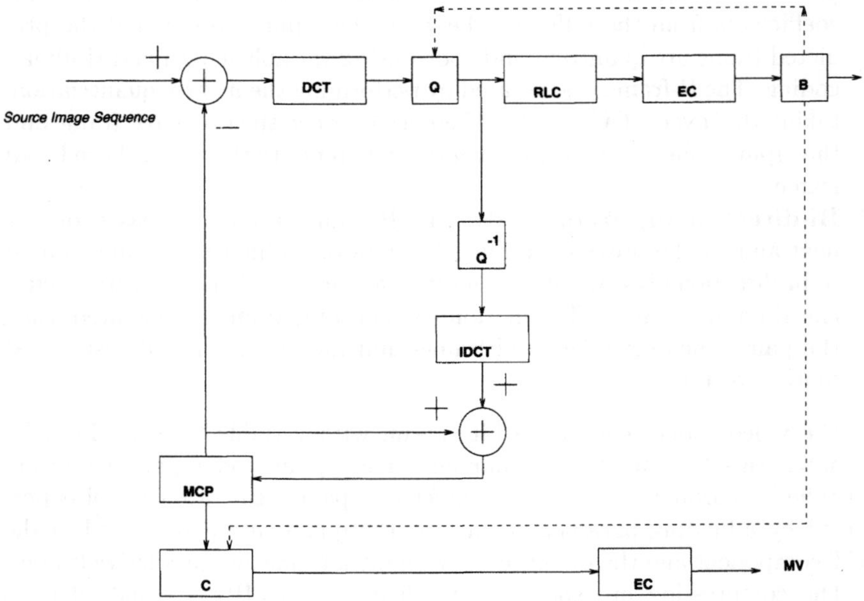

Fig. 7.16. MPEG-1: Coding module

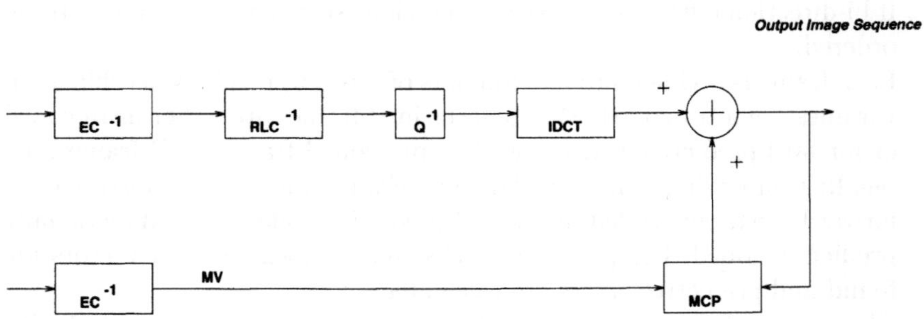

Fig. 7.17. MPEG-1: Decoding module

The promising results obtained with object-based coding techniques for still images motivated its extension to video sequences. Objects in a video stream can be defined as regions specified by color, shape, textural content and motion. The methods used for motion estimation and texture coding are extensions of those used in the block-based methodologies. However, since actual objects and not flat rigid blocks are tracked, the motion-compensated prediction is more exact therefore reducing the amount of information needed to encode the residual prediction error signal.

MPEG-4 is a new multimedia standard which specifies coding of audio and video objects, both natural and synthetic, a multiplexed representation of many such simultaneous objects, as well as the description and dynamics of the scene containing the objects. The video portion of the MPEG-4 standard, the so-called MPEG-4 visual part, deals with the coding of natural and synthetic visual data, such as facial animation and mesh-based coding. Central to the MPEG-4 visual part is the concept of video object and its temporal instance, the so-called video object planes (VOP). A VOP can be fully described by shape and/or variations in the luminance and chrominance values. In natural images, VOPs are obtained by interactive or automatic segmentation and the resulting shape information can be represented as a binary shape mask. The segmented sequences contains a number of well defined VOPs. Each of the VOPs are coded separately and multiplexed to form a bitstream that users can access and manipulate. The encoder sends together with video objects information about scene composition to indicate where and when VOPs of video objects are to be displayed. MPEG-4 extends the concept of I-frames, P-frames and B-frames of MPEG-1 and MPEG-2 to VOPs, therefore the standard defines I-VOP, as well as P-VOP and B-VOP based on forward and backward prediction. The encoder used to code the video objects of the scene has three main components: (i) motion coder which uses macroblock and block motion estimation and compensation similar to that of MPEG-1 but modified to work with arbitrary shapes, (ii) the texture coder that uses block DCT coding adapted to work with arbitrary shapes, and (iii) shape coder that deals with shape. A rectangular bounding box enclosing the shape to be coded is formed such that its horizontal and vertical dimensions are multiples of 16 pixels. The pixels on the boundaries or inside the object are assigned a value of 255 and are considered opaque while the pixels outside the object but inside the bounding box are considered transparent and are assigned a value of 0. Coding of each (16×16) block representing shape can be performed either lossy or losslessly. The degree of lossiness of coding the shape is controlled by a threshold that can take values of $0, 16, 32, ..., 256$. The higher the value of the threshold, the more lossy the same representation. In addition, each shape block can be coded in intra-mode or in inter-mode. In intra-mode, no explicit prediction is performed. In inter-mode, shape information is differenced with respect to the prediction obtained using a motion vector and the resulting error may be coded. Decoding is the inverse sequence of operations with the expection of encoder specific functions.

The object-based description of MPEG-4 allows increased interactivity and scalability both in the temporal and the spatial domain. Scalable coding offers a means of scaling the decoder if resources are limited or vary with time. Scalable coding also allows graceful degradation of quality when bandwidth resources are limited or vary with time. Spatial scalability encoding means that the decoder can either offer the base layer or display an enchancent layer output based on problem constraints and user defined specifications. On the other hand, temporal scalable coding refers to a decoder that can increase temporal resolution of decoded video using enhancement VOPs in conjunction with decoded base layer VOPs. Therefore, the new standard is better suited to address variable Quality-of-Service requests and can accommodate high levels of user interaction. It is anticipated that in full development MPEG-4 will offer increased flexibility in coding quality control, channel bandwidth adaptation and decoder processing resource variations.

7.9 Conclusion

In this chapter many coding schemes were reviewed. To achieve a high compression ratio at a certain image quality, a combination of these techniques is used in practical systems. The choice of the appropriate method heavily depends on the application on hand. With the maturing of the area, international standards have become available. These standards include the JPEG standard, a generic scheme for compressing still color images, the MPEG suite of standards for video coding applications, and the H.261/H.263 standards for video conferencing and mobile communications. It is anticipated that these standards will be widely used in the next few years and will facilitate the development of emerging applications.

The tremendous advances in both software and hardware have brought about the integration of multiple media types within a unified framework. This has allowed the merging of video, audio, text, and graphics with enormous possibilities for new applications. This integration is at the forefront of the convergence of the computer, telecommunications and broadcast industries. The realization of these new technologies and applications, however, demands new methods of processing visual information. Interest has shifted from pixel based models, such as pulse code modulation, to statistically dependent pixel models, such as transform coding to object-based approaches. Therefore, in view of the requirements of future applications, the future direction of image coding techniques is to further develop model-based schemes as well as perceptually motivated techniques.

Visual information is an integral part of many newly emerging multimedia applications. Recent advances in the area of mobile communications and the tremendous growth of the Internet have placed even greater demands on the need for more effective video coding schemes. However, future coding techniques must focus on providing better ways to represent, integrate and exchange visual information in addition to efficient compression methods. These efforts aim to provide the user with greater flexibility for content-based access and manipulation of multimedia data. Numerous video applications, such as portable video phones, video conferencing, multimedia databases, and video-on-demand can greatly benefit from better compression schemes and this added content-based functionality. International video coding standards, such as the H.261, and more recently the H.263, are widely used for very low bit rate applications such as those described above. These existing standards, including MPEG-1 and MPEG-2, are all based on the same framework, that is, they employ a block-based motion compensation scheme and the discrete cosine transform for intra-frame encoding. However, this block-based approach introduces blocking and motion artifacts in the reconstructed sequences. Furthermore, the existing standards deal with video exclusively at the frame level, thereby preventing the manipulation of individual objects within the bit stream. Second generation coding algorithms have focused on representing a scene in terms of objects rather than square blocks. This approach not only improves the coding efficiency and alleviates the blocking artifacts, but it can also support the content-based functionalities mentioned previously by allowing interactivity and manipulation of specific objects within the video stream. These are some of the objectives and issues addressed within the framework of the MPEG-4 and future MPEG-7 standards.

High compression ratios and very good image quality, in fact perceptually lossless image quality, can be achieved by incorporating the characteristics of the human visual system into traditional image compression schemes, or using second generation techniques which are specifically designed to account for the HVS characteristics. While these techniques are successful in addressing the current need for both efficiency and image quality, the on-going development and evolution of video applications might render the current state of these techniques unsatisfactory in a few years.

It has become evident that in order to keep up with the growing sophistication of multimedia applications, the focus of still image compression research should not only be on finding new or improving existing techniques, primarily second generation techniques, but also on improving our understanding of the human visual system, and refining the existing models. Indeed, existing models are capable of accounting for only a few of the many behavioral attributes of the HVS. A perceptually motivated scheme is only as good as the perceptual model it uses. With a more general and complete perceptual model image compression techniques will be able to further eliminate visual information that is of no importance to the human visual system, thus achieving a better performance.

References

1. Raghavan, S. V., Tripathi, S. K. (1998): Networked Multimedia Systems: Concepts, Architecture and Design. Prentice Hall, Upper Sandle River, New Jersey.

2. Netravali, A. N., Haskell, B. G. (1995): Digital Pictures: Representation, Compression and Standards. 2nd edition, Plenum Press, New York, N. Y.
3. Joint Photographic Expertc Group (1998): JPEG Home Page. *www.disc.org.uk/public/jpeghomepage.htm.*
4. ISO/IEC, JTC1/SC29/WG1 N505 (ITU-T SG8) (1997): Coding of still images. Electronic Preprint.
5. Pennebaker, W. B., Mitchell J. L. (1993): JPEG Still Image Data Compression Standard. Van Nostrand Reinhold, New York, NY.
6. Chiarglione, L. (1997): MPEG and multimedia communications. IEEE Transactions on Circuits and Systems for Video Technology, **7**:5-18.
7. Chiariglione, L. (1995): MPEG: A technological basis for multimedia applications. IEEE Multimedia, **2**(1): 85-89.
8. Jayant, N., Johnston, J. D., Safranek, R. J. (1993): Signal compression based on models of the human perception. Proceedings of the IEEE, **81**(10): 1385-1422.
9. Glenn, W. E. (1993): Digital image compression based on visual perception and scene properties. Society of Motion Picture and Television Engineers Journal, 392-397.
10. Tong, H. (1997): A Perceptually Adaptive JPEG Coder. M.A. Sc. Thesis, Department of Electrical and Computer Engineering, University of Toronto.
11. Gersho, A., Ramamurthi, B. (1982): Image coding using vector quantization. Proceedings of the IEEE Conference on Acoustic Speech and Signal Processing, **1**:428-431.
12. Clarke, R. J. (1985): Transform Coding of Images. Academic Press, New York, N.Y.
13. Rao K. R., Yip, P. (1990): Discrete Cosine Transform: Algorithms, Advances, Applications. Academic Press, London, U.K.
14. Woods, J. W. (1991); Subband Image Coding. Kluwer, Boston, MA.
15. Shapiro, J. M. (1993): Embedded image coding using zerotrees of wavelet coefficients. IEEE Transactions on Signal Processing, **41**: 3445-3462.
16. Davis, G., Danskin, J., Heasman, R. (1997): Wavelet image compression construction kit. On line report. *www.cs.dartmouth.edu/~gdavis/wavelet/wavelet.html*
17. Kunt, M., Ikonomopoulos, A., Kocher, M. (1985): Second generation image coding techniques. Proceedings of the IEEE, **73**(4): 549-574.
18. Ebrahimi, T., kunt, M. (1998): Visual data compression for multimedia applications. Proceedings of the IEEE, **86**(6): 1109-1125.
19. Pearson, D. (1995): Developments in model-based video coding. Proceedings of the IEEE, **83**: 892-906.
20. Fisher, Y. (ed.) (1995): Fractal Image Compression: Theory and Application to Digital Images. Springer Verlag, New York, N.Y.
21. Jayant, N (1992): Signal compression: Technology targets and research directions. IEEE Journal on Selected Areas in Communications, **10**:796-818.
22. Domanski, M., Bartkowiak, M. (1998): Compression. in Sangwine, S.J., Horne, R.E.N. (eds.), The Colour Image Processing Handbook, 242-304, Chapman & Hall, Cambridge, Great Britain.
23. Penney, W. (1988): Processing pictures in HSI space. The Electronic System Design Magazine, 61-66.
24. Moroney, N. M., Fairchild, M. D. (1995): Color space selection for JPEG image compression. Journal of Electronic Imaging, **4**(4): 373-381.
25. Kuduvalli, G. R., Rangayyan, R. M. (1992): Performance analysis of reversible image compression techniques for high resolution digital tele radiology. IEEE Transactions on Medical Imaging, **11**: 430-445.

26. Gonzales, R.C., Wood, R. E. (1992): Digital Image Processing. Addison-Wesley, Massachusetts.
27. Roger, R. E., Arnold, J. F., Reversible image compression bounded by noise. IEEE Transactions on Geoscience and Remote Sensing, **32**: 19-24.
28. Provine, J. A., Rangayyan, R. M. (1994): Lossless compression of Peano scanned images. Journal of Electronic Imaging, **3**(2): 176-180.
29. Witten, I. H., Moffat, A., Bell, T. C. (1994): Managing Gigabytes, Compressing and Indexing Documents and Images. Van Nostrand Reinhold.
30. Boncelet Jr., C. G., Cobbs, J. R., Moser, A. R. (1988): Error free compression of medical X-ray images. Proceedings of Visual Communications and Image Processing '88, **1001**: 269-276.
31. Wallace, G. K. (1991): The JPEG still picture compression standard. Communications of ACM, **34**(4): 30-44.
32. Ahmed, N., Natarajan, T., Rao, K. R. (1974): Discrete cosine transform. IEEE Transactions on Computers, **23**: 90-93.
33. Bhaskaran, V., Konstantinides, K. (1995): Image and Video Compression Standards. Kluwer, Boston, MA.
34. Léger, A., Omachi, T., Wallace, C. K. (1991): JPEG still picture compression algorithm. Optical Engineering, **30**: 947-954.
35. Egger, O., Li., W. (1995): Subband coding of images using symmetrical filter banks. IEEE Transactions on Image Processing, **4**(4): 478-485.
36. Van Dyk, R. E., Rajala, S. A. (1994): Subband/VQ coding of color images with perceptually optimal bit allocation. IEEE Transaction on Circuits and Systems for Video Technology, **4**(1): 68-82.
37. Lewis, A. S., Knowles, G. (1992): Image compression using the 2-D wavelet transform. IEEE Transactions on Image Processing, **1**(2): 244-250.
38. Chen, D., Bovik, A. C. (1990): Visual pattern image coding. IEEE Transactions on Communications, **38**(12): 2137-2145.
39. Barnsley, M. F. (1988): Fractals Everywhere. Academic Press, N. Y.
40. Jacquin, A. E. (1992): Image coding based on a fractal theory of iterated contractive image transformation. IEEE Transactions on Image Processing, **1**: 18-30.
41. Lu, G. (1993): Fractal image compression. Signal Processing: Image Communications, **4**(4): 327-343.
42. Jayant, N. Johnston, J., Safranek, R. (1993): Perceptual coding of images. SPIE Proceedings, **1913**: 168-178.
43. Klein, S. A., Silverstein, A. D., Carney, T. (1992): Relevance of human vision to JPEG-DCT compression. SPIE Proceedings **1666**: 200-215.
44. Nill, N. B. (1985): A visual model weighted cosine transform for image compression and quality assessment. IEEE Transactions on Communications, **33**: 551-557.
45. Rosenholtz, R., Watson, A. B. (1996): Perceptual adaptive JPEG coding. Proceedings, IEEE International Conference on Image Processing, **I**: 901-904.
46. Eom, I. K., Kim, H. S., Son, K. S., Kim, Y. S., Kim, J. H. (1995): Image coding using wavelet transform and human visual system. SPIE Proceedings, **2418**: 176-183.
47. Kocher, M., Leonardi, R. (1986): Adaptive region growing technique using polynomial functions for image approximations. Signal Processing, **11**(1): 47-60.
48. Mitchell, J., Pennebaker, W., Fogg, C. E., Legall, D. J. (1997): MPEG Video Compression Standard. Chapman and Hall, N.Y.
49. Fleury, P., Bhattacharjee, S., Piron, L., Ebrahimi, T., Kunt, M. (1998): MPEG-4 video verification model: A solution for interactive multimedia applications. Journal of Electronic Imaging, **7**(3): 502-515.

50. Ramos, M. G. (1998): Perceptually based scalable image coding for packet networks. Journal of Electronic Imaging, **7**(3): 453-463.
51. Strang, G., Nguyen, T. (1996): Wavelets and Filter Banks. Wellesley-Cambridge Press, Wellesley, MA.
52. Chow, C. H., Li, Y. C. (1996): A perceptually tuned subband image coder based on the measure of just noticable distortion profile. IEEE Transaction on Circuits and Systems for Video Technology, **5**(6): 467-476.

8. Emerging Applications

Multimedia data processing refers to a combined processing of multiple data streams of various types. Recent advances in hardware, software and digital signal processing allow for the integration of different data streams which may include voice, digital video, graphics and text within a single platform. A simple example may be the simultaneous use of audio, video and closed-caption data for content-based searching and browsing of multimedia databases or the merging of vector graphics, text, and digital video. This rapid development is the driving force behind the convergence of the computing, telecommunications, broadcast, and entertainment technologies. The field is developing rapidly and emerging multimedia applications, such as intelligent visual search engines, multimedia databases, Internet/mobile audio-visual communication, and desktop video-conferencing will all have a profound impact on modern professional life, health care, education, and entertainment.

The full development and consumer acceptance of multimedia will create a host of new products and services including new business opportunities for innovative companies. However, in order for these possibilities to be realized, a number of technological problems must be considered. Some of these include, but are not limited to the following:

1. **Novel methods to process multimedia signals in order to meet quality of service requirements.** In the majority of multimedia applications, the devices used to capture and display information vary considerably. Data acquired by optical, electro-optical or electronic means are likely to be degraded by the sensing environment. For example, a typical photograph may have excessive film grain noise, suffer from various types of blurring, such as motion or focus blur, or have unnatural shifts in hue, saturation or brightness. Noise introduced by the recording media degrades the quality of the resulting images. It is anticipated that the use of digital processing techniques, such as filtering and signal enhancement will improve the performance of the system.
2. **Efficient compression and coding of multimedia signals.** In particular, visual signals with an emphasis on negotiable quality of service contracts must be considered. Rich data types such as digital images and video signals have enormous storage and bandwidth requirements. Techniques that allow images to be stored and transmitted in more compact

formats are of great importance. Multimedia applications are putting higher demands on both the achieved image quality and compression ratios.

Quality is the primary consideration in applications such as DVD drives, interactive HDTV, and digital libraries. Existing techniques achieve compression ratios of 10:1 to 15:1, while maintaining reasonable image quality. However, higher compression ratios can reduce the high cost of storage and transmission, and also lead to the advent of new applications, such as future display terminals with photo quality resolution, or the simultaneous broadcast of a larger number of visual programs.

3. **Innovative techniques for indexing and searching multimedia data.** Multimedia information is difficult to handle both in terms of its size and the scarcity of tools available for navigation and retrieval. A key problem is the effective representation of this data in an environment in which users from different backgrounds can retrieve and handle information without specialized training. Unlike alphanumeric data, multimedia information does not have any semantic structure. Thus, conventional information management systems cannot be directly used to manage multimedia data. Content-based approaches seem to be a natural choice where audio information along with visual indices of color, shape, and motion are more appropriate descriptions. A set of effective quality measures are also necessary in order to measure the success of different techniques and algorithms.

In each of these areas, a great deal of progress has been made in the past few years driven in part by the availability of increased computing power and the introduction of new standards for multimedia services. For example, the emergence of the MPEG-7 multimedia standard demands an increased level of intelligence that will allow the efficient processing of raw information; recognition of dominant features; extraction of objects of interest; and the interpretation and interaction of multimedia data. Thus, effective multimedia signal processing techniques can offer promising solutions in all of the aforementioned areas.

Digital video is an integral part of many newly emerging multimedia applications. Recent advances in the area of mobile communications and the tremendous growth of the Internet have placed even greater demands on the need for more effective video coding schemes. However, future coding techniques must focus on providing better ways to represent, integrate and exchange visual information in addition to efficient compression methods. These efforts aim to provide the user with greater flexibility for "content-based" access and manipulation of multimedia data. Numerous video applications such as portable videophones, video-conferencing, multimedia databases, and video-on-demand can greatly benefit from better compression schemes and this added "content-based" functionality.

The next generation of coding algorithms have focused on representing a scene in terms of "objects" rather than square blocks [1],[2], [3]. This approach not only improves the coding efficiency and alleviates the blocking artifacts, but it can also support the content-based functionalities mentioned previously by allowing interactivity and manipulation of specific objects within the video stream. These are some of the objectives and issues addressed within the framework of the MPEG-4 and future MPEG-7 standards [4].

In order to obtain an object-based representation, an input video sequence must first be segmented into an appropriate set of arbitrarily shaped regions. In a videophone-type application for example, an accurate segmentation of the facial region can serve two purposes: (i) it can allow the encoder to place more emphasis on the facial region since this area, the eyes and mouth in particular, is the focus of attention to the human visual system of an observer, and (ii) it can also be used to extract features, such as personal characteristics, facial expressions, and composition information so that higher level descriptions can be generated. In a similar fashion, the contents within a video database can be segmented into individual objects, where the following features can be supported: (i) sophisticated query and retrieval functions, (ii) advanced editing and compositing, and (iii) better compression ratios.

A method to automatically locate and track a facial region of a head-and-shoulders videophone type sequence using color and shape information is reviewed here. The face localization method consists of essentially two components, namely: (i) a color processing unit, and (ii) a fuzzy-based shape and color analysis module. The color processing component utilizes the distribution of skin-tones in the HSV color space to obtain an initial set of candidate regions or objects. The latter shape and color analysis module is used to correctly identify the facial regions when falsely detected objects are extracted. A number of fuzzy membership functions are devised to provide information about each object's shape, orientation, location, and average hue. An aggregation operator, similar to the one used in Chap. 3, combines these measures and correctly selects the facial area. The methodology presented here is robust with regards to different skin types, and various types of object or background motion within the scene. Furthermore, the algorithm can be implemented at a low computational complexity due to the binary nature of the operations performed.

8.1 Input Analysis Using Color Information

The detection and automatic location of the human face is important and vital in numerous applications including human recognition for security purposes, human-computer interfaces, and more recently, for video coding, and content-based storage/retrieval in image and video databases. Several techniques based on shape and motion information have recently been proposed for the automatic location of the facial region [5],[6]. In [5] the technique is based on fitting an ellipse to a thresholded binary edge image while in [6] the approach utilizes the shape of the thresholded frame differences. In the approach presented color is used as the primary tool in detecting and locating the facial areas in a scene with a complex or moving background.

Color is a key feature used to understand and recollect the contents within a scene. It is also found to be a highly reliable attribute for image retrieval as it is generally invariant to translation, rotation, and scale changes [7]. The segmentation of a color image is the process of classifying the pixels within the image into a set of regions with a uniform color characteristic. The objective in our approach is to detect and isolate the color regions that correspond to the skin areas of the facial region. However, the shape or distribution of the regions that are formed depend on the chosen color space [8]. Therefore, the most advantageous color space must first be selected in order to obtain the most effective results in the segmentation process. It has been found that the skin clusters are well partitioned in the HSV (hue, saturation, value) space, and the segmentation can be performed by a simple thresholding scheme in one dimension rather than a more expensive multidimensional clustering technique. Furthermore, this color model is very intuitive in describing the color/intensity content within a scene. Analogous results have been found in the similar HSI space [9].

It was mentioned in Chap. 1 that color information is commonly represented in the widely used RGB coordinate system. This color space is hardware oriented and is suitable for acquisition or display devices but not particularly applicable in describing the perception of colors. On the other hand, the HSV color model corresponds more closely to the human perception of color.

The HSV color space is conveniently represented by the hexcone model shown in Chap. 1 [10]. The hue (H) is measured by the angle around the vertical axis and has a range of values between 0 and 360 degrees beginning with red at 0°. It gives a measure of the spectral composition of a color. The saturation (S) is a ratio that ranges from 0 (on the V axis), extending radially outwards to a maximum value of 1 on the triangular sides of the hexcone. This component refers to the proportion of pure light of the dominant wavelength. The value (V) also ranges between 0 and 1 and is a measure of the relative brightness. A fast algorithm [10] is used here to convert the set of RGB values to the HSV color space.

Certain steps of the proposed segmentation scheme require the comparison of color features. For example, during clustering color regions are compared with one another to test for similarity. As mentioned in Sect. 6.8.1 when comparing the colors of two regions or pixels, a problem is encountered when one or both of the regions or objects have no or very little chromatic information. That is, a gray scale object can not successfully be compared to an object that has substantial chromatic information. As done in the segmentation scheme in Sect. 6.8, all the pixels in the image are classified as either chromatic or achromatic pixels. This is done by considering the discontinuities of the hue color channel.

Classifying the pixels as either chromatic or achromatic can be considered a crude form of segmentation since the image is segmented into two groups. Although this form of segmentation does have an affect in the face localization algorithm there is no change in the pixel colors. The chromatic/achromatic information is used, in the algorithm, as an indication of whether two colors should be considered similar.

The segmentation of the skin areas within an image is most effective when a suitable color space is selected for the task, as mentioned earlier. This is the case when the skin clusters are compact, distinct, and easy to extract from the color coordinate system. The complexity of the algorithm must also be low to facilitate real-time applications. The HSV color space was found to be the most suitable as it produced clusters that were clearly separated, allowing them to be detected and readily extracted. Three color spaces were compared during experimentation: the HSV, RGB and L*a*b* color spaces. These three coordinate systems cover the different color space groups (hardware-based, perceptually uniform, and hue-oriented) and are frequently selected color models for testing the performance of many proposed color image segmentation algorithms. The RGB and L*a*b* spaces showed ambiguity in the partitioning of the regions.

Data from two different skin-colored regions , as well as the lip area from a different set of images were manually extracted and plotted in each of the aforementioned coordinate systems in order to observe the clusters formed. The results obtained from the RGB space are shown in Fig. 8.1.

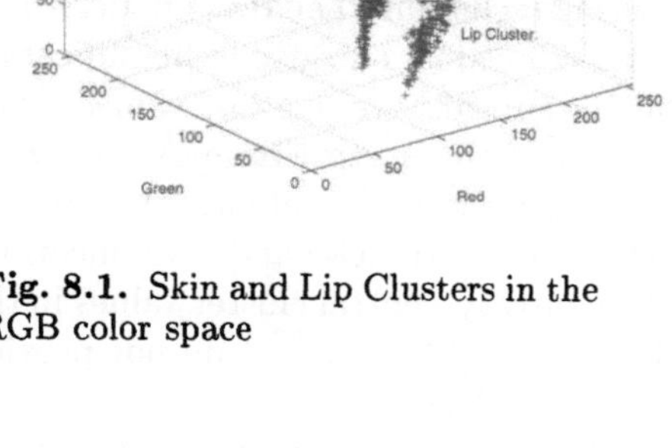

Fig. 8.1. Skin and Lip Clusters in the RGB color space

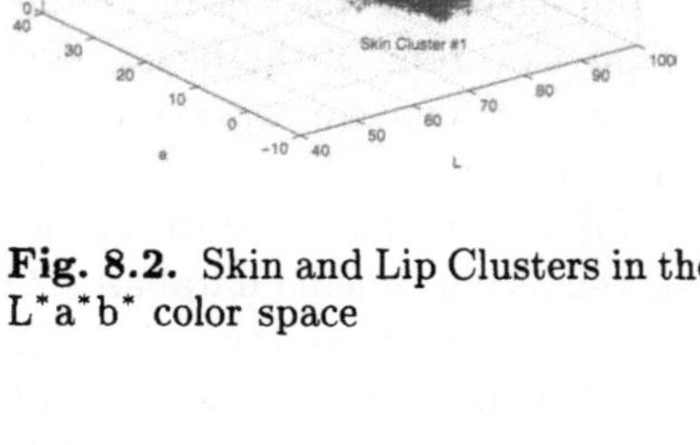

Fig. 8.2. Skin and Lip Clusters in the L*a*b* color space

In the figures above, it can be seen that the skin clusters are positioned relatively close to one another, however, the individual clusters are not compact.

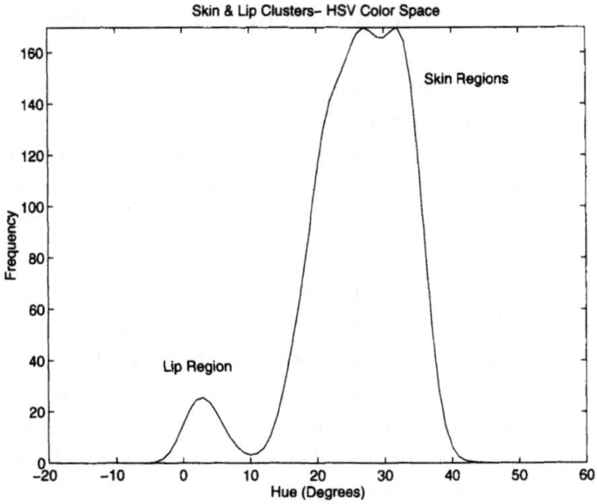

Fig. 8.3. Skin and Lip hue Distributions in the HSV color space

Each forms a diagonal, elongated shape that makes the extraction process difficult. In Fig. 8.2, the skin and lip clusters are displayed in the L*a*b* color space. In this case, the individual clusters are more compact but are spaced quite a distance apart. In fact, the Euclidean distance from skin cluster #1 to the lip cluster is roughly equivalent to that from skin cluster #1 to #2. Thus, the skin clusters do not have a global compactness which once again makes them difficult to isolate and extract. The L*a*b* space is also computationally expensive due to the cube-root expressions in the transformation equations. Finally, in Fig. 8.3, the hue component of the skin and lip clusters from the HSV space are shown. The graph illustrates that the spectral composition of the skin and lip areas are distinct and compact. Skin clusters #1 and #2 are contained between the hue range of 10° and 40° while the lip region lies at a mean hue value of about 2° (i.e. close to the red hue value at 0°).

Thus, the skin clusters are well partitioned allowing the segmentation to be performed by a thresholding scheme in the hue axis rather than a more expensive multidimensional clustering technique. The HSV model is also advantageous in that the mean hue of the skin values can give us an indication of the skin tone of the facial region in the image. Average hue values closer towards 0° contain a greater amount of reddish spectral composition while those towards 60° contain greater yellowish spectral content. This can be useful for content-based storage and retrieval for MPEG-4 and -7 applications as well as multimedia databases. On the contrary, central cluster values in the other coordinate systems, (i.e. $[R_c\ G_c\ B_c]^T$ or $[L^*_c\ a^*_c\ b^*_c]^T$) do not provide the same meaningful description to a human observer.

Having defined the selected HSV color space, a technique to determine and extract the color clusters that correspond to the facial skin regions must be devised. This requires an understanding of where these clusters form in the space just outlined in the previous section.

The identification and tracking of the facial region is determined by utilizing the a priori knowledge of the skin-tone distributions in the HSV color space outlined above. It has been found that skin-colored clusters form within a rather well defined region in chromaticity space [11], and also within the HSV hexcone model [12], for a variety of different skin types. In the HSV space in particular, the skin distribution was found to lie predominantly within the limited hue range between $0° - 50°$ (Red-Yellow), and in certain cases within $340° - 360°$ (Magenta-Red) for darker skin types [13]. The saturation component suggests that skin colors are somewhat saturated, but not deeply saturated, with varying levels of intensity.

The hue component is the most significant feature in defining the characteristics of the skin clusters. However, the hue can be unreliable when: 1) the level of brightness (e.g. value) in the scene is low, or 2) the regions under consideration have low saturation values. The first condition can occur in areas of the image where there are shadows, or generally, under low lighting levels. In the second case, low values of saturation are found in the achromatic regions of a scene. Thus, appropriate thresholds must be defined for the value, and saturation components where the hue attribute is reliable. The following polyhedron that corresponds to skin colored clusters has been defined with well defined saturation and value components, based on a large sample set [13]:

$$T_{hue1} = 340° \leq \mathrm{H} \leq T_{hue2} = 360° \tag{8.1}$$

$$T_{hue3} = 0° \leq \mathrm{H} \leq T_{hue4} = 50° \tag{8.2}$$

$$\mathrm{S} \geq T_{sat1} = 20\% \tag{8.3}$$

$$\mathrm{V} \geq T_{val} = 35\% \tag{8.4}$$

The extent of the above hue range is purposely designed to be quite wide so that a variety of different skin-types can be modeled. As a result of this, however, other objects in the scene with skin-like colors may also be extracted. Nevertheless, these objects can be separated by analyzing the hue histogram of the extracted pixels. The valleys between the peaks are used to identify the various objects that possess different hue ranges (e.g. facial region and different colored objects). scale-space filtering [14] is used to smoothen the histogram and obtain the meaningful peaks and valleys. This process is carried out by convolving the original hue histogram, $f_h(x)$, with a Gaussian function, $g(x, \tau)$ of zero mean and standard deviation τ as follows:

$$F_h(x,\tau) = f_h(x) * g(x,\tau) \tag{8.5}$$

$$F_h(x,\tau) = \int_{-\infty}^{\infty} f_h(u) \frac{1}{\sqrt{2\pi}\,\tau} exp[\frac{-(x-u)^2}{2\tau^2}]\, du \tag{8.6}$$

where $F_h(x,\tau)$ represents the smooth histogram. The peaks and valleys are determined by examining the first and second derivatives of F_h above. In the remote case that another object matches the skin color of the facial area (i.e. separation is not possible by the scale-space filter), then the shape analysis module that follows provides the necessary discriminatory functionality.

A series of post-processing operations which include median filtering, and region filling/removal are subsequently used to refine the regions obtained from the initial extraction stage.

Median filtering is the first of two post-processing operations that are performed after the initial color extraction stage. The median operation is introduced in order to smoothen the segmented object silhouettes and also eliminate any isolated misclassified pixels that may appear as impulsive-type noise. Square filter windows of size (5×5) and (7×7) provide a good balance between adequate noise suppression, and sufficient detail preservation. This operation is computationally inexpensive since it is carried out on the bi-level images, e.g. object silhouettes.

The result of the median operation is successful in removing any misclassified *noise-like* pixels, however, small isolated regions and small holes within object areas may still remain after this step. Thus, the application of median filtering by region filling and removal is followed. This second post-processing operation fills in small holes within objects which may occur due to color differences, e.g. eyes and mouth of the facial skin region, extreme shadows, or any unusual lighting effects (specular reflection). At the same time, any erroneous small regions are also eliminated as candidate object areas.

It has been found that the hue attribute is reliable when the saturation component is greater than 20% and meaningless when it is less than 10% [13]. Similar results have also been confirmed in the cylindrical $L^*u^*v^*$ color model [15]. Saturation values between 0% and 10% correspond to the achromatic areas within a scene while those greater than 20% to the chromatic ones. The range between 10% and 20% represents a sort of transition region from the achromatic to the chromatic areas. It has been observed, that in certain cases, the addition of a select number of pixels within this 10-20% range can improve the results of the initial extraction process. In particular, the initial segmentation may not capture smaller areas of the face when the saturation component is decreased due to the lighting conditions. Thus, pixels within this transition region are selected accordingly [13], and merged with the initially extracted objects. A pixel within the transitional region is added to a particular object if its distance is within a threshold of the closest object. A reasonable selection can be made if the threshold is set to a factor between 1.0-1.5 of the distance from the centroid of the object to its most distant point. The results from this step are once again refined by the two post-processing operations described earlier.

At this point, one or more of the extracted objects correspond to the facial regions. In certain video sequences however, gaps or holes have been found around the eyes of the segmented facial area. This occurs in sequences where the forehead is covered by hair and as a result, the eyes fail to be included in the segmentation. Two morphological operators are utilized to overcome this problem and at the same time smoothen the facial contours. A morphological closing operation is first used to fill in small holes and gaps, followed by a morphological opening operation which is used to remove small spurs and thin channels [16]. Both of these operations maintain the original shapes and sizes of the objects. A compact structuring element, such as a circle or square without holes can be used to implement these operations and also help to smoothen the object contours. Furthermore, these binary morphological operations can be implemented by low complexity *hit or miss* transformations [16].

The morphological stage is the final step involved prior to any analysis of the extracted objects. The results at this point contain one or more objects that correspond to the facial areas within the scene. The block diagram in Fig. 8.4 summarizes the proposed face localization procedure. The shape and color analysis unit, described next, provides the mechanism to correctly identify the facial regions.

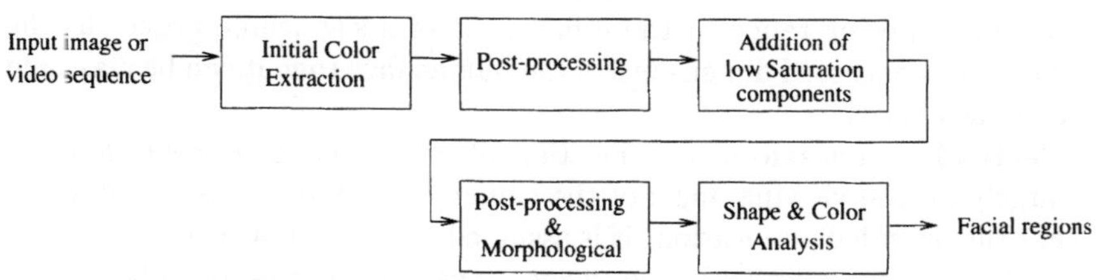

Fig. 8.4. Overall scheme to extract the facial regions within a scene

8.2 Shape and Color Analysis

The input to the shape and color analysis module may contain objects other than the facial areas. Thus, the function of this module is to identify the actual facial regions from the set of candidate objects. In order to achieve this, a number of expected facial characteristics such as shape, color, symmetry, and location are used in the selection process. Fuzzy membership functions are constructed in order to quantify the expected values of each characteristic. Thus, the value of a particular membership function gives an indication of the 'goodness of fit' of the object under consideration with the corresponding feature. An overall 'goodness of fit' value can finally be derived for each object by combining the measures obtained from the individual primitives.

For the segmentation and localization scheme a set of features that are suitable for our application purposes are utilized. In facial image databases, such as employees databases or videophone-type sequences, such as video archives of newscasts and interviews, the scene consists of predominantly upright faces which are contained within the image. Thus, features such as the location of the face, its orientation from the vertical axis, and its aspect ratio can be utilized to assist with the recognition task. These features can be determined in a simple and fast manner as opposed to measurements based on facial features, such as the eyes, nose, and mouth which may be difficult to compute due to the fact that these features may be small or occluded in certain images. More specifically, the following four primitives are considered in the face localization system [17], [18]:

1. **Deviation from the average hue value of the different skin-type categories.** The average hue value for different skin-types varies amongst humans and depends on the race, gender, and the age of the person. However, the average hue of different skin-types falls within a more restricted range than the wider one defined by the HSV model [13]. The deviation of an object's expected hue value from this restricted range gives an indication of its similarity to skin-tone colors.
2. **Face aspect ratio.** Given the geometry and the shape of the human face, it is reasonable to expect that the ratio of height to width falls within a specific range. If the dimensions of a segmented object fit the commonly accepted dimensions of the human face then it can be classified as a facial area.
3. **Vertical orientation.** The location of an object in a scene depends largely on the viewing angle of the camera, and the acquisition devices. For the intended applications it is assumed that only reasonable rotations of the head are allowed in the image plane. This corresponds to a small deviation of the facial symmetry axis from the vertical direction.
4. **Relative position of the facial region in the image plane.** By similar reasoning to (3) above, it is more probable that the face will not be located right at the edges of the image but more likely within a central window of the image.

8.2.1 Fuzzy Membership Functions

A number of membership function models can be constructed and empirically evaluated. A trapezoidal function model is utilized here for each primitive in order to keep the complexity of the overall scheme to a minimum. This type of membership function attains the maximum value only over a limited range of input values. Symmetric or asymmetrical trapezoidal shapes can be obtained depending on the selected parameter values. As in Chap. 3, the membership function can assume any value in the interval $[0,1]$, including both of the extreme values. A value of 0 in the function above indicates that the event is impossible. On the contrary, the maximum membership value of 1 represents total certainty. The intermediate values are used to quantify variable degrees of uncertainty. The estimates for the four membership functions are obtained by a collection of physical measurements of each primitive from a database of facial images and sequences [13].

The hue characteristics of the facial region (for different skin-type categories) were used to form the first membership function. This function is built using the discrete universe of discourse $[-20°, 50°]$ (e. g. $-20° = 340°$). The lower bound of the average hue observed in the image database is approximately $8°$ (African-American distribution) while the upper bound average value is around $30°$ (Asian distribution) [13]. A range is formed using these values, where an object is accepted as a skin-tone color with probability 1 if its average hue value falls within these bounds. Thus, the membership function associated with the first primitive is defined as follows:

$$\mu(x) = \begin{cases} \frac{(x+20)}{28} & \text{if } -20° \leq x \leq 8° \\ 1 & \text{if } 8° \leq x \leq 30° \\ \frac{(50-x)}{20} & \text{if } 30° \leq x \leq 50° \end{cases} \tag{8.7}$$

Experimentation with a wide variety of facial images has led to the conclusion that the aspect ratio (height/width) of the human face has a nominal value of approximately 1.5. This finding confirms previous results reported in the open literature [9]. However, in certain images compensation for the inclusion of the neck area which has similar skin-tone characteristics to the facial region must also be considered. This has the effect of slightly increasing the aspect ratio. Using this information along with the observed aspect ratios from the database, the parameters of the trapezoidal function for this second primitive can be tuned. The final form of the function is given by:

$$\mu(x) = \begin{cases} \frac{(x-0.75)}{0.5} & \text{if } 0.75 \leq x \leq 1.25 \\ 1 & \text{if } 1.25 \leq x \leq 1.75 \\ \frac{(2.25-x)}{0.5} & \text{if } 1.75 \leq x \leq 2.25 \\ 0 & \text{otherwise} \end{cases} \tag{8.8}$$

The vertical orientation of the face in the image is the third primitive used in the shape recognition system. As mentioned previously, the orientation of the facial area (i.e. deviation of the facial symmetry axis from the vertical axis) is more likely to be aligned towards the vertical due to the type of applications considered. A reasonable threshold selection of $30°$ can be made for valid head rotations also observed within our database. Thus, a membership value of 1 is returned if the orientation angle is less than this threshold. The membership function for this primitive is defined as follows:

$$\mu(x) = \begin{cases} 1 & \text{if } 0° \leq x \leq 30° \\ \frac{(90-x)}{60} & \text{if } 30° \leq x \leq 90° \end{cases} \tag{8.9}$$

The last primitive used in the knowledge-based system refers to the relative position of the face in the image. Due to the nature of the applications considered, a smaller weighting is assigned to objects that appear closer to the edges and corners of the images. For this purpose, two membership functions are constructed. The first one returns a confidence value for the location of the segmented object with respect to the X-axis. Similarly, the second one quantifies our knowledge about the location of the object with respect to the Y-axis. The following membership function has been defined for the position of a candidate object with respect to either the X or Y-axis:

$$\mu(x) = \begin{cases} \frac{(x-(d))}{\frac{d}{2}} & \text{if } d \leq x \leq \frac{3d}{2} \\ 1 & \text{if } \frac{3d}{2} \leq x \leq \frac{5d}{2} \\ \frac{((3d)-x)}{\frac{d}{2}} & \text{if } \frac{5d}{2} \leq x \leq 3d \\ 0 & \text{otherwise} \end{cases} \tag{8.10}$$

The membership function for the X-axis is determined by letting $d = \frac{D_x}{4}$, where D_x represents the horizontal dimensions of the image (i.e. in the X-direction). In a similar way, the Y-axis membership function is found by letting $d = \frac{D_y}{4}$, where D_y represents the vertical dimensions of the image (e. g. in the Y-direction).

8.2.2 Aggregation Operators

The individual membership functions expressed above must be appropriately combined to form an overall decision. A number of fuzzy operators can be used to combine or fuse together the various sources of information. Conjunctive type of operators weigh the criterion with the smallest membership value more heavily while disjunctive ones assign the most weight to the criterion with the largest membership value. Here, a compensative operator which offers a compromise between conjunctive and disjunctive behavior is utilized. This type of operator is defined as the weighted mean of a *logical* AND and a *logical* OR operator:

$$A \bigodot_{\gamma} B = (A \bigcap B)^{1-\gamma} \cdot (A \bigcup B)^{\gamma} \tag{8.11}$$

where A, and B are sets defined on the same space and represented by their membership functions [19]. If the product of membership functions is utilized to determine the intersection (*logical* AND) and the possibilistic sum for the union (*logical* OR), then the form of the operator becomes as follows [19]:

$$\mu_c = \prod_{j=1}^{m} \mu_j^{(1-\gamma)} \left(1 - \prod_{j=1}^{m} (1 - \mu_j)\right)^{\gamma} \tag{8.12}$$

where μ_c is the overall membership function which combines all the knowledge primitives for a particular object, and μ_j is the jth elemental membership value associated with the jth primitive. The weighting parameter γ is interpreted as the *grade of compensation* taking values in the range of [0, 1] [19]. The product and the possibilistic sum however, are not the only operators that may be used [20]. A simple and useful t-norm function is the *min* operator while the corresponding one for the t-conorm is the *max* operator. These operators were selected to model the compensative operator, which assumes the form a weighted product as follows:

$$\mu_c = ((\min_{j=1}^{m} \mu_j)(\max_{j=1}^{m} \mu_j))^{0.5} \tag{8.13}$$

where the grade of compensation $\gamma = 0.5$ provides a good compromise of conjunctive and disjunctive behavior [20]. The aggregation operator defined in (8.13) is used to form the final decision based on the designed primitives.

Multimedia databases are comprised of a number of different media types, such as images and video that are binary by nature, and hence unstructured. An appropriate set of interpretations must be derived for these media objects in order to allow for content-based functionalities which include storage and retrieval. These interpretations, or 'metadata' are generated by applying a set of feature extracting functions on the contained media objects [21]. These functions are media dependent and are unique even within each media type. The following four steps are necessary in extracting the features from image object types: (i) object locator design, (ii) feature selection, (iii) classifier design, and (iv) classifier training. The function of the object locator is to isolate the individual objects of interest within the image through a suitable segmentation algorithm. In the second step, specific features are selected to identify the different types of objects that might occur within the images of interest. The classifier design stage is then used to establish a mathematical basis for distinguishing the different objects based on the designed features. Finally, the last step is used to train and update the classifier module by adjusting various parameters. In the previous section the object locator to automatically isolate and track the facial area within a facial image database or a videophone-type sequence was described. Now, the use of a set of features that may be used in constructing a metadata feature vector for the classifier design and training stages is proposed.

Having determined the facial regions within the image, an n-dimensional feature vector, $\mathbf{f} = (f_1, f_2, \ldots, f_n)$, that may be used for content-based storage and retrieval purposes can be constructed. Several features that may be incorporated within a more detailed metadata feature vector are presented here. More specifically, the use of hair and skin color, and face location and size are proposed as a preliminary set.

Hair color is a significant human characteristic that can be effectively employed in user queries to retrieve particular facial images. A scheme to categorize black, gray/white, brown, and blonde hair colors within the HSV space has been determined. First, the H, S, and V component histograms of the hair regions are formed and smoothened using the scale-space filter defined earlier. The peak values from each histogram are subsequently determined and used to form the appropriate classification. The following regions were suitably found from the large sample set for the various categories of hair color:

1. *Black* $V_p < 15\%$
2. *Gray* $S_p < 20\% \cap V_p > 50\%$
3. *Brown* $S_p \geq 20\% \cap 15 \leq V_p < !40\%$
4. *Blonde* $20° < H_p < 50° \cap S_p \geq 20\% \cap V_p \geq 40\%$

where H_p, S_p, and V_p denote the peaks of the corresponding histograms. Thus, dark or black hair is characterized by low intensity values. Gray or white hair is characterized by low saturation and high intensity values. On the other hand, brown or blonde hair colors are typically well saturated but differ in their intensity values. The expected value component of dark brown hair lies at approximately $V_p \approx 20\%$, lighter brown at around $V_p \approx 35\%$, and blonde hair at higher values, $V_p \geq 40\%$. Therefore, this information can be used to appropriately categorize the facial regions extracted earlier. A suitably sized template is used above each facial area for the classification process as shown in Fig. 8.5. The template consists of regions, $R_1 + R_2 + R_3$. This provides a fast yet good approximation to the overall description [22].

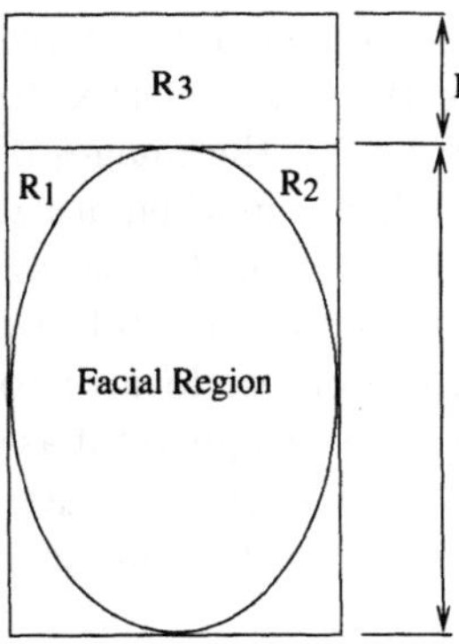

Fig. 8.5. Template for hair color classification = $R_1 + R_2 + R_3$.

The next feature we propose to use is the average hue value of the facial area. We have found that darker skin-types tend to shift towards 0° (e.g. average hue = 8° for our darker skin-type sample set) while lighter colored skin-types towards 30° [13]. In certain cases, however, lighter skin-types with a reddish appearance may also have a slightly reduced average hue value (i.e. 15°). Nevertheless, the hue sector can be partitioned to discriminate between lighter and darker skin-types as follows: (i) darker colored skin, $H < 15°$, and (ii) lighter, $H \geq 15°$. This can give a reasonable approximation, however, it is believed that the saturation and value components can improve upon these results.

Table 8.1. Miss America (Width×Height=360×288):Shape & Color Analysis.

Attributes	Objects		
	O_1	O_2	O_3
Centroid Location			
X	177	245	244
μ_1	1	0	0
Y	188	120	269
μ_2	1	1	0.02
Orientation			
θ°	4.92	47.74	44
μ_3	1	0.7	0.77
Object Ratio			
r	1.61	1.16	1.32
μ_4	1	0.82	1
Mean Hue			
H_m ($^\circ$)	20	-6	-5
μ_5	1	0.5	0.54
Aggregation	1.0	0.0	0.0

Finally, the location and size of each facial area (e.g. centroid location and size relative to the image, respectively) can provide very useful information in a retrieval system. These combined features can give an indication of whether the face is a portrait shot or if perhaps the body is included. In addition to this, it can also provide information about the spatial relationships of a particular facial region with other objects or faces within the scene.

8.3 Experimental Results

The scheme outlined in Fig. 8.4 was used to locate and track the facial region in a number of still images and video sequences. The results from videophone type sequences (i.e. newscast or interview-type sequences) are presented below: (i) 'Carphone', (ii) 'Miss America' and (iii) 'Akiyo' . The segmentation results in Fig. 8.7, 8.10 and 8.12 illustrate the robustness of the technique to the various cases of object/background motion, lighting, and scale variations. A parameter selection of $\tau = 2$ was made in the Gaussian function in order to smoothen the histograms. This provided adequate smoothing, and was found to be appropriate for the skin-tone distribution models [13]. A similar value [15] has also been suggested in the HVC space. The shape and color analysis module was used to identify the facial regions from the set of candidate objects. An object was classified as a facial region if its overall membership function, μ_c exceeded a predefined threshold of 0.75. In the CIF 'Miss America' sequence of Fig. 8.10 three candidate regions were extracted by the localization procedure in Fig. 8.4, and the results of these are summarized in Table 8.1. Only the first object was selected based on the aggregation of the membership function values. The objects C_2 and C_3 were rejected as they scored poorly in their mean hue value and location, and had reduced membership values in the Orientation primitive. In Fig. 8.12, the facial region was successfully identified and tracked for the 'Akiyo' sequence. Two candidate objects were extracted in this case, and once again, the face was correctly selected based on the aggregation values.

Fig. 8.6. Carphone: Frame 80

Fig. 8.7. Segmented frame

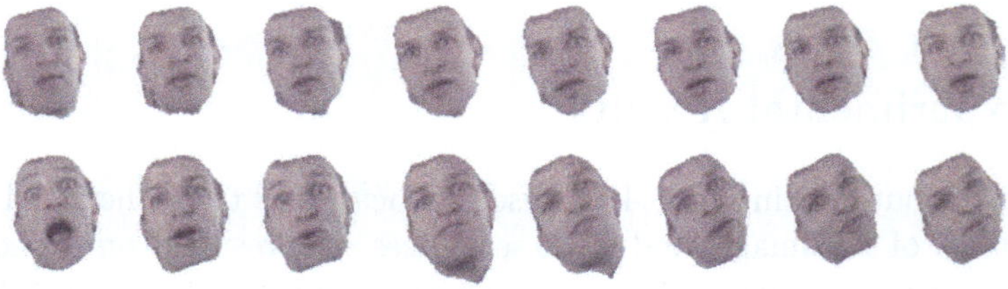

Fig. 8.8. Frames 20-95

Once the facial region is identified, then the proposed metadata features can be computed according to the methodology provided in the previous section. The feature values for each of the image sequences are summarized in Table 8.1. The average hue value of the facial area (e.g. skin) is, in all three cases, greater than 20° which puts them in the lighter skin category, as expected. Next, the S_p and V_p values of the hair region obtained from our constructed template are observed. According to the classification scheme, the tabulated values indicate that the facial image in the 'Carphone' sequence has brown hair while the other two, black. These *fuzzy* descriptions are ap-

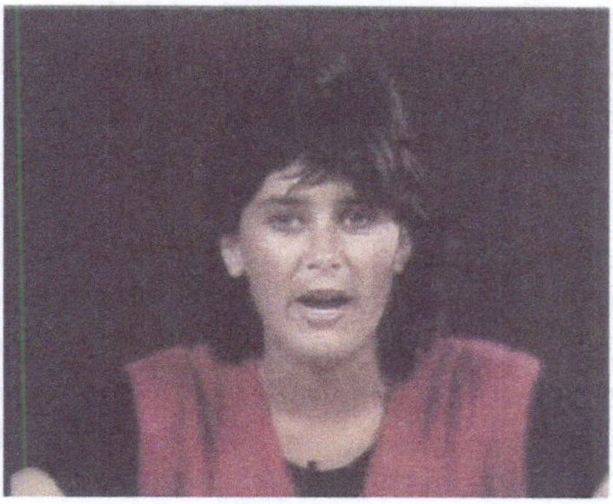

Fig. 8.9. Miss America: Frame 20

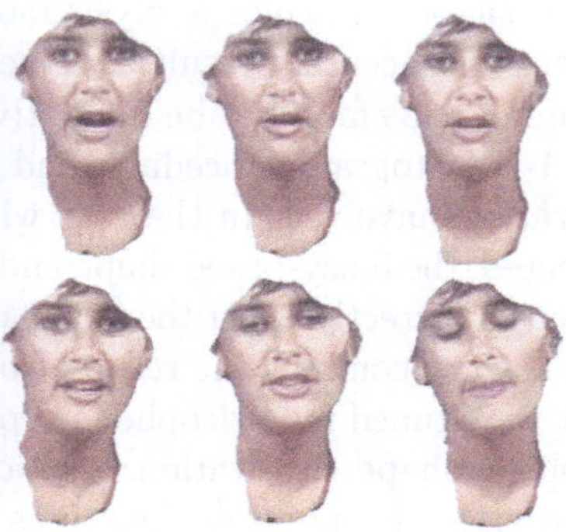

Fig. 8.10. Frames 20-120

Fig. 8.11. Akiyo: Frame 20

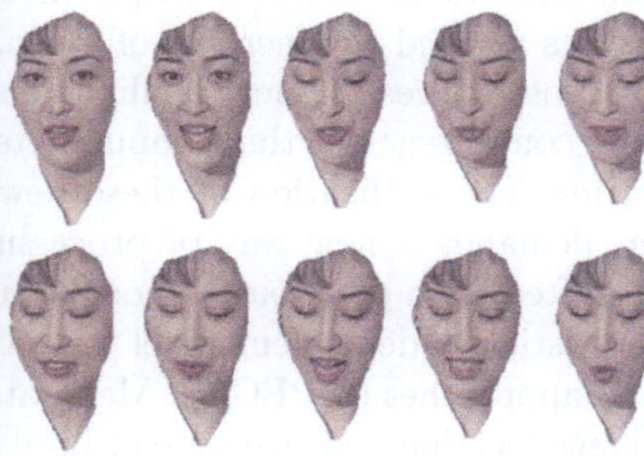

Fig. 8.12. Frames 20-110

propriate representations of the images shown in Fig. 8.8. Finally, the last two features give us an indication of the location and size of the face within the scene. In all cases, the facial region is relatively close to the center of the image (location is with respect to the top left corner), and is of significant size (e.g. closeup).

8.4 Conclusions

The automatic extraction of facial images in digital pictures is vital in numerous multimedia applications, including multimedia databases, video-on-demand, human-computer interfaces, and video coding. A method to locate and track the facial areas within videophone type sequences was presented. The attributes of color and shape were utilized in devising a two-stage segmentation scheme which consisted of a color processing unit, and a fuzzy-based shape and color analysis module. The suggested method led to consistent and accurate results for the intended applications. Furthermore, the technique was found to be of relatively low computational complexity due to the 1-D histogram procedure, and the binary nature of the post-processing operations involved. In the case where more than one candidate object was detected, the fuzzy-based shape and color analysis module provided the mechanism to correctly select the facial area. A compensative aggregation operator was used to combine the results from a series of fuzzy membership functions that were tuned for videophone-type applications. A number of features such as object shape, orientation, location, and average hue were used to form the appropriate membership functions. The proposed fuzzy-based face tracking scheme appears to be quite promising and can be used with an additional feature extraction stage to provide higher level descriptions in future video coding environments.

The tremendous advances in both software and hardware have brought about the integration of multiple media types within a unified framework. This has allowed the merging of video, audio, text, and graphics with enormous possibilities for new applications. This integration is at the forefront in the convergence of the computer, telecommunications, and broadcast industries. The realization of these new technologies and applications, however, demands a new way of processing audio/visual information. Interest has shifted from pixel-based models, such as pulse code modulation (PCM), to statistically dependent pixel models, to the current audio/visual object-based approaches (MPEG-7). Metadata features such as hair and skin color, and face location and size were utilized as a preliminary set. The results of the findings were encouraging in extracting vital information from facial images. Efforts for content-based video description is an active research topic. It is highly desirable to index multimedia data using visual features such as color, texture, shape; sound features such as audio, and speech; and textual features of script and closed-caption. It is also of great interest to have the capabilities to browse and search for this content using compressed data since most video data will likely be stored in compressed formats. Other areas of interest are in the area of temporal segmentation where it is of importance to extract shots, scenes, or objects. Furthermore, higher level descriptions for the direction and magnitude of dominant object motion, and the entry and exit instances of objects of interest are highly desirable. These are all future research areas to be investigated and fueled with the upcoming MPEG-7 standard. In this chapter certain aspects of color based multimedia data processing have been examined. However, further analysis is warranted to address issues of real-time architectures and realizations, modularity, software portability, and system robustness.

References

1. Musmann, H.G., Hotter, M., Ostermann, J. (1989): Object-oriented analysis-synthesis coding of moving objects. Signal Processing: Image Communications, **1**(2), 117-138.
2. Hotter, M. (1990): Object-oriented analysis-synthesis coding based on moving two-dimensional objects. Signal Processing: Image Communications, **2**(4), 409-428.
3. Herodotou, N., Plataniotis, K.N., Venetsanopoulos, A.N., (1998): A color segmentation scheme for object-based video coding. in Proceedings, IEEE Symposium on Advances in Signal Filtering and Signal Processing, **I**, 25-30.
4. Chiariglione, L. (1997): MPEG and multimedia communications. IEEE Transactions on Circuits and Systems for Video Technology, **7**(1), 5-18.
5. Eleftheriadis, A., Jacquin, A. (1995): Automatic face location detection for model-assisted rate control in H.261-compatible coding of video. Signal Processing: Image Communication, **7**(4), 435-455.
6. Reinders, M.J.T., van Beek, P.J.L., Sankur, B., van der Lubbe, J.C.A. (1995): Facial feature localization and adaptation of a generic face model for model-based coding. Signal Processing: Image Communication, **7**(1), 57-74.
7. Jain, A.K., Vailaya, A. (1996): Image retrieval using color and shape. Pattern Recognition, **29**(8), 1233-1244.
8. Uchiyama, T., Arbib, M.A. (1994): Color image segmentation using competitive learning. IEEE Transactions on Pattern Analysis and Machine Intelligence, **16**(12), 1197-1206.
9. Lee, C.H., Kim, J.S., Park, K.H. (1996): Automatic human face location in a complex background using motion and color information. Pattern Recognition, **29**(11), 1877-1889.
10. Foley, J., van Dam, A., Feiner, S., Hughes, J. (1990): Computer Graphics, Principles and Applications. Addison-Wesley, N.Y.
11. Chang, T.C., Huang, T.S., Novak, C. (1994): Facial feature extraction from color images. in Proceedings, 12th International Conference on Pattern Recognition, **3**, 39-43.
12. Herodotou, N., Venetsanopoulos, A.N. (1997): Image segmentation for facial image coding of videophone sequences. in Proceedings, 13th International Conference on Digital Signal Processing, **1**, 233-236.
13. N. Herodotou, N., Plataniotis, K.N., Venetsanopoulos, A.N. (1999): Automatic location and tracking of the facial region in color video sequences. Signal Processing: Image Communications, **14**(5), 359-388.
14. Carlotto, M.J. (1987): Histogram analysis using a scale-space approach. IEEE Transactions on Pattern Analysis and Machine Intelligence, **9**(1), 121-129.
15. Gong, Y., Sakauchi, M. (1995): Detection of regions matching specified chromatic features. Computer Vision and Image Understanding, **61**(2), 263-269.
16. Serra, J. (1982): Image Analysis and Mathematical Morphology. Academic Press, New York, N.Y.
17. Herodotou, N., Plataniotis, K.N., Venetsanopoulos, A.N., (1999): A fuzzy based face tracking scheme. Computational Intelligence and Applications, Mastorakis N. (ed.), 272-276, Word Scientific Press.
18. Herodotou, N., Plataniotis, K.N., Venetsanopoulos, A.N., (1999): Automatic location and tracking of the facial region in color video sequeneces. Signal Processing: Image Communications, **14**(5), 359-388.
19. Zimmermann, H.J., P. Zysno, P. (1980): Latent connectives in human decision making. Fuzzy Sets and Systems, **4**, 37-51.

20. Plataniotis, K.N., Androutsos, D., Venetsanopoulos, A.N. (1997): Multichannel filters for image processing, Signal Processing: Image Communications, **9**(2), 143-158.
21. Prabhakaran, B. (1997): Multimedia Database Management Systems, Kluwer Academic Publishers, Massachusetts.
22. Herodotou, N., Plataniotis, K.N., Venetsanopoulos, A.N., (1998): A color segmentation and classification scheme for facial image and video retrieval. in Proceedings, IX European Signal Processing Conference, **3**, 1721-1724.

A. Companion Image Processing Software

CIPAView, the companion software which complements this book was written exclusively using *J*ava[1]. This choice was made on the basis of two key characteristics of this particular language. First, Java is architecture independent, meaning that the companion software can be run on any platform (*I*ntel, *S*un, etc) which has a Java interpreter installed. Secondly, the Java Developer's Kit (JDK) provides an extremely practical, and convenient way by which applications can be developed quickly and easily. This is due in part to Java's object-oriented nature, but mostly because of the extensive libraries of commonly used methods (routines) and objects (e.g. image format readers, file access support, image filters, etc). These libraries also include a set of streamlined user-interface (UI) development tools which facilitate the development of intuitive and easy-to-use interactive programs. In contrast to the standard technique, this book *d*oes not append its companion software at the end. Instead, the networking availability inspired an idea to provide this software via an Internet browser. The relevant code can be found at the book's web page that can be accessed through *S*pringer Science Online **http://www.springer.de**.

Fig. A.1 shows the main CIPAView window which contains a menubar by which the user is able to access the various filters, and image processing routines.

CIPAView is capable of processing images in a number of standard formats. These are JPG, GIF, PPM, PGM, and RAW. Image files can be opened from the 'File' menu option using either the 'Open' or 'Open As' command for accessing JPG and GIF files or PPM, PGM and RAW files respectively. Once a desired image is loaded, the user is free to perform a wide range of operations on the image. These are:

- Image filtering.
- Image analysis
- Image transforms
- Noise generation
- Image histogram determination

[1] *J*ava is a registered trademark of *S*un Microsystems

Fig. A.1. Screenshot of the main CIPAView window at startup.

A.1 Image Filtering

CIPAView's support for image filtering includes the following:

- Vector Median Filter
- Basic Vector Directional Filter
- Generalized Vector Directional Filter
- Adaptive Nearest Neighbor Filter
- Double Window Adaptive Nearest Neighbor Filter
- Fuzzy Vector Directional Filter
- Adaptive Non-parametric Filter (with *E*xponential Kernel)
- Adaptive Non-parametric Filter (with *G*aussian Kernel)

A.2 Image Analysis

The image analyses which can be performed using CIPAView can be split into two major categories. These are *I*mage Segmentation and *E*dge Detection and routines such as those shown below are included. The screenshot in Fig. A.2 depicts the result of a Difference Vector Mean edge detection.

- Segmentation
  - Region Growing
  - Seed Selection
  - Histogram Thresholding
  - Hybrid
- Edge Detection

- Difference Vector Mean
- Difference Vector α-Trimmed
- Difference Vector Median
- Minimum Vector Distance
- Nearest Neighbor Minimum Vector Distance
- Vector Range

Fig. A.2. Screenshot of Difference Vector Mean edge detector being applied

A.3 Image Transforms

CIPAView has the ability to perform simple image transforms. Examples include color to gray scale conversion, color space transformations, quantization and channel separation. Fig. A.3 depicts a simple quantization operation being performed on a gray scale input image.

A.4 Noise Generation

Since it is to be able to explore the performance of a particular routine or filter in a non-ideal environment, noise generation algorithms have been included in this software package. These are:

- Gaussian Noise
- Impulsive Noise

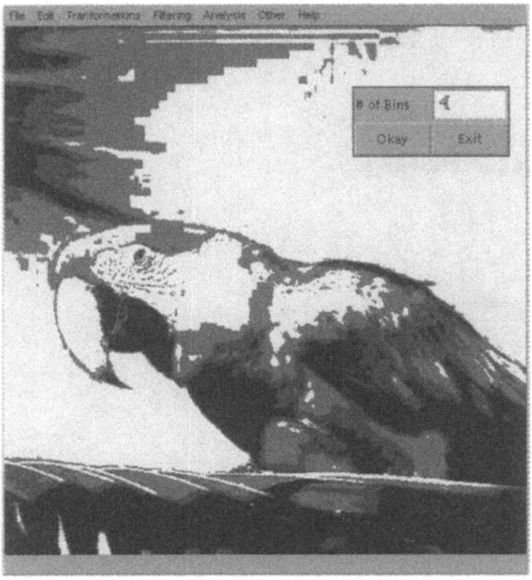

Fig. A.3. Gray scale image quantized to 4 levels

- Mixed Noise

Fig. A.4 shows a screenshot of an input image which is corrupted by Impulsive noise.

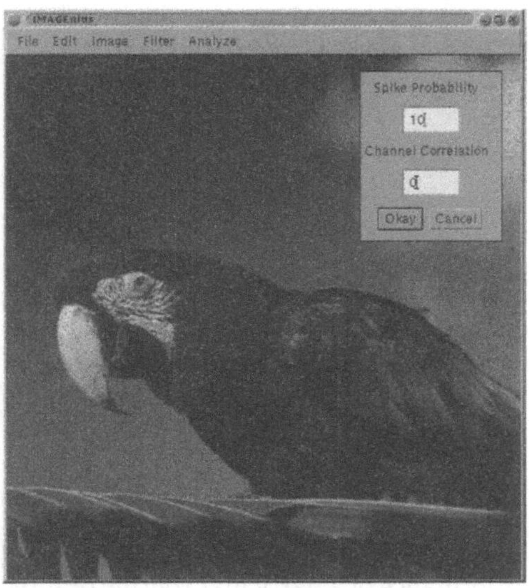

Fig. A.4. Screenshot of an image being corrupted by Impulsive Noise.

Index

MIX
Papier aus verantwortungsvollen Quellen
Paper from responsible sources
FSC® C105338

If you have any concerns about our products,
you can contact us on
ProductSafety@springernature.com

In case Publisher is established outside the EU,
the EU authorized representative is:
Springer Nature Customer Service Center GmbH
Europaplatz 3, 69115 Heidelberg, Germany

Printed by Libri Plureos GmbH
in Hamburg, Germany